Veterinary Clinical Procedures in Large Animal Practice

Jody Rockett, DVM
Susanna Bosted, DVM

DELMAR
CENGAGE Learning

Australia • Brazil • Japan • Korea • Mexico • Singapore • Spain • United Kingdom • United States

Veterinary Clinical Procedures in Large Animal Practice
Jody Rockett, Susanna Bosted

Vice President, Career Education Strategic Business Unit: Dawn Gerrain

Director of Learning Solutions: Sherry Dickinson

Managing Editor: Robert L. Serenka, Jr.

Acquisitions Editor: David Rosenbaum

Product Manager: Christina Gifford

Editorial Assistant: Scott Royael

Director of Production: Wendy A. Troeger

Production Manager: JP Henkel

Senior Content Project Manager: Kathryn B. Kucharek

Director of Marketing: Wendy Mapstone

Channel Manager: Gerard McAvey

Cover Images: Getty Images, Inc.

Cover Design: Mike Egan

© 2007 Delmar, Cengage Learning

ALL RIGHTS RESERVED. No part of this work covered by the copyright herein may be reproduced, transmitted, stored or used in any form or by any means graphic, electronic, or mechanical, including but not limited to photocopying, recording, scanning, digitizing, taping, Web distribution, information networks, or information storage and retrieval systems, except as permitted under Section 107 or 108 of the 1976 United States Copyright Act, without the prior written permission of the publisher.

> For product information and technology assistance, contact us at
> **Cengage Learning Customer & Sales Support, 1-800-354-9706**
> For permission to use material from this text or product,
> submit all requests online at **www.cengage.com/permissions**
> Further permissions questions can be emailed to
> **permissionrequest@cengage.com**

Library of Congress Control Number: 2006003495

ISBN-13: 978-1-4018-5787-5

ISBN-10: 1-4018-5787-6

Delmar
Executive Woods
5 Maxwell Drive
Clifton Park, NY 12065
USA

Cengage Learning is a leading provider of customized learning solutions with office locations around the globe, including Singapore, the United Kingdom, Australia, Mexico, Brazil, and Japan. Locate your local office at **www.cengage.com/global**

Cengage Learning products are represented in Canada by Nelson Education, Ltd.

To learn more about Delmar, visit **www.cengage.com/delmar**

Purchase any of our products at your local bookstore or at our preferred online store **www.CengageBrain.com**

Notice to the Reader
Publisher does not warrant or guarantee any of the products described herein or perform any independent analysis in connection with any of the product information contained herein. Publisher does not assume, and expressly disclaims, any obligation to obtain and include information other than that provided to it by the manufacturer. The reader is expressly warned to consider and adopt all safety precautions that might be indicated by the activities described herein and to avoid all potential hazards. By following the instructions contained herein, the reader willingly assumes all risks in connection with such instructions. The publisher makes no representations or warranties of any kind, including but not limited to, the warranties of fitness for particular purpose or merchantability, nor are any such representations implied with respect to the material set forth herein, and the publisher takes no responsibility with respect to such material. The publisher shall not be liable for any special, consequential, or exemplary damages resulting, in whole or part, from the readers' use of, or reliance upon, this material.

Printed in Canada
3 4 5 6 7 14 13 12

To the animals that have made me laugh, and to those that didn't kill me when given either opportunity or cause. And to my husband—for the same reasons. Live well.

J.R.

To the students of veterinary medicine, surgery, and technology who are devoting their lives to the medical care of our large animal species.

To my husband—thank you for all the encouragement, advice, and endless hours of proofreading which made the writing possible. Without your help, many of the photographs demonstrating real-life situations would never have been taken. And to Paul, who is too young to understand, thanks for bringing me a smile every day.

S.B.

Table of Contents

SECTION 1 — **CARE AND RESTRAINT TECHNIQUES** 1

CHAPTER 1
ROPES AND KNOTS 2
   Key Terms 2
   Objectives 2
   Ropes 3
   Finishing the End of a Rope 3
      Procedure for Finishing or Securing the End of a Rope 3
   Quick-release Knot 5
      Procedure for Tying a Quick-release Knot 5
   Bowline 6
      Procedure for Tying a Bowline 6
   Tomfool Knot 7
      Procedure for Tying a Tomfool Knot 7
   Double Half Hitch 8
      Procedure for Tying a Double Half Hitch 9
   Tail Tie 9
      Procedure for Placing a Tail Tie 10
   Braiding an Eye Splice 10
      Procedure for Braiding an Eye Splice 11
   Rope Halters 12
      Procedure for Building a Rope Halter for Cattle or Sheep .. 12
      Procedure for Building a Temporary Rope Halter
      for a Horse 14
   Review Questions 15
   Bibliography 15

CHAPTER 2

RESTRAINT TOOLS AND TECHNIQUES 16
- Key Terms 16
- Objectives 16
- Complications of Restraint 17
- Restraint of the Horse 17
  - Guidelines for Restraint of the Horse 17
  - Rules of Tying 18
  - Stock 18
    - Procedure for Using Stock 18
  - Haltering and Leading 19
    - Procedure for Haltering the Horse 20
    - Procedure for Leading the Horse 21
  - Applying Chains 22
    - Procedure for Applying Chains 22
  - Twitches 23
    - Procedure for Hand Twitching 24
    - Procedure for Applying Mechanical Twitches 25
  - Loading Horses in Trailers 26
    - Procedure for Loading Horses in Trailers 26
  - Special Handling Scenarios 28
    - Foals 28
    - Leading Foals 28
    - Procedure for Cradling Foals 29
    - Stallions 29
- Restraint of Cattle 30
  - Guidelines for Restraint of Cattle 30
  - Processing Facilities 31
  - Operating Chutes 32
    - Procedure for Operating Chutes 32
  - Haltering 33
    - Procedure for Haltering Cattle 33
  - Tailing-up Cattle 34
    - Procedure for Tailing-up Cattle 34
  - Casting Cattle 35
    - Procedure for Casting Cattle 35
  - Flanking 36
    - Procedure for Flanking Calves 36
  - Securing Cow Feet for Examination 37
    - Procedure for Securing Feet for Examination 38
  - Miscellaneous Equipment 39
    - Hot Shots 39
    - Procedure for Use of a Hot Shot 40
    - Nose Tongs 40
    - Procedure for Placing Nose Tongs 40
    - Nose Rings 42
    - Procedure for Placing a Nose Ring 42
- Restraint of the Goat 43
  - Guidelines for Restraint of the Goat 43
  - Collaring and Leading Goats 43
    - Procedure for Collaring and Leading Goats 43

Stanchion 44
    Procedure for Placing Goat in Stanchion 44
Restraint of the Pig 45
  Guidelines for Restraint of the Pig 45
  Pig Boards 45
    Procedure for Using a Pig Board 46
  Castration Restraint 46
    Procedure for Holding Pig in Castration Position 46
  Snout Snare 47
    Procedure for Applying a Snout Snare 47
Restraint of the Llama 48
  Guidelines for Restraint of the Llama 48
  Haltering and Leading 48
    Procedure for Haltering and Leading the Llama 49
  Stock 49
    Procedure for Placing Llama in a Stock 50
Review Questions 51
Bibliography 51

CHAPTER 3
GROOMING AND STALL MAINTENANCE 52

Key Terms 52
Objectives 52
Basic Grooming 53
  Procedure for Grooming the Horse 54
  Procedure for Grooming the Cow 56
  Procedure for Grooming the Llama 57
Blankets and Fly Masks 57
  Blanketing Procedure 58
  Types of Blankets 59
  Fly Masks 59
  Fly Mask Application Procedure 59
General Husbandry Procedures 60
  Daily Stall Cleaning 60
  Disinfecting and Stripping Stalls 62
  Dirt Floor Stalls 62
  Cement Floors with Rubber Mats 62
  Llama-specific Stall Care and Dung Piles 63
Water Buckets, Feed Tubs, Salt Blocks, and Hay Nets 63
  Daily Care of Water Buckets, Tubs, and Salt Blocks 63
  Care of Water Buckets and Salt Blocks Between Animals ...64
  Hay Nets 64
Identifying Feed 65
  Hay Types 65
  Hay Quality 67
  Hay Quantity 67
  Grains 67
Review Questions 67
Bibliography 68

SECTION 2

PHYSICAL EXAMINATION 69

CHAPTER 4
PHYSICAL EXAMINATION 70
- Key Terms 70
- Objectives 70
- Basic History Taking 71
  - Procedures for History Taking 71
    - General History Questions 71
    - Vaccination History Questions 72
    - Dairy Animal Production Questions 73
    - Hoof Care and Health Questions 73
    - Horse-specific Questions 74
    - Beef Cattle-specific Questions 76
    - Dairy-specific Questions 77
    - Dairy Calf-specific Questions 79
    - Llama-specific Questions 80
    - Pig-specific Questions 80
- Patient Observation 81
- Physical Examination 82
  - Procedures Used in Physical Examinations 83
  - Physical Examination of the Horse 83
  - Physical Examination of the Dairy Cow 87
  - Physical Examination of Beef Cattle 91
  - Physical Examination of the Llama 95
  - Physical Examination of the Pig 98
- Female Reproductive Examination 101
  - Procedures Used in Reproductive Examination of the Female 102
  - Mare Reproductive Examination 102
  - Cow Reproductive Examination 105
  - Small Ruminant Reproductive Examination 107
  - Sow Reproductive Examination 109
- Reproductive Examination of the Male 109
  - Procedures for Reproductive Examination of the Male 110
  - Stallion Breeding Soundness Evaluation 110
  - Bull Breeding Soundness Evaluation 113
  - Small Ruminant Breeding Soundness Evaluation 118
- Examination for Legal Purposes 119
  - Health Certificate 120
  - Insurance Examinations 121
  - Prepurchase Examination of Horses 121
- Review Questions 123
- Bibliography 123

SECTION SAMPLE COLLECTION AND CLINICAL PROCEDURES ..125

CHAPTER 5
SAMPLE COLLECTION 126
  Key Terms 126
  Objectives 126
  Blood Collection 127
    Venipuncture 127
      Procedure for Jugular Venipuncture of the Horse 128
      Procedure for Facial Sinus Venipuncture of the Horse 130
      Procedure for Jugular Venipuncture of the Cow 131
      Procedure for Coccygeal Venipuncture of the Cow 132
      Procedure for High-neck Jugular Venipuncture of
      the Llama 133
      Procedure for Low-neck Jugular Venipuncture of
      the Llama 134
      Procedure for Jugular Venipuncture of the Goat 135
      Procedure for Jugular Venipuncture of the Small Pig 136
      Procedure for Jugular Venipuncture of the Large Pig 137
      Procedure for Venipuncture of the Marginal Ear Vein
      of the Pig 139
      Alternative Collection Sites 140
    Arterial Puncture 140
      Procedure for Arterial Puncture of the Carotid Artery
      of the Horse 140
      Procedure for Arterial Puncture of the Facial, Transverse
      Facial, and Greater Metatarsal Arteries 141
      Procedure for Arterial Puncture of the Coccygeal Artery
      of the Cow 142
      Procedure for Arterial Puncture of the Median
      Auricular Artery 142
    Coagulation Study Blood Collection 143
      Procedure for Coagulation Study Blood Collection 143
  Fecal Sample Collection 144
    Procedure for Fecal Sample Collection from the Horse . . .144
    Procedure for Fecal Sample Collection from the Cow 145
    Procedure for Fecal Sample Collection from the Llama . . .145
  Urine Collection 146
    Procedure for Urinary Catheterization of the Horse 146
    Procedure for Free-catch Urine Collection in the Horse . .147
    Procedure for Urine Collection Using the Digital Technique
    in Cows 148
    Procedure for Llama Urine Free-catch Technique 149
    Procedure for Llama Urinary Catheterization 149
  Centesis 150
    Abdominocentesis 150
      Preparation for Abdominocentesis
      of the Horse: For Both Needle and Teat Cannula
      Method 151

        Procedure for Abdominocentesis of the Horse: Needle Method Continued 152
        Procedure for Abdominocentesis of the Horse: Teat Cannula Method Continued 152
        Procedure for Abdominocentesis of the Cow 153
        Procedure for Abdominocentesis of the Llama 154
        Complication Management 155
    Thoracocentesis 155
        Procedure for Thoracocentesis of the Horse and Cow 156
        Procedure for Thoracocentesis of the Llama 157
    Rumenocentesis 158
        Procedure for Rumenocentesis 158
Respiratory Sampling 159
    Nasal Swabs 160
        Procedure for Nasal Swabs 160
    Transtracheal Wash 161
        Procedure for Transtracheal Wash 162
Biopsy, Aspiration, Scrapes, and Smears 163
    Uterine Biopsy 163
        Procedure for Uterine Biopsy of the Horse 164
    Liver Biopsy 165
        Procedure for Liver Biopsy of the Cow 166
        Procedure for Liver Biopsy of the Horse 167
        Procedure for Liver Biopsy of the Llama 169
    Dermal Biopsy 170
        Procedure for Dermal Biopsy 170
    Fine-needle Aspiration 171
        Procedure for Fine-needle Aspiration Using Technique One 172
        Procedure for Fine-needle Aspiration Using Technique Two 173
    Skin Scrapes 174
        Procedure for Skin Scrapes 174
    Impression Smears 175
        Procedure for Impression Smears 175
Cultures and Testing 176
    Uterine Culture 176
        Procedure for Uterine Culture of the Horse 177
        Procedure for Uterine Culture of the Cow 177
    Skin Cultures 179
        Procedure for Bacterial Culture of the Skin 179
        Procedure for Dermatophyte Culture 180
    Milk Cultures 181
        Procedure for Milk Culture 182
    California Mastitis Test (CMT) 183
        Procedure for California Mastitis Test 183
Necropsy Sampling 184
        Procedure for Necropsy 185
Review Questions 191
Bibliography 191

CHAPTER 6
CLINICAL PROCEDURES 193

- Key Terms 193
- Objectives 193
- Administering Oral Medications 194
  - Administering Pastes 194
    - Procedure for Administering Oral Pastes to the Horse .194
    - Procedure for Administering Oral Pastes to the Goat ..195
    - Procedure for Administering Oral Pastes to the Llama .196
  - Balling Guns 196
    - Procedure for Using a Balling Gun 196
- Parenteral Administration of Drugs 197
  - Common Needle Sizes 198
  - Administering Intramuscular Injections 199
    - Procedure for an Intramuscular Injection in the Horse .199
    - Procedure for Intramuscular Injection in the Cow 202
    - Beef Quality Assurance Recommendations 205
    - Care of Automatic Syringe Guns 205
    - Procedure for Intramuscular Injection in the Goat 206
    - Procedure for Intramuscular Injection in the Pig 208
    - Procedure for Intramuscular Injection in the Llama ..209
  - Administering Intravenous Injections 211
    - Procedure for Intravenous Injection in the Horse 211
    - Procedure for Intravenous Injection in the Cow 213
    - Procedure for Intravenous Injection in the Goat 214
    - Procedure for Intravenous Injection in the Pig 215
    - Procedure for Intravenous Injection in the Llama 216
  - Administering Subcutaneous Injections 217
    - Procedure for Subcutaneous Injection in the Cow 218
    - Procedure for Subcutaneous Injection in the Goat and Llama 219
    - Procedure for Subcutaneous Injection in the Pig 219
  - Growth Hormone Implants 220
    - Procedure for Placing Growth Hormone Implants 221
  - Intubation 221
    - Nasogastric Intubation 222
      - Procedure for Nasogastric Intubation of the Horse 222
    - Orogastric Intubation 224
      - Procedure for Orogastric Intubation of Cows, Llamas, and Goats 225
  - Catheters 228
    - Catheter Types 228
    - Catheter Sizes for Various Animal Species 229
    - General Intravenous Catheter Placement and Care Recommendations 230
    - Intravenous Catheters 231
      - Procedure for Placing an Over-the-needle Intravenous Catheter in the Horse 232
      - Procedure for Placing a Through-the-needle Intravenous Catheter in the Horse 236

Procedure for Placing an Intravenous Catheter
in Cows and Llamas 237
Procedure for Placing an Intravenous Catheter
in the Goat 240
Procedure for Placing an Intravenous Catheter
in the Pig 242
Removal of Intravenous Catheters 244
Intraarterial Catheters 244
Procedure for Placing an Intraarterial Catheter 245
Bandaging 247
Basic Lower Limb Wraps 247
Procedure for Applying a Lower Limb Wrap 248
Tail Wraps 250
Procedure for Placing a Tail Wrap 250
Abdominal Wraps 251
Procedure for Applying an Abdominal Wrap 252
Reproductive Treatments 252
Sheath Cleaning 252
Procedure for Cleaning the Sheath 253
Intramammary Infusion 255
Procedure for Performing an Intramammary Infusion 255
Ocular Procedures 256
Administering Ocular Medication 256
Procedure for Administering Ocular Medication 256
Fluorescence Staining 257
Procedure for Performing a Fluorescein Stain 257
Nasolacrimal Flushing 258
Procedure for Flushing the Nasolacrimal Duct 258
Nasolacrimal Cannulation or Indwelling Infusion Tube 259
Procedure for Cannulating the Nasolacrimal Duct 259
Identification Techniques 261
Hot Branding 261
Procedure for Hot Branding 262
Freeze Branding 262
Procedure for Freeze Branding 263
Ear Tattooing 264
Procedure for Tattooing the Ear 265
Ear Tags 266
Procedure for Attaching Ear Tags 266
Ear Notching 267
Procedure for Notching a Pig Ear 268
Review Questions 269
Bibliography 270

CHAPTER 7
NEONATAL CLINICAL PROCEDURES 271
Key Terms 271
Objectives 271
Bottle Feeding 272
    Procedure for Bottle Feeding a Foal 272
    Procedure for Bottle Feeding a Calf 273
    Procedure for Bottle Feeding a Kid 275
    Procedure for Bottle Feeding a Pig 276
    Procedure for Bottle Feeding a Cria 276
Milk Replacers and Colostrum 277
    Common Milk Replacer Components 278
    Colostrum Recommendations 278
Intubation 279
    Nasogastric Intubation of Foals 279
        Procedure for Nasogastric Intubation of Foals 279
    Orogastric Intubation 281
        Procedure for Orogastric Intubation of Calves 281
        Procedure for Orogastric Intubation of Kids 282
        Procedure for Orogastric Intubation of Cria 283
Early Neonatal Care Checklist 284
    Procedure for Immediate Postpartum Care of All Species 285
    Procedure for Postpartum Care of the Foal 286
    Procedure for Postpartum Care of the Calf 288
    Procedure for Postpartum Care of the Kid 289
    Procedure for Postpartum Care of the Piglet 289
    Procedure for Postpartum Care of the Cria 290
Parenteral Nutrition of Foals 291
    Procedure for Administering Parenteral Nutrition to the Foal 291
Nasal Oxygen 292
    Procedure for Administering Nasal Oxygen to the Foal 293
Enemas 293
    Procedure for Administering an Enema to a Foal 294
Routine Neonatal Processing Procedures 295
    Clipping Needle Teeth 295
        Procedure for Clipping Needle Teeth 296
    Disbudding Kids 297
        Procedure for Disbudding Kids 297
    Lamb Tail Docking 298
        Procedure for Docking Lambs' Tails 298
Review Questions 300
Bibliography 300

SECTION 4: SURGICAL, RADIOGRAPHIC, AND ANESTHETIC PREPARATION 301

CHAPTER 8
SURGICAL PREPARATION 302
- Key Terms 302
- Objectives 302
- Surgery Team 303
  - Scrub Nurse 303
  - Circulator 303
- Personal Preparation for Surgery 304
  - Procedure for Personal Preparation for Surgery 305
- Instrument Pack Preparation and Sterilization 307
  - Procedure for Surgery Pack Preparation 308
- Sterilizing Instruments 310
  - Procedure for Sterilizing Instruments 311
- Packing and Operating an Autoclave 313
  - Procedure for Packing and Operating an Autoclave 313
- Preparation of the Patient for Surgery 316
  - Three-step Surgical Scrub 316
    - Procedure for Three-step Surgical Scrub 316
- Draping the Patient Using Aseptic Technique 318
  - Procedure for Draping the Patient 318
- Cushioning the Recumbent Patient 320
  - Procedure for Cushioning the Recumbent Patient 321
- Sutures, Needles, and Suturing Techniques 322
- Tying and Cutting a Suture Knot 323
  - Procedure for Tying and Cutting a Suture Knot 325
- Suture Patterns 327
  - Procedure for How to Sew Four Suture Patterns 327
- Surgical Castration of the Large Animal 329
  - Surgical Castration of Calves 330
    - Procedure for Castration of Calves 331
  - Recumbent Surgical Castration of Horses 332
    - Procedure for Recumbent Castration of Horses 333
  - Surgical Castration of Llamas 335
    - Procedure for Castration of Llamas 336
  - Surgical Castration of Pigs 337
    - Procedure for Castration of Pigs 337
- Preparation of the Patient for Laparotomy 339
  - Preparation for a Flank Laparotomy 339
    - Procedure for Standing Flank Laparotomy of the Horse 340
    - Procedure for Flank Laparotomy of the Ruminant 343
  - Ventral Laparotomy 345
    - Procedure for Ventral Laparotomy in Ruminants 346
- Obstetrical Procedures 349
  - Assisted Vaginal Delivery of a Fetus 349
    - Procedure for Assisted Vaginal Delivery of Horses 350
    - Procedure for Assisted Vaginal Delivery of Ruminants 351
    - Procedure for Assisted Vaginal Delivery of Pigs 352

Cesarean Section 353
    Comparison of Cesarean Section in Large Animals 353
Vaginal Prolapse Repair 353
    Procedure for Vaginal Prolapse Repair of the Cow 355
Uterine Prolapse 357
    Procedure for Uterine Prolapse Repair 358
Miscellaneous Common Large Animal Procedures 360
    Dehorning Calves 360
        Procedure for Dehorning Calves 361
    Dehorning Goats 364
        Procedure for Dehorning Goats 365
    Bovine Eye Enucleation 367
        Procedure for Bovine Eye Enucleation 368
    Rectal Prolapse Reduction 370
        Procedure for Rectal Prolapse Reduction 371
    Preparation for Endoscopy in the Horse 373
        Procedure for Preparation for Endoscopy in the Horse ...373
Review Questions 376
Bibliography 376

CHAPTER 9

SELECTED LOWER LIMB RADIOGRAPHIC PROCEDURES 378

Key Terms 378
Objectives 378
Basic Radiographic Equipment 379
Basic Radiographic Safety 380
    Safety Guidelines 380
Hoof Preparation for Radiographs 381
    Procedure for Hoof Preparation 382
Lower Limb Radiographs 383
    Procedure Used for All Distal Limb Radiographs 384
    Radiographic Views of Coffin 385
        Procedure for Lateral View of Coffin 386
        Procedure for 45-degree Dorsopalmar View of Coffin 387
        Procedure for Horizontal Zero-degree Dorsopalmar
        View of Coffin 388
        Procedure for Oblique Views of Coffin 389
    Radiographic Views of Navicular 390
        Procedure for Palmar-proximal or Skyline
        View of Navicular 390
        Procedure for Lateral View of Navicular 392
        Procedure for 65-degree Dorsopalmar or Upright
        Pedal View of Navicular 392
        Procedure for Oblique Views of Navicular 394
    Radiographic Views of Pastern 395
        Procedure for Dorsopalmar View of Pastern 395
        Procedure for Lateral View of Pastern 396
        Procedure for Oblique Views of Pastern 397
    Radiographic Views of Fetlock 399
        Procedure for Dorsopalmar View of Fetlock 399
        Procedure for Extended Lateral View of Fetlock 401

Procedure for Flexed Lateral View of Fetlock 402
Procedure for Oblique Views of Fetlock 403
Procedure for Flexor Surface or Skyline View of Fetlock . . .404
Radiographic Views of Metacarpals and Metatarsals 405
Procedure for Dorsopalmar View of Metacarpals
and Metatarsals 406
Procedure for Lateral View of Metacarpals
and Metatarsals 407
Procedure for Oblique Views of Metacarpals
and Metatarsals 408
Radiographic Views of Carpus 410
Procedure for Dorsopalmar View of Carpus 410
Procedure for Extended and Flexed Lateral Views
of Carpus 411
Procedure for Oblique Views of Carpus 413
Procedure for Skyline Views of Carpus 415
Radiographic Views of Tarsus 417
Procedure for Dorsoplantar View of Tarsus 417
Procedure for Lateral and Flexed Lateral Views of Tarsus .418
Procedure for Oblique Views of Tarsus 420
Procedure for Flexed Dorsoplantar View of Tarsus 422
Preparation of the Limb for Ultrasonographic Examination 423
Procedure for Ultrasonographic Preparation 423
Review Questions 424
Bibliography 425

CHAPTER 10
ANESTHESIA 426
Key Terms 426
Objectives 426
Veterinary Anesthesia in Large Animal Practice 427
Roles of the Veterinary Anesthetist 427
Preanesthetic Period 427
Minimum Data Base 427
Preparation of Horses for General Anesthesia
or Heavy Sedation 428
Procedure for Preparing Horses for General Anesthesia . .430
Preparation of Ruminants for General Anesthesia or
Heavy Sedation 431
Procedure for Preparing Ruminants for Anesthesia 432
Preparation of Llamas for General Anesthesia
or Heavy Sedation 433
Procedure for Preparing Llamas for Anesthesia 434
Preparation of Pigs for General Anesthesia or
Heavy Sedation 435
Procedure for Preparing Pigs for Anesthesia 435
Stages of Anesthesia 436
Description of Stages of Anesthesia 436

Parameters to Monitor in All Patients under Anesthesia 439
    Procedure for Monitoring Horses 439
    Procedure for Monitoring Ruminants .445
    Procedure for Monitoring Pigs 449
Monitoring Equipment 451
    Blood Pressure Monitors 452
    Doppler Blood Pressure Monitor 452
        Procedure for Doppler Blood Pressure Monitoring 453
    Oscillometric Blood Pressure Monitoring 454
        Procedure for Oscillometric Blood Pressure Monitoring . .455
    Direct Blood Pressure Monitoring 456
        Procedure for Direct Blood Pressure Monitoring 457
    Electrocardiograph 459
        Procedure for Setting Up the ECG 460
    Pulse Oximetry 461
        Procedure for Pulse Oximetry 462
Sedatives and Tranquilizers Used in Large Animal Anesthesia . . .463
    Phenothiazine Tranquilizers 464
    Alpha-2 Receptor Agonists 465
    Alpha-2 Receptor Antagonists 468
    Benzodiazepines 469
    Opioids 470
Anesthetic Agents 471
    Dissociative Anesthetics 471
    Other Injectable Anesthetics 473
    Inhalant Anesthetics 474
Sample General Anesthetic Regimens 477
    Induction and Maintenance of Anesthesia in Horses 477
    Induction and Maintenance of Anesthesia in Ruminants 479
    Induction and Maintenance of Anesthesia in Llamas 481
    Induction and Maintenance of Anesthesia in Pigs 482
Field Anesthesia 483
    Induction of Field Anesthesia in the Horse 483
        Procedure for Method One: Horse Field Induction
        and Recovery 484
        Procedure for Method Two: Horse Field Induction
        and Recovery 487
    Induction of Field Anesthesia in the Ruminant 489
        Procedure for Induction of Ruminants 489
Inhalant Anesthesia in Large Animal Surgery 490
    Parts of the Anesthetic Delivery System 491
    Operation of Anesthesia Machine 494
    Intubation: Acquiring Access to the Airways 495
        Procedure for Nasotracheal Intubation 497
        Procedure for Orotracheal Intubation 498
Local and Regional Anesthetic Blocks 500
    Mechanism of Action 501
    Systemic and Toxic Effects of Local Anesthetic Agents 501
    Local Anesthetic Agents Commonly Used in Large Animal
    Medicine and Surgery 502
    Methods of Producing Local Anesthesia 503

Intrasynovial Blocks 504
　　　　　Procedure for Intrasynovial Blocks 506
　　　Horse Lower Limb Blocks 508
　　　　　Procedure for Administering Diagnostic Lower
　　　　　　Limb Blocks to Horses 509
　　　Ring Block 512
　　　　　Procedure for Teat Block 512
　　　　　Procedure for Digit Block in Ruminants 514
　　　Intravenous Regional Block (Bier Block) 516
　　　　　Procedure for Intravenous Limb Block 517
　　　Infiltrative Blocks 518
　　　　　Procedure for Performing an Inverted L Regional Block . .519
　　　Paravertebral Block 520
　　　　　Procedure for Paravertebral Block 521
　　　Caudal Epidural Anesthesia 523
　　　　　Procedure for Caudal Epidural in the Ruminant 523
　Review Questions 525
　Bibliography 525

SECTION **APPENDIX**

　　　Conversion Chart 527
　　　Normal Temperature, Pulse, and Respiration 527
　　　Aging by Eruption of Permanent Incisors and Canines 528
　　　Basic Husbandry Periods 528
　　　Peritoneal Fluid Analysis of Cattle 529
　　　Peritoneal Fluid Analysis of Horses 530
　　　Complete Blood Count Normal Values 531
　　　Blood Chemistry Normal Values 532
　　　Milk Composition Chart 532
　　　Common Names of Large Animals 533
　　　Common Names for Leg Anatomy 533
　　　Common Names for Food Animal Anatomy 534
　　　Horse Vaccines 535
　　　Cattle Vaccines 536
　　　Goat Vaccines 538
　　　Llama Vaccines 539
　　　Pig Vaccines 540
Bibliography **541**
Glossary **542**
Index **547**

Join us on the web at
agriculture.delmar.com

Preface

When it comes to working with large animals, a wide and varied range of skills must be mastered by the potential veterinary technician or future DVM. From restraint to specimen collection, physical examination to anesthesia, veterinary medical team members are expected to exhibit current and best practices. *Veterinary Clinical Procedures in Large Animal Practice* addresses these issues in a clear, concise approach.

Care has been taken in each chapter to present the material in a uniform and easily followed format. We have intentionally departed from the standard paragraph prose format. Our intent? To concisely answer the critical questions everyone has when learning a new procedure. "What do I need, what do I do, and what can go wrong?" Ultimately our goal was to provide these answers in a clinically accessible format eliminating the need to wade through more traditional texts.

With this goal in mind, all chapters are constructed such that the purpose for each procedure is clearly stated and followed by a list of potential complications. Equipment requirement lists precede the step-by-step instructions for performing each procedure. Finally, procedural instructions are augmented by short explanatory statements or comments.

We believe the students, technicians, and practioners using this text will find it concise, reliable, and informative. It has been our sincere pleasure to contribute a resource to the talented, compassionate people who have dedicated their lives to the betterment of animals.

Acknowledgments

The authors would like to thank Dr. Rick Parker for his contributions to this text. Additionally, the authors and Delmar, Cengage Learning would like to express their appreciation to those individuals who reviewed the manuscript and offered helpful suggestions:

Khursheed Mama, DVM, Dipl ACVA
Assistant Professor of Anesthesia
Colorado State University
College of Veterinary Medicine
Fort Collins, Colorado

Clyde Gillespie, DVM
Animal Medical Clinic
Heyburn, Idaho

Other reviewers as selected by Delmar:

Betsy Krieger, DVM
Veterinary Technology Program
Front Range Community College
Fort Collins, CO

Bonnie Ballard, DVM
Gwinnett Technical College
Lawrenceville, GA

About the Authors

After receiving her bachelors degree in Microbiology from the University of Wyoming, Jody Rockett, DVM, attended the University of Missouri, College of Veterinary Medicine. She enjoyed several years practicing as an associate veterinarian and subsequently founded the Veterinary Technology program at the College of Southern Idaho. Dr. Rockett currently directs the veterinary technology program at CSI and enjoys spending time with her husband and two children.

Susanna Bosted, DVM, attended the University of Idaho, where she graduated with a bachelors degree in Microbiology. She then attended Washington State University, College of Veterinary Medicine, graduating in 1989. In addition to her roles as president of the Idaho Veterinary Medical Association and Instructor at the College of Southern Idaho, Dr. Bosted has owned and operated a mixed animal practice for the past 13 years. She, her veterinarian husband, and their son enjoy their free time at home with their collection of farm animals and other critters.

SECTION ONE

Care and Restraint Techniques

Chapter 1
Ropes and Knots

Chapter 2
Restraint Tools and Techniques

Chapter 3
Grooming and Stall Maintenance

CHAPTER 1

Ropes and Knots

> *How do you catch a loose horse? Make a noise like a carrot.*
> BRITISH CAVALRY JOKE

KEY TERMS

bight
hondo
lariat
loop
sisal
tensile strength

OBJECTIVES

- Identify the types of ropes available and explain how to maintain them.
- List and describe the types of ropes used in large animal medicine.
- Describe how to tie basic knots used in large animal medicine.
- Describe how to build a rope halter for cattle and sheep.

ROPES

For years ropes were made out of hemp, and although very durable and weather resistant, their roughness caused rope burns and irritation. Ropes these days are generally made of synthetic materials, cotton, or **sisal**. Synthetic materials are also weather resistant and difficult to break, but just like hemp, they tend to burn either the patient or the handler if pulled rapidly across the skin. An added benefit is that these ropes tend to be lighter weight than cotton or sisal. Cotton ropes most often are used for positioning large animals, because they are softer and less apt to burn. They should not be left out in the sun or inclement weather, however, because they will break down quickly and lose **tensile strength.** Specialized ropes, such as **lariats** (commonly used to rope a calf), may be coated with wax to help them slide through the **hondo** (or **loop**) more effectively. The diameter and coating of lariats tend to cause rope burns when used for restraint purposes. All ropes can be purchased in almost any length and thickness at the local hardware or ranch supply store.

SAFETY ALERT

- All ropes, halters, and lead ropes used for restraint should be checked often for signs of fraying or weakness.
- Nylon halters and ropes should be washed regularly to lessen the spread of disease.
- Washing cotton ropes speeds their degradation.
- Lariats are not meant to be washed and will lose their special handling characteristics if that is done.

FINISHING THE END OF A ROPE

Whenever a rope is cut, it is important to do something with the raw end to prevent it from unraveling. The simplest method is to tie a knot in the end, but that makes the end bulky and difficult to manipulate. Whatever method you choose, make sure it is appropriate for the intended use of the rope.

Procedure for Finishing or Securing the End of a Rope

TECHNICAL ACTION	RATIONALE/AMPLIFICATION
1. Tape the end.	1a. Using black electrician's tape, start wrapping the rope about 1 cm (½ inch) from the end with the tape.
	1b. Pull very hard, hard enough to compress the strands of the rope.
	1c. Continue up the rope for approximately 5 cm (2 inches).

SECTION 1 Care and Restraint Techniques

TECHNICAL ACTION	RATIONALE/AMPLIFICATION
2. Whip the end.	2a. Select a thin piece of nylon string or cord about 30 cm (12 inches) long.
	2b. Lay the cord on the rope in a U shape, with the bottom of the U about 5 cm (2 inches) from the end of the rope and the arms of the U heading off the end of the rope (Fig. 1-1A).
	2c. Hold the loop of cord against the rope.
	2d. Using the long arm of the cord, and starting about 1 cm from the end of the rope, tightly wrap the rope with the cord (Fig. 1-1B).
	2e. When at least 2.5 cm (1 inch) of the rope is covered, pass the long end of the cord through the loop (Fig. 1-1C).
	2f. Pull the short end of the cord until the loop disappears and the long end of the cord is sucked up tight to the wraps (Fig. 1-1D).
	2g. Cut both ends of the cord close to the wrap.

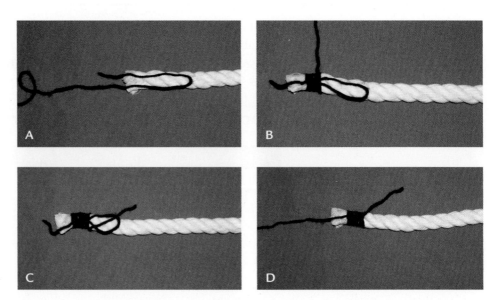

FIG. 1-1 (A) Step one to whipping the end of a rope. (B) Tightly wrapping the end of the rope with the cord. (C) Passing the end of the whipping cord through the loop. (D) Pull the end of the cord in the direction shown.

CHAPTER 1 Ropes and Knots

TECHNICAL ACTION	RATIONALE/AMPLIFICATION
3. Burn the end.	3a. Done with a nylon rope.
	3b. Hold a lighter flame to the raw end of the rope until the strands melt and curl back.
	3c. Rotate the rope to get even coverage.
	3d. Takes roughly 30 seconds to get complete and even melting up to 1 cm up the rope.
	3e. Tape may then be applied as above for a finished appearance.

QUICK-RELEASE KNOT

This simple, easy-to-tie knot is used most frequently by horse people. The main drawback to this knot is that it continues to tighten when the long end of the rope is pulled on, which can sometimes make it difficult to release.

Purpose

- To tie a horse or llama to a hitching post, rail or hook
- Allows the lead rope to be untied quickly and easily by simply pulling on the free end
- To secure any restraint rope that may need to be released quickly

Procedure for Tying a Quick-release Knot

TECHNICAL ACTION	RATIONALE/AMPLIFICATION
1. Pass the rope over the hook or rail from left to right.	1a. Make the short end at least 45-60 cm (18-24 inches) long.
2. Make a loop in the short end and pass it over the long end (Fig. 1–2A).	2a. The loop should be about 10-15 cm (4-6 inches) in diameter.
3. Pass the short end behind the long end and the loop and push a bight (a fold of rope) into the loop (Fig. 1–2B).	3a. Make the bight at least 25 cm (10 inches) long.
4. Pull the bight to tighten the knot.	4a. Animals learn quickly to untie this knot by pulling the short end, so you can pass the short end through the bight.

FIG. 1-2 (A) Loop in the short end held over the long end or the rope attached to the animal's halter. (B) Placing the bight within the loop.

BOWLINE

The bowline is probably the first knot learned by those who work around boats and water, because no matter how much force is pulled against it, it can still be untied. It's named such because this knot has been used for centuries to secure the tug line to the bow of a boat.

Purpose

- Used to make a fixed-diameter loop
- Easily released even when pulled extremely tight, so used for securing an animal to a post or dragging a dead animal

Procedure for Tying a Bowline

TECHNICAL ACTION	RATIONALE/AMPLIFICATION
1. Make a loop in the long end of the rope such that the short end of the rope overlaps the long end (Fig. 1-3A).	—
2. Pass the short end of the rope up through the loop.	2a. This is the rabbit coming out of the hole.
3. Reach under the long end of the rope and grasp the short end such that it wraps around the long end (Fig. 1-3B).	3a. This is the rabbit running around the tree.

CHAPTER 1 Ropes and Knots

TECHNICAL ACTION	RATIONALE/AMPLIFICATION
4. Pass the short end of the rope back through the loop in the opposite direction of the first pass (Fig. 1–3C).	4a. This is the rabbit running back down into the hole.
5. Tighten the knot by pulling on both the long and short ends.	—

FIG. 1–3 (A) Note carefully how this loop was formed. It is important that the short end cross over the long end. (B) Wrapping the short end of rope around the long end. (C) Pass the short end through the loop and pull in the direction indicated by the arrow.

TOMFOOL KNOT

Named for the way the knot disappears if the free ends are pulled, the Tomfool knot is handy to know when you need to secure or hobble the feet of an animal. Tied correctly, the knot will hold fast without cutting off the circulation, even if the animal struggles against it.

Purpose

- Used to tie two legs together

Procedure for Tying a Tomfool Knot

TECHNICAL ACTION	RATIONALE/AMPLIFICATION
1. Grasp the center of the rope in both hands.	1a. The right hand should be positioned thumb up and the left hand thumb down.

TECHNICAL ACTION	RATIONALE/AMPLIFICATION
2. Rotate both hands counterclockwise to form two loops.	2a. Now both thumbs are in the middle facing each other (Fig. 1–4A).
3. Bring the loops together so that the right loop overlaps the left loop.	3a. Overlap by half so that one side of the right loop is in the middle of the left loop.
4. Pull the side of the right loop through the left loop and the near side of the left loop through and over the right loop.	4a. Fig. 1–4B.
5. Pull both hands apart to create two loops knotted together in the middle. See Fig. 1–4C.	5a. If you pull on the ends of the rope when the loops are not over something, the whole thing will come undone.
6. Place each loop over a limb and pull each end very tight to snug the loops around the legs.	6a. Finish off with a simple overhand or slip knot to keep the animal from wiggling loose.

FIG. 1–4 (A) Note how hands are positioned in relation to each other and to the rope. (B) Passing the two loops through each other. (C) Pull in direction indicated by arrows.

DOUBLE HALF HITCH

Also known as a clove hitch. People who work regularly around livestock use this hitch all the time; often tying it with little or no thought.

Purpose

- To secure a rope to a post, rail, or hook when it does not need to be released quickly
- May be used to tie a horse, cow, or llama to an object, although this is not recommended for safety reasons

Procedure for Tying a Double Half Hitch

TECHNICAL ACTION	RATIONALE/AMPLIFICATION
1. Pass the rope around the post.	1a. May be a rail, hook, or whatever you wish to tie to.
2. Pass the short end under the long end and then back over the top.	2a. In effect, you are creating a closed loop around the post.
3. Continue down between the post and the loop you just formed.	3a. Fig. 1–5A.
4. Pull it tight.	—
5. Pass the short end over and under the long end, forming a loop.	—
6. Pass the short end up through the loop and pull it tight.	6a. Fig. 1–5B.

FIG. 1–5 (A) Pull in direction shown by arrow. (B) Second half of the double half hitch.

TAIL TIE

Ropes often are tied to a horse's tail to help us maneuver the animal more effectively. Horses have very strong tails that can support their body weight. Cattle on the other hand, have very weak tails that may be broken or even pulled off if tied by the tail.

Purpose

- Used to lift or move the back end of a recumbent or ataxic horse
- Used in horses to link one horse to another, head to tail (as in a pack string)

- Used in horses to tie the tail out of the way
- May be used in cattle *only* to hold the tail out of the way, *never* to lift or move the animal

Procedure for Placing a Tail Tie

TECHNICAL ACTION	RATIONALE/AMPLIFICATION
1. Lay a rope over the tail at the tip of the tail bone.	1a. Make the short end about 18 inches long.
2. Fold all the tail hairs up over the rope.	2a. This can be difficult in a horse with a very short, thin tail.
3. Pass the short end of the rope behind the tail and make a fold or bight in it.	—
4. Pass the fold or bight over the folded tail and under the rope which is looped around the tail.	4a. Fig. 1–6A.
5. Pull tight.	5a. Fig. 1–6B.

 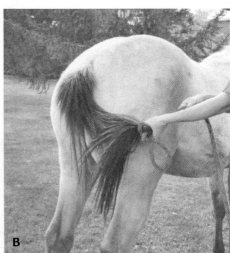

FIG. 1-6 (A) Passing the bight through the tail loop. (B) Finished tail tie.

BRAIDING AN EYE SPLICE

Placing an eye splice at the end of a rope makes it useful for many situations requiring a slip loop. Once having accomplished the braiding technique, you will be able to splice ropes together or finish off the end of the rope by braiding back.

Purpose

- To create a permanent loop in the end of a rope that can withstand a great deal of force
- Basis for creating a hondo in a lariat

Procedure for Braiding an Eye Splice

TECHNICAL ACTION	RATIONALE/AMPLIFICATION
1. Unravel 8-10 inches of the rope.	—
2. Make a bight or loop such that the unraveled ends are at right angles to the axis of the still-braided strands of the rope.	2a. Make the bight (eye) whatever size suits your needs (Fig. 1–7A).
3. Lift one strand of the intact rope, and pass the center strand of the loose ends beneath it. This will be strand 1.	3a. Fig. 1–7B.
4. Raise the strand in the side of the rope next to where strand 1 exits.	—
5. Pass loose strand 2 beneath this strand, so that strand 2 enters the rope where strand 1 exits.	5a. Fig. 1–7C.
6. Turn the ropes over and pass loose strand 3 under the strand where strand 2 exits, and it exits the rope where strand 1 entered.	—
7. Continue passing the loose strands over each other and through the rope until their ends are reached or cut.	7a. Fig. 1–7 D.

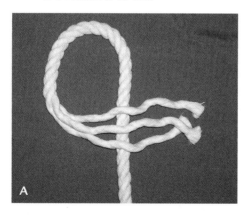

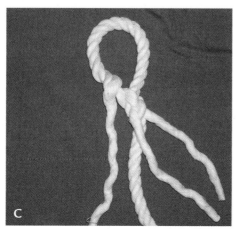
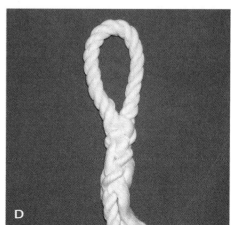

FIG. 1-7 (A) First step in creating an eye splice. (B) Passing the first strand of rope through the rope. (C) Passing the second strand of rope through the rope. (D) Finished product. Ends may be trimmed to smooth the appearance of the rope.

ROPE HALTERS

In large animal medicine, it is common to make your own halters, especially for cattle. Building your own halter not only saves you money but makes it possible to have on hand halters that are a variety of weights and sizes. Horse halters are more complicated to build and, in this author's experience, not suitable for most of the restraint procedures. Knowing how to build a temporary rope halter, however, is extremely useful when you need to restrain a horse that has been injured and has no halter immediately available.

Purpose

- Restraint or control of the heads of sheep, goats, or cattle
- Leading cattle, sheep, goats, and horses
- Tying cattle, sheep, or goats to a fixed object

Equipment

- 12–14 feet of three-strand cotton or nylon rope $\frac{1}{2}$-inch thick for adult cattle, $\frac{3}{8}$-inch thick for sheep, goats, and small calves
- 10–14 feet of $\frac{1}{2}$-inch cotton or nylon rope for a horse
- Electrical tape, hog rings, or cord to secure the end of the rope

Procedure for Building a Rope Halter for Cattle or Sheep

TECHNICAL ACTION	RATIONALE/AMPLIFICATION
1. Finish one end of a 12-foot to 14-foot three-strand cotton or nylon rope.	1a. Wrap end tightly with electrician's tape or whip the end (refer to Procedure for Finishing or Securing the End of a Rope).
2. Make the nose piece for a cow.	2a. Measure 18 inches from the finished end of the rope for a large breed of cow (12-14 inches for a smaller breed or a calf).
	2b. At this 18-inch point, separate the strands enough to insert the long end (Fig. 1–8A).
	2c. Have two strands on top and one underneath.
	2d. Insert the long (unfinished) end so that the short (finished) end points down and the long end points up.
3. To make the nosepiece for a sheep: Complete the directions above for this step.	3a. Measure 6-8 inches using $\frac{3}{8}$-inch rope.

CHAPTER 1 Ropes and Knots **13**

TECHNICAL ACTION	RATIONALE/AMPLIFICATION
4. About 4 inches from the loop, separate the strands of the long end. Place the short, finished end through the long end.	4a. Two strands on top and one underneath (Fig. 1–8B).
5. Pull the short end all the way through, so that a single loop is formed with four strands on top and two strands underneath where the ends intertwine.	5a. Fig. 1–8C.
6. To form the crown, follow steps 7 through 10.	6a. The piece that goes behind the ears of the cow.
7. Take the short end of the rope in your left hand. Place your right hand about 3 inches from your left hand.	—
8. Turn your left hand toward you and your right hand away from you so that the rope twists open.	—
9. Continue to twist the rope until three small loops are formed by the strands.	9a. Fig. 1–8D.

FIG. 1–8 (A) First step in creating the loop for the nose piece. (B) Passing the short end of rope through the strands separated on the long end. (C) Pull the short end tightly to secure the loop. (D) These loops were created simply by twisting the rope in a direction opposite to its existing twist.

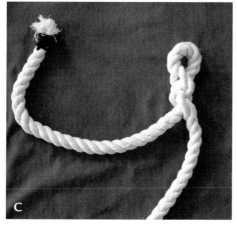
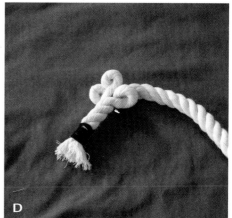

Continues

TECHNICAL ACTION	RATIONALE/AMPLIFICATION
10. Place the long end of the rope through these three loops.	**10a.** Pull all but about 2 feet of the long end of the rope through the loops.
	10b. For sheep, the crown needs to be only about 12 inches long.
11. Now place the long end through the loop completed in steps 1 through 4.	**11a.** This will go under the jaw of the sheep or cow and tightens when you pull on the lead.
12. Adjust the halter so it looks like the one in Fig. 1–8E and finish the lead-rope end as described in Procedure for Finishing or Securing the End of a Rope.	—

FIG. 1–8 Continued (E) Finished product. The crown length is adjustable to fit a variety of head sizes.

Procedure for Building a Temporary Rope Halter for a Horse

TECHNICAL ACTION	RATIONALE/AMPLIFICATION
1. Loop the rope around the horse's neck.	**1a.** Rope should be loose enough to fit your hand perpendicularly with ease between the rope and the horse's throat latch.
2. Tie a bowline to secure the loop.	**2a.** Make sure it is not a slip knot (Fig. 1–9A).
3. Fold the long end of the rope up through the neck loop.	**3a.** Fig. 1–9B.
4. Pass the bight or loop over the bridge of the horse's nose.	**4a.** Fig. 1–9C.
5. Secure the second loop by tying a second knot at the throat latch.	**5a.** For a temporary halter, just tie a simple overhand knot.

FIG. 1-9 (A) Bowline secured around horse's neck. (B) Passing a bight or fold of rope through the neck loop. (C) Finished product.

REVIEW QUESTIONS

1. What are the advantages and disadvantages of synthetic rope?
2. How does a lariat differ from other restraint ropes?
3. Why is it important to check ropes frequently?
4. What is a quick-release knot most often used for?
5. How can a quick-release knot be modified so that the animal doesn't release itself?
6. What are the advantages of using a bowline?
7. Why does a tail tie have limited use in cattle?
8. What is a Tomfool knot and what is it used for?
9. Name two ways to finish off the end of a rope.
10. What else can the braiding back technique be used for?

BIBLIOGRAPHY

Heersche, G. (1982). *How to make a rope halter.* Lexington, KY University of Kentucky, College of Agriculture Cooperative Extension Service, 4 AA-0-200.

Leahy, J., & Barrow, P. (1953). *Restraint of animals* (2nd ed). Ithaca, NY: Cornell Campus Store.

CHAPTER 2

Restraint Tools and Techniques

A dog looks up to you, a cat looks down upon you, but a pig will look you right in the eye.
WINSTON CHURCHILL

KEY TERMS

casting
chute
halter
hot shot

hyperthermia
nose ring
pig board
snout snare

stock
sweep tub
tailing-up
twitch

OBJECTIVES

- Identify behavioral characteristics unique to each species.
- Describe restraint techniques used on large animals.
- Compare and contrast restraint procedures used in each species.
- Identify various facilities, tools, and equipment used in large animal restraint.

COMPLICATIONS OF RESTRAINT

Defined as forcible confinement, restraint is required for proper transportation, examination, and treatment of any animal species. The degree of restraint required reflects the species, the animal's familiarity with handling, anticipated invasiveness, and the duration of the procedure. It is the handler's responsibility to use techniques that facilitate the success and safety of all humans and animals involved in a procedure. Unfortunately, despite all attempts to minimize complications, restraint can adversely affect some animals. Undesirable effects that can be associated with restraint include:

- Trauma including contusions, bruising, lacerations, or nerve paralysis
- Metabolic disturbances such as acidosis, hypoxia, hypocalcemia, hyperglycemia, hypoglycemia
- Hyperthermia
- Regurgitation
- Emotional stress

RESTRAINT OF THE HORSE

Horses are the large animal most accustomed to handling. Whether it involves haltering, grooming, or transportation, frequent handling is a component of all equine husbandry protocols.

GUIDELINES FOR RESTRAINT OF THE HORSE

The ability to work cooperatively with horses takes many years to develop. The following guidelines, however, will help the inexperienced handler avoid many problems.

- Horses are large herd animals. Their general approach is to run first and ask questions later. Thus, most accidents involving equines are the result of nervous, not aggressive, behavior on the horse's part. Having said that, horses that choose to do so can inflict extensive damage via biting, kicking (horses kick caudally), and striking (using the front limbs).
- Always let horses know that you are in the area. Horses do not appreciate surprises. Speak quietly; avoid loud noises and sudden movement.
- When standing near a horse, place your hand gently on the animal. This will alert you to any pending movement on the horse's part.
- Never go under a horse.
- Never stand directly behind in the blind spot of a horse.
- If you must walk behind the horse, walk very close to the horse with your body touching the animal. Alternatively you can walk behind at a distance of 15 feet from the hind end.
- Do not permit yourself to be pinned or squeezed between the horse and a solid object. For example, do not stand between the horse and the wall. Change positions such that the horse is between you and the wall.

- Horses occasionally will cow kick (kick cranially with a hind limb) or strike using a forelimb.
- Most handler accidents occur when the horse is placed in a confined area (stock, trailer) or is within physical contact distance of another horse. Pay close attention in these situations.

RULES OF TYING

To facilitate safety of both horses and handlers, the following guidelines should be adhered to when tying horses.

- Always use a quick-release knot.
- Secure to solid post or hitching rack. Never tie to gates, fence rails, or any object that a frightened animal could pull apart.
- Lead ropes should be tied short and high enough to prevent animal from lifting a leg over the rope. This is usually at the level of the animal's head.
- Ensure a minimum of 12 feet between animals.

STOCK

Purpose

- Inhibit horse's movement through bodily confinement

Complications

- Injury to horse
- Bodily injury to personnel

Equipment

- Stock
- Halter and lead rope

Procedure for Using Stock

TECHNICAL ACTION	RATIONALE/AMPLIFICATION
1. Place halter and lead rope on horse.	—
2. Open front and back gates of stock.	2a. Fig. 2–1.
3. Walk horse into stock while you remain outside just to the left of the stock.	3a. If the horse refuses to enter, the handler can precede the horse into the stock. Be certain that the front of the stock is open before attempting this.
	3b. Do not look at the horse as you approach the stock. Keep looking forward.

TECHNICAL ACTION	RATIONALE/AMPLIFICATION
	3c. Many horses will halt 3 feet from the stock. Allow them to quietly assess the situation for 2–3 minutes, then cluck to encourage forward movement.
4. Close front gate, close hind gate. Tie the horse at this time if warranted.	**4a.** Horses should never be placed in cattle chutes.
	4b. Never leave a horse unattended in a stock.

FIG. 2–1   Equine stock.

HALTERING AND LEADING

Purpose

- Provide fundamental restraint for horse

Complications

- Injury to handler if stepped on by the horse

Equipment

- Halter
- Lead rope

Procedure for Haltering the Horse

TECHNICAL ACTION	RATIONALE/AMPLIFICATION
1. Approach horse slowly from the left side in the area of the shoulder.	1a. The shoulder is the safest area of the horse.
	1b. If the horse is not aware of your presence, make a soft sound that will alert him.
	1c. Never approach a horse directly from behind.
	1d. Most horses are accustomed to being handled from the left side.
2. Place lead rope around neck.	2a. The lead rope should be as close to the head as possible, not low on the neck.
	2b. The lead rope should be attached to the ventral D-ring of the halter.
3. Place nose band over horse's nose and buckle strap behind ears.	3a. Some halters have a ventral clasp that is buckled instead of the crown piece, which goes behind the ears.
	3b. Try to minimize contact with the ears, because many horses are sensitive about their ears being touched.
	3c. Fig. 2–2.

FIG. 2–2   Haltering the horse.

Procedure for Leading the Horse

TECHNICAL ACTION	RATIONALE/AMPLIFICATION
1. The horse should be haltered with the lead rope attached.	1a. Fig. 2–3.
2. Stand on the left side of the horse.	2a. The left side is also referred to as the near side; the right side is called the off side.
3. Hold lead rope in right hand 12 inches from the halter. Coil the excess rope and place in left hand.	3a. Never wrap any portion of a lead rope around yourself. 3b. If needed, the handler can always grasp the cheek piece or chin strap of the halter.
4. Walk forward at a brisk pace.	4a. Do not look at the horse. He knows that he should be following you.
5. Should the horse become unruly, circle to the left. This will allow you to act as a pivot while the horse circles around you.	—

FIG. 2–3 Leading the horse.

APPLYING CHAINS

Purpose

- Increase the amount of restraint through increased pressure

Complications

- Head tossing
- Tissue trauma (buccal or oral)

Equipment

- Halter
- Lead rope with shank chain

Procedure for Applying Chains

TECHNICAL ACTION	RATIONALE/AMPLIFICATION
1. Halter the horse.	1a. Chains can be used over the nose, under the chin, in the mouth, or under the lip, if necessary. All of the methods increase the amount of restraint, with varying degrees of discomfort.
	1b. Horses should never be tied using a chain lead rope, because it could result in severe trauma if the horse were to pull back.
2. **Chain over nose.** Disconnect chain from ventral halter D-ring and pass chain through left side ring, over nose, and attach to right side ring.	2a. If the chain is long enough, it can be passed through the right side D-ring and clipped to the ventral ring, or it can be passed through the right side-D ring and clasped to right cheek D-ring.
	2b. Looping the chain one time around the halter's nose band will help prevent the chain from slipping down the nose (Fig. 2–4).
3. **Chain under chin:** Pass chain through left side D-ring under chin and attach to right side D-ring.	3a. Fig. 2–5.
4. **Chain in mouth:**	
a. Pass chain through left side D-ring under chin and attach to right side D-ring. Loosen chain such that there is enough slack to reach horse's lips.	4a. The chain should rest comfortably, just touching the commissure on the lips. If the chain is too tight or too low in the mouth, it will cause unnecessary discomfort.

TECHNICAL ACTION	RATIONALE/AMPLIFICATION
b. Place right arm ventral to head and grasp head in the area of nose band with right hand.	**4b.** Placing a chain in the mouth is rarely done in veterinary practice. Care should be taken to prevent harm to the delicate tissues of the lips.
c. Insert left thumb in commissure of the lips. As horse opens mouth, slide chain in (as if placing a bit in mouth).	**4c.** Fig. 2–6.
d. Apply slight pressure to chain to remove slack.	
5. Chain under lip:	**5a.** This can be extremely painful when pressure is applied. Do not maintain pressure when the horse is cooperative. The chain should rest snugly, without causing any discomfort.
a. See steps 4a and 4b above.	
b. Lift upper lip and slide chain over gum under the lip.	
c. Apply slight pressure to remove slack.	

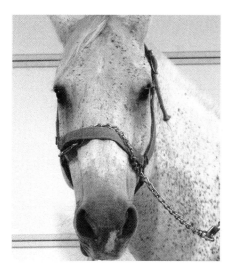

FIG. 2–4 Chain over the nose.

FIG. 2–5 Chain under the chin.

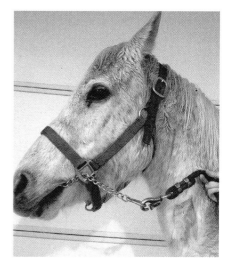

FIG. 2–6 Chain in the mouth.

TWITCHES

Purpose

- Application of pressure with the intent to distract attention or induce endorphin release

Complications

- Personnel injury
- Trauma to lip

Equipment

- Halter
- Lead rope
- Twitch

Procedure for Hand Twitching

TECHNICAL ACTION	RATIONALE/AMPLIFICATION
1. Apply halter and lead rope.	1a. Never attempt to apply a twitch without haltering the horse.
	1b. Twitches provide minor pain for the horse, thereby creating a distraction during a clinical procedure.
	1c. If applied to the nose, twitches are thought to cause endorphin release, thereby suppressing pain in the horse.
2. Turn head toward you and grasp the loose skin in the neck area just cranial to the shoulder. Grasp a large amount of skin, twist slightly, and hold firmly.	2a. Hand twitching is performed most commonly in the neck area. A twitch should never be applied to a horse's ear. Holding onto the nose manually is difficult and better accomplished using a mechanical twitch.
	2b. Turning the head toward you loosens the skin, making it easier to grasp.
	2c. Fig. 2–7.

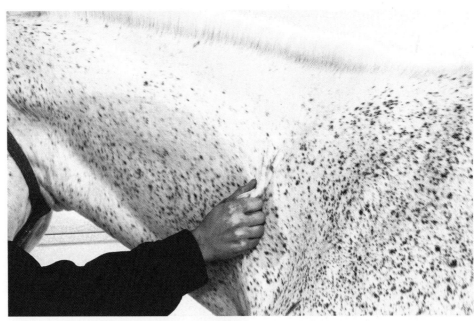

FIG. 2–7 Applying a hand twitch to the neck.

CHAPTER 2 Restraint Tools and Techniques 25

Procedure for Applying Mechanical Twitches

TECHNICAL ACTION	RATIONALE/AMPLIFICATION
1. Apply halter and lead rope.	1a. Never attempt to apply a twitch without haltering the horse.
	1b. Twitches provide minor pain for the horse, thereby creating a distraction during a clinical procedure.
	1c. If applied to the nose, twitches are thought to cause endorphin release, thereby suppressing pain in the horse.
2. Select twitch type.	2a. Mechanical twitches include the Kendal humane twitch, chain twitch, or rope twitch.
	2b. The Kendal humane twitch is hinged, which prevents over-tightening. It can also be made self-retaining.
3. Place hand through twitch and firmly grasp upper lip.	3a. The horse will most likely resist by raising or flipping its head up and down.
	3b. Grasp as much of the upper lip as possible.
	3c. Fig. 2–8.

FIG. 2–8 Applying a mechanical twitch.

TECHNICAL ACTION	RATIONALE/AMPLIFICATION
4. Secure rope and chain twitches by rapidly twisting wooden handle.	
a. The humane twitch is secured by pressing arms together (like a nutcracker), then winding the attached string around bottom of arms and clipping to side or ventral D-ring of halter.	4a. To keep twitch in place, the handler must not pull it downward. Think of pushing the twitch into the nose.
	4b. The most common reason for twitches coming off at inappropriate times is that the handler pulls down.
	4c. If a self-retaining twitch is not needed, the humane twitch does not need to be clipped to the halter.
5. To remove twitch, untwist and then rub horse's nose with the palm of your hand to stimulate circulation.	5a. Twitches should not remain in place more than 20 minutes without loosening temporarily to facilitate circulation.

LOADING HORSES IN TRAILERS

Purpose

- Place horse safely in trailer for transport

Complications

- Injury to personnel (crushing, rope burn)
- Trauma to horse (head or legs most commonly)

Equipment

- Halter
- Lead rope
- Protective equipment such as wraps, bell boots, or head bumpers

Procedure for Loading Horses in Trailers

TECHNICAL ACTION	RATIONALE/AMPLIFICATION
1. Familiarize yourself with the trailer, noting specifically where the horse is tied, method of closing door, and presence of any emergency exits.	1a. Many types and models of trailers are available. These can include stock trailers, designed to carry both horses and cattle, 2-horse and 4-horse regular-load trailers, and 2-horse to 4-horse slant-load trailers.

TECHNICAL ACTION	RATIONALE/AMPLIFICATION
2. Apply any protective equipment owner elects to use.	2a. Protective equipment helps prevent injury to the horse during transit.
	2b. Equipment commonly includes leg wraps, bell boots, head bumpers, tail wraps, and blankets.
3. Lead horse onto trailer.	3a. Do not look at horse when loading. Look toward where you want to go.
	3b. Fig. 2–9.
	3c. If the horse initially refuses to enter, allow it to look at the trailer for 1–2 minutes, then begin clucking to encourage forward motion.
4. Secure butt rope, internal gate, or boom arms, if present. Tie horse using quick-release knot.	4a. Always use a quick-release knot. This knot can be released with a quick pull on the rope end.
	4b. Chapter 1.
5. Secure all exterior doors.	—

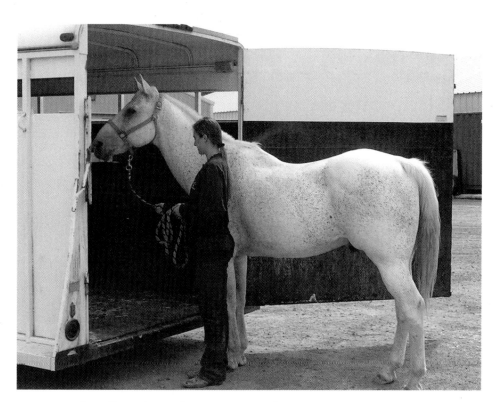

FIG. 2–9 Loading a horse in a step-up trailer.

SPECIAL HANDLING SCENARIOS

Foals

Foals are very delicate and often lack the handling experience of older animals. The handler should take care to assess the individual foal's temperament and degree of training. Make every attempt to keep the mare and foal within sight of each other. In general, the closer the mare is to the foal the better the result of the procedure.

Leading Foals

Technique is dependent upon the amount of prior training.

- If the foal is not halter broken, the best method is to cradle the foal in your arms and push it up against the mare's side.
- If the foal has some experience with the halter, a rump rope can be used. Attach lead rope to the ventral halter D-ring. Use this lead rope as the standard lead rope, then place a second lead rope around foal's rump and back through the halter. Apply more pressure to the butt rope than to the halter lead rope (approximately 70-to-30 percent of pressure) (Fig. 2–10).
- If the foal is halter broken, use the techniques described for adult horses.

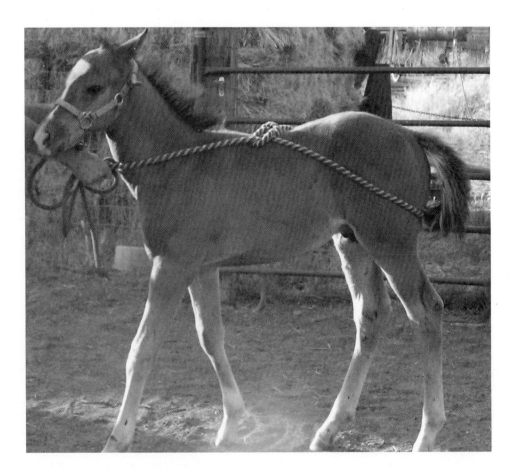

FIG. 2–10 Application of a rump rope to a foal.

Procedure for Cradling Foals

TECHNICAL ACTION	RATIONALE/AMPLIFICATION
1. Place one arm around foal's chest and the other around the rump. Cradle foal in arms.	1a. Some foals will slump in the handler's arms. If this happens, loosen hold and foal will begin bearing its own weight.
	1b. Fig. 2–11.
2. If a foal is accepting the procedure being performed, shift so that foal is against the side of the mare or wall.	2a. Mare should also be very calm.
3. With exuberant foals, the handler can grasp the base of the tail.	3a. Do not pull tail upward or foal will attempt to sit down.

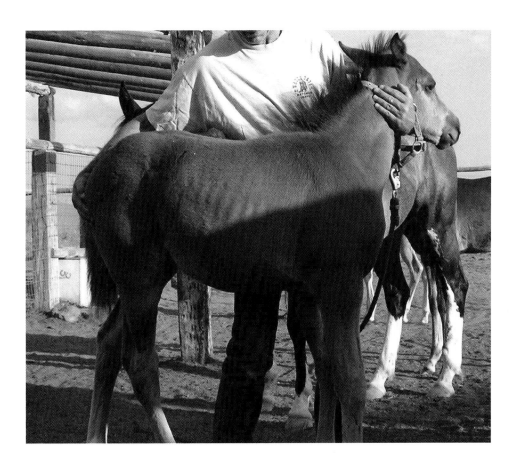

FIG. 2-11 Cradling a foal.

Stallions

Although most stallions are well behaved, they are much more exuberant and volatile than geldings and mares. In general they are more likely to strike, bite, or rear. Knowing this, special care should be taken to maintain adequate physical distance between stallions and other horses. A minimum of 20 feet should be maintained between stallions and other animals. This distance is especially important if the stallion must be in the presence of mares in heat (estrus). Novice horseman would be advised to avoid handling stallions until they are very competent handling mares and geldings.

RESTRAINT OF CATTLE

Cattle exhibit marked variance in their response to handling. Dairy cows or 4-H animals accustomed to intimate daily contact tolerate restraint more readily than range cattle. The selection of restraining facilities and equipment directly reflects this variance. Wise handlers will always assess breed, sex, and production use prior to unloading any cattle.

GUIDELINES FOR RESTRAINT OF CATTLE

- Cattle are not typically haltered or led. They are pushed or driven via arm waving and similar measures. Handlers must remember to avoid standing where they want the cattle to go. For example, do not stand in front of the **chute** if you want the cow to exit the chute through the front or head gate. Care should be taken not to overly agitate the cattle when driving them.
- Cattle are very herd oriented. Hence, they are much easier to manipulate as a group than as individuals.
- The fight-flight distance for cattle is approximately 15–20 feet, although this can vary dramatically from breed to breed.
- Always inspect facilities prior to processing. Familiarize yourself with the chute, and ensure that fences are sturdy and gates are locked. Loose cattle are extremely difficult to catch.
- Cattle breeds vary dramatically with regard to temperament. In general, dairy breeds are much more docile than beef breeds. (This statement is true for cows, not bulls.)
- Caution should always be exercised when working with bulls. This is especially relevant when working with dairy bulls.
- Avoid the use of dogs unless the dog is extremely well trained.
- Although most cattle are not overtly aggressive, the following actions can cause damage to personnel.
  1. Butting: Cattle will swing their heads and use them as battering rams. This is especially dangerous in horned animals.
  2. Kicking: The cow kick is usually forward and to the side. Restrained cattle are less likely to kick caudally, although this can happen. Cattle almost never kick with both hind limbs simultaneously.
  3. Trampling: Cattle will often run directly over individuals who are blocking their path of escape.
  4. Biting: Cows rarely, if ever, bite.

Dramatic advances have been made in recent years regarding humane handling and processing of cattle. Many individuals have contributed to the welfare of cattle and revolutionized our understanding of bovine restraint. Review of such material is advised.

FIG. 2-12 (A) Cow in a squeeze chute. (B) Alley way. (C) Sweep tub. (D) Chute with palpation cage.

PROCESSING FACILITIES

Cattle typically are processed using facilities that have the following equipment (Fig. 2-12).

- Chute: This piece of equipment provides immobilization via a head catch and squeezable sides. The side panels can be dropped to examine feet and legs, while the side bars can be dropped to examine the dorsum of the animal. Head plates with halters can be attached to the head catch to facilitate dehorning. Many models and brands are available. Technicians and veterinarians should familiarize themselves with individual chute operation prior to use.

- Stock: A stock is similar to a chute but does not possess squeezable sides. It is not meant for processing large numbers of cattle and should never be used with fractious animals.

- Palpation cage: These gates are placed directly behind the chute. They facilitate entrance into the tailgate of the chute by prohibiting cattle in the alley way from approaching the chute.

- Alley way: The alley way is the narrow passage area that prevents cattle from turning around as they approach the chute. Poles or back gates can be used to prevent cattle from backing up in the alley way as they approach the chute.

- Sweep tub: Sweep tubs permit a small group of cattle to be squeezed together to facilitate passage in the alley way. Minimizing excess room decreases the opportunity for cattle to turn around and refuse to enter the alley way.

OPERATING CHUTES

Purpose

- Provide the most effective method of restraint for cattle

Complications

- Injury to cattle
- Injury to personnel

Equipment

- Chute

Procedure for Operating Chutes

TECHNICAL ACTION	RATIONALE/AMPLIFICATION
1. Ready chute for operation by opening head catch, releasing squeeze, and opening tailgate.	1a. Familiarize yourself with the individual chute operation before running cattle. There are many models available, each with slight variations in operation.
2. Allow cow to enter.	—
3. Catch head.	3a. Experienced chute operators will often close the tailgate before catching the head. This prevents the cow from backing out of the chute before her head is caught. Slow operators will miss cattle using this technique.
4. Close tailgate.	—
5. Apply squeeze.	5a. Chutes can accommodate various sizes of animals by manually moving the side panels inward or outward. If the cow can't be squeezed sufficiently, check the position of the side panel. If needed, move the panel inward by adjusting the pins located on the ventral corners.
6. To release cow, release squeeze and then release head catch.	6a. If a dehorning plate is attached, the cow will need to be released from the side of the chute, instead of through the head catch.

HALTERING

Purpose

- Provide restraint of head
- Permit leading of cattle that have been halter broken

Complications

- Trauma to personnel

Equipment

- Cow halter

Procedure for Haltering Cattle

TECHNICAL ACTION	RATIONALE/AMPLIFICATION
1. Place cow in chute or stock.	1a. If cow has been broken to a halter, it can be applied in the stall or pen.
2. Place crown piece of halter over ears, then slip nose through nose piece. Adjust halter such that nose band crosses over bridge of nose halfway between the nostrils and eyes.	2a. Cattle halters typically are made of a single piece of rope that forms both the halter and lead rope. The halter is adjustable and should be fitted individually to each cow.
	2b. The adjustable portion of the nose band should always go under the chin, not across the bridge of the nose. The standing end or lead rope portion should be on the left side of the cow.
	2c. Fig. 2–13.

FIG. 2–13 Haltering a cow.

TAILING-UP CATTLE

Purpose
- Provide restraint for examination or minor surgical procedures

Complication
- Injury to coccygeal vertebrae
- Injury to personnel

Equipment
- None

Procedure for Tailing-up Cattle

TECHNICAL ACTION	RATIONALE/AMPLIFICATION
1. Place cow in chute or stock and catch head.	1a. Fig. 2–14.
2. Stand caudal to cow and use both hands to grasp the base of the tail, 5–6 inches from the anus. Push the tail straight over the dorsum. Do not permit tail to curl or to shift left to right.	2a. Movement or deviation of the tail dramatically decreases the effectiveness of tailing-up. A cow is very likely to kick if inappropriate technique is used.
	2b. This technique should not be confused with tail twisting, which involves bending the tail to make a cow move forward.
	2c. Grasping the tail base decreases the likelihood of damaging coccygeal vertebrae.
3. Maintain pressure until directed to release.	3a. This hold can result in arm fatigue.

FIG. 2–14 Tailing-up a cow.

CASTING CATTLE

Purpose

- Place cow in lateral or sternal recumbency

Complications

- Tissue trauma
- Cow resists recumbency

Equipment

- Forty feet of heavy cotton rope
- Halter

Procedure for Casting Cattle

TECHNICAL ACTION	RATIONALE/AMPLIFICATION
1. Place halter on cow.	—
2. Ensure that ground surface is suitable for recumbency.	—
3. Secure cow to fence post. Tie knot low to the ground.	3a. The cow must be secured to prevent it from backing up. 3b. Knot must be low to prevent cow from hanging in recumbency.
4. Locate midpoint of rope and place over dorsum of neck.	—
5. Run rope medial to forelimbs, crisscross over back, then run medial to hind limbs.	5a. Fig. 2–15. 5b. The Burley method of **casting** is superior to the half-hitch method in that it does not place pressure on the trachea, penis, or mammary veins. The rope ends can also be used to tie the hind limbs in a flexed position, thus eliminating the need for an assistant to hold the ropes. 5c. If the cow kicks, gently toss the rope between the hind limbs instead of passing it through with your hand.
6. Standing caudal to cow, pull on rope ends until cow drops into recumbency.	6a. This technique is often more successful if two people work together, with one holding each rope end.

FIG. 2-15 Rope placement for casting a cow.

FLANKING

Purpose

- Place calf or goat in lateral recumbency

Complications

- Tissue injury to animal
- Back pain in restraining personnel

Equipment

- None

Procedure for Flanking Calves

TECHNICAL ACTION	RATIONALE/AMPLIFICATION
1. Bring animal to area suitable for recumbency.	1a. Animals that weigh over 200 lbs should not be flanked.
2. Stand on left side of animal.	2a. Fig. 2–16.
3. Place left arm over neck and grasp ventral neck area. Reach over animal and use right hand to grasp right flank area.	—
4. Push right knee into animal's flank. As calf jumps forward, use this momentum to lift animal off its feet.	4a. It is critical to use the animal's own momentum, otherwise you will not be able to lift the calf.

TECHNICAL ACTION	RATIONALE/AMPLIFICATION
5. Bend knee to push the left side underneath the calf, and quickly press animal down onto its left side.	5a. Maintain position by placing your left knee over the animal's neck. 5b. Holding the cannon bone of the down forelimb will inhibit the calf from rising.

FIG. 2-16 Flanking a calf to lay it on the ground.

SECURING CATTLE FEET FOR EXAMINATION

Purpose

- Provide immobilization of foot to facilitate examination of hoof, application of bandage, hoof trimming, and other procedures

Complications

- Cow becomes recumbent in chute
- Trauma to personnel secondary to kicking

Equipment

- Cotton rope 15–30 feet
- Chute or stock

Procedure for Securing Feet for Examination

TECHNICAL ACTION	RATIONALE/AMPLIFICATION
1. Place cow in chute or stock.	1a. Tilt tables can be used to examine feet and are far superior to tying up the feet.
2. Remove side panel.	—
3. Tie square knot or clove hitch around cannon area of leg. Apply half hitch distal to square knot.	3a. Take care to avoid being kicked during this maneuver.
4. Slip running end of rope over beam and pull to lift leg off ground.	4a. At this point the leg is off the ground and can be examined visually, but the cow can still kick.
5. Apply a second half hitch over fetlock and pull limb tight against the corner post of the stock or chute.	5a. Fig. 2–17.
6. Apply third or fourth half hitches as needed to minimize motion of foot.	—

FIG. 2–17 Securing a hind foot for examination or treatment.

MISCELLANEOUS EQUIPMENT

HOT SHOT

Encouraging cattle to move through processing facilities often requires ancillary equipment such as the **hot shot**, prod, or paddle (Table 2–1, Fig. 2–18). This equipment offers varying degrees of encouragement to the livestock and should be used with care and compassion.

Purpose

- Equipment used to stimulate cattle to move through a processing area or stop movement toward a handler

Complications

- Pain and distress

TABLE 2–1 EQUIPMENT DESCRIPTION

EQUIPMENT NAME	COMPOSITION OR APPEARANCE	COMMENTS
Hot shot	Battery-powered electrical device	Device causes intense pain. Use with caution. Avoid use if possible.
Prod	Graphite rod	—
Paddle	Plastic and looks similar to a boat paddle	They are typically filled with beads to make a rattling sound.
Pole	Pole that is attached to the nose ring of a bull	A pole prevents the bull from moving toward a handler.

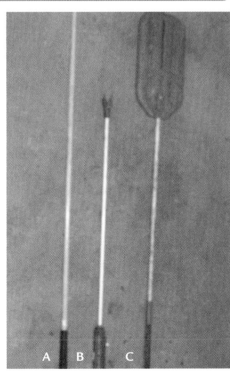

FIG. 2–18 Cow processing equipment including a prod (A), a hot shot (B), and a paddle (C).

Procedure for Use of a Hot Shot

TECHNICAL ACTION	RATIONALE/AMPLIFICATION
1. Cow must be in a contained alley way or chute area.	1a. Never apply a hot shot to an animal in an open pen or pasture.
2. Remove safety clip, if present.	2a. The safety clip prevents accidental discharge of the hot shot.
3. Confirm that battery is operational by depressing button. A distinct buzzing sound should be heard.	—
4. Touch cow with hot shot.	—
5. Before putting hotshot down, touch the end against a metal surface. This will discharge the residual electricity.	5a. Replace safety clip, if present. 5b. Hot shots should be used only as a last resort. They cause extreme pain. Anyone who uses a hot shot frequently should probably experience the procedure; it will help one to develop empathy.

NOSE TONGS

Purpose

- Restraint device used to control head

Complications

- Damage to nasal septum
- Personnel trauma
- Pain and discomfort to animal

Equipment

- Tongs
- Lead rope
- Halter

Procedure for Placing Nose Tongs

TECHNICAL ACTION	RATIONALE/AMPLIFICATION
1. Place cow in chute or stock and catch head.	—

TECHNICAL ACTION	RATIONALE/AMPLIFICATION
2. Apply halter if available.	2a. Many times a halter or head plate is not available, which is why the tongs are used.
3. If halter is not used, restrain by grasping head and pin between left arm and thigh.	3a. Restraint of the head is difficult. Cattle will often swing their heads about in an attempt to avoid tongs placement.
4. Place tongs using a rotating motion. Insert one side of the tongs, then rotate across the nostril and place other on opposite side.	4a. Use only tongs which have a space between the balls. Tongs which leave no room for the nasal septum are considered inhumane.
	4b. Before placement, always examine balls to ensure that surfaces are smooth.
5. Apply tension to maintain tongs in position. A lead rope can be attached to the end of most tongs.	5a. Never leave an animal unattended when nose tongs are in place.
	5b. Fig. 2–19.

FIG. 2–19 Application of nose tongs.

NOSE RINGS

Purpose

- A permanent restraint device used to control head of bulls

Complications

- Torn nasal septum

Equipment

- Self-piercing **nose ring** with screw and Allen wrench
- Cloth rag

Procedure for Placing a Nose Ring

TECHNICAL ACTION	RATIONALE/AMPLIFICATION
1. Place bull in chute.	1a. Proper restraint is critical for attaining correct placement.
2. Place halter or secure head in head plate.	2a. Head plates are much preferred over halters, because they immobilize the head better.
3. Select appropriate size ring.	3a. Rings typically are placed when bulls are 1–2 years of age. Do not use large rings on these small bulls, because they will enlarge the nasal septum hole inappropriately.
	3b. Ring sizes are small: 2.5 inches, medium: 3 inches, and large: 3.5 inches.
	3c. Rings that are too large or placed too close to the end of the septum will hang out and catch on objects. A torn nasal septum can be catastrophic to a bull's breeding career.
4. Apply topical anesthetic or inject 5.0 cc lidocaine into the septum. Injection should be made 1–2 inches from the end of septum.	4a. Nose ring placement is extremely painful for the animal. Postsurgical analgesics should be used.
5. Open ring and use sharpened end to pierce through nasal septum. Once through, use rag to wipe screw area of the ring free of blood.	—
6. Place screw in ring and tighten using Allen wrench.	6a. Be careful with the screw. They are easy to drop and lose.

RESTRAINT OF THE GOAT

Goats are unique in their dual role as both production and pet animals. Discerning individual animal status is best accomplished through consultation and observation of the animal with the owner. Goats fulfilling pet roles should be referred to by name. Interactions with this category of goat should parallel relationships to pets such as dogs. Alternatively, interactions with production goats will mirror interactions with cattle.

GUIDELINES FOR RESTRAINT OF THE GOAT

- Goats are gregarious, fun-loving animals that do not possess a strong herd instinct. They typically respond well to gentle handling and willingly accompany handlers away from the herd. Goats are not typically haltered but are led using a collar around the neck.
- Distraught or angry goats will vocalize and stamp their forefeet. Although they do not usually charge, extremely stressed goats may attempt to jump over a handler. This typically results in the feet of the goat being placed in the center of the handler's chest as the animal attempts to spring over.
- Horned goats may attempt to butt the handler. Goats do not bite or kick.
- Scent glands contribute to the unpleasant odor of intact males. This foul smell is especially noticeable during breeding season. Bucks will also urinate on their beards and forelimbs to increase their attractiveness to the does. Wise handlers do not stand unaware in front of bucks during the breeding season.

COLLARING AND LEADING GOATS

Purpose

- General restraint

Complications

- None

Equipment

- Collar
- Lead rope

Procedure for Collaring and Leading Goats

TECHNICAL ACTION	RATIONALE/AMPLIFICATION
1. Place collar on neck, leaving 2 inches of space between collar and neck.	1a. Most goats receiving veterinary care are considered companion animals. As such they typically

TECHNICAL ACTION	RATIONALE/AMPLIFICATION
	are handled in a manner consistent with that of other large animals of companion status (horse, llama). They are not handled in the same way as are production animals (cow, pig). This distinction is very important to goat owners.
2. Attach lead rope and lead animal from left side.	2a. Most goats are accustomed to handling and will follow readily.
	2b. If a lead rope is not available, it is acceptable to grasp collar directly.

STANCHION

Purpose

- Restrain for hoof trim, milking, artificial insemination or examination

Complications

- Goat jumps off stanchion

Equipment

- Stanchion
- Collar

Procedure for Placing Goat in Stanchion

TECHNICAL ACTION	RATIONALE/AMPLIFICATION
1. Lead goat to stanchion using collar.	1a. A stanchion is essentially a raised platform with a head catch. It is not suitable for fractious animals.
2. Encourage goat to jump onto platform by clucking and lifting up on collar. Secure head in head catch.	2a. Many stanchions have attached grain boxes. Placing a handful of grain in the box rewards the goat for good behavior.
	2b. Milk goats readily accept stanchions; however, production and pet goats may require encouragement (Fig. 2–20A).
	2c. If a stanchion is unavailable, the animal can be straddled or held by the collar (Figs. 2–20B and 2–20C).

 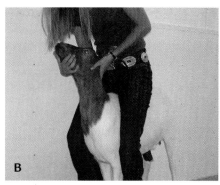 

FIG. 2-20A (A) Goat secured in stanchion. (B) Straddling the goat to facilitate jugular access. (C) Lateral restraint hold for jugular access.

RESTRAINT OF THE PIG

Intelligent and independent animals such as pigs do not appreciate restraint, and they never hesitate to voice their displeasure when subjected to this insult. Use of appropriate processing facilities can help to minimize this disruption; however, many pig owners do not have these facilities available.

GUIDELINES FOR RESTRAINT OF THE PIG

- Pigs are independent animals with little herd instinct. Although they do not seek comfort from a herd, pigs show great concern when fellow pigs are distressed. Thus, handlers should take care when working within a confined group.
- Most pigs are not aggressive. When stressed, however, pigs can inflict extensive damage using their teeth. This is especially true with boars, which have elongated canine teeth (tusks).
- Sows with litters are very protective. Care should be taken to confine the sow when handling the piglets.

PIG BOARDS

Purpose

- General restraint for examination or moving pigs from one area to another.

Complications

- Pig escapes under board
- Handler knocked over

Equipment

- **Pig board** (3-foot by 4-foot piece of plywood with hand holds)

Procedure for Using Pig Board

TECHNICAL ACTION	RATIONALE/AMPLIFICATION
1. Hold board parallel to pig.	1a. Keep board close to the ground.
2. Use board to push pig gently toward corner of enclosure.	—

CASTRATION RESTRAINT

Purpose

- Restrain pig for castration

Complications

- Handler bitten on leg or knocked to ground

Equipment

- None

Procedure for Holding Pig in Castration Position

TECHNICAL ACTION	RATIONALE/AMPLIFICATION
1. Grasp both hind limbs proximal to hocks.	1a. This hold is also called the pig handstand.
	1b. Use of suitable ear plugs for the protection of the handler and surgeon is advised.
2. Position pig such that handler is standing over pig, with one leg on either side. The pig's head should be between the handler's legs, facing the opposite direction.	2a. Fig. 2–21.
	2b. Do not twist the hind limbs, because this can dislocate the hips.
	2c. Small pigs can be lifted so that forefeet do not touch the ground.

FIG. 2–21 Pig castration restraint.

SNOUT SNARE

Purpose

- Used as a restraint device on larger pigs for sample collection or other clinical procedures

Complications

- Damage to entire snout as result of tourniquet effect
- Damage to nasal cartilage

Equipment

- Snout snare

Procedure for Applying a Snout Snare

TECHNICAL ACTION	RATIONALE/AMPLIFICATION
1. Approach pig from left or right side caudal to head.	1a. Fig. 2–22.
2. Place loop over snout, working caudally until loop is caudal to canines.	2a. Snares that are placed too far cranially will cause pain from pressure on nasal cartilage.
3. Tighten cable and apply pressure toward the pig.	3a. Pigs will pull backward against the snare and squeal loudly.
	3b. Do not use snares for longer than 20-minute intervals, because the tight cable will act as a tourniquet.

FIG. 2–22 Application of a snout (hog) snare.

TECHNICAL ACTION	RATIONALE/AMPLIFICATION
4. To remove snare, step toward pig, loosen cable, and remove from mouth.	3c. Most pigs will permit a snare application 2 or 3 times. After this they will use evasive maneuvers, making it impossible to apply the snare.

RESTRAINT OF THE LLAMA

A gradual rise in llama popularity has brought these animals out of the exotic designation into a routine large animal arena. Technicians working in mixed or large animal practices will most likely encounter llamas with some frequency. Unique in their own right, llamas are restrained using techniques common to both horses and cattle. For example, llamas are haltered routinely and led, yet they are also restrained with head catches. Ultimately familiarity with handling will be the most influential factor in determining the restraint techniques used on each animal.

GUIDELINES FOR RESTRAINT OF THE LLAMA

- Llamas are social and curious animals. They vary dramatically in their response to restraint, and this is most likely the result of prior handling experiences. Some llamas are extremely well trained, while others may never have been touched. Therefore it is advisable to talk with the owner to determine proper restraint protocol.
- Although most llamas are very docile, they can cause harm in a number of ways. These include biting (especially dangerous in intact males with canine teeth), spitting, regurgitation, and kicking. Most llamas will kick like a cow does, but occasionally they will kick directly behind with one foot.

HALTERING AND LEADING

Purpose

- Secure for examination or to move animal

Complications

- Unable to catch llama
- Spitting or biting

Equipment

- Halter and lead rope

Procedure for Haltering and Leading the Llama

TECHNICAL ACTION	RATIONALE/AMPLIFICATION
1. Approach llama quietly from the left side. Do not make direct eye contact.	1a. Llamas are accustomed to being handled from the left side.
	1b. Direct eye contact can be interpreted as confrontational behavior.
2. Place lead rope around neck.	—
3. Place nose band over nose and secure crown piece behind ears.	3a. Limit contact with ears if possible.
4. Lead llama from the left side.	4a. Optimally, the handler remains between the llama's head and shoulder while leading.
	4b. Fig. 2–23.

FIG. 2-23 Leading a llama.

STOCK

Purpose

- Inhibit llama movement through bodily confinement

Complications

- Lying down in the stock

Equipment

- Stock
- Halter and lead rope

Procedure for Placing Llama in a Stock

TECHNICAL ACTION	RATIONALE/AMPLIFICATION
1. Halter the llama.	—
2. Open chute or stock head catch, tailgate, and release squeeze.	2a. Llamas can be placed in chutes or stocks.
	2b. Stocks designed specifically for llamas are available. Most veterinary clinics, however, will use stocks designed for horses and cattle.
3. Walk llama into chute or stock.	3a. If the handler must walk the llama through a chute, two people are required. One will operate the chute and the other will handle the llama. Stocks can be managed by an individual.
4. Secure head catch.	4a. Fig. 2–24.
5. Close tail gate.	—
6. Apply squeeze if using chute.	6a. Do not apply a snug squeeze, because many llamas will lie down in response to squeezing.

FIG. 2-24 Llama stock.

REVIEW QUESTIONS

1. State three rules that should be followed to ensure safety when tying horses.
2. Identify the primary difference between a chute and a stock.
3. Describe the procedure for cradling a foal.
4. List the standard facilities used to process cattle.
5. Diagram the placement of the ropes used for casting cattle.
6. Compare and contrast nose rings and nose tongs.
7. Name the equipment used to restrain goats for milking.
8. Identify the animal species most likely to regurgitate when stressed.
9. State two complications associated with the use of a pig snout snare.

BIBLIOGRAPHY

Fowler, M. (1995). *Restraint and handling of wild and domestic animals* (2nd ed). Ames: Iowa State University Press.

Fowler, M. (1998). *Medicine and surgery of South American camelids* (2nd ed.). Ames: Iowa State University Press.

Mackenzie, S. (1998). *Equine safety.* Clifton Park, NY: Thomson Delmar Learning.

McCurnin, D. (1998). *Clinical textbook for veterinary technicians* (4th ed.). Philadelphia: W. B. Saunders.

Noordsy, J. (1989). *Food animal surgery* (2nd ed.). Lenexa, KS: Veterinary Medicine Publishing.

Pratt, P. (1998). *Principles and practice of veterinary technology.* St. Louis: Mosby.

CHAPTER 3

Grooming and Stall Maintenance

Why is it that cleaning stalls is far more enjoyable than cleaning houses?

J.R.

KEY TERMS

curry
fly mask
hay flake
hay net
legume
nosocomial infection

rug
sheet
stable blanket
stall stripping
sweet feed

OBJECTIVES

- Describe routine husbandry procedures required to maintain animal health.
- Identify purpose and complications associated with routine husbandry procedures.
- List the basic types of grains and hays.

Basic Grooming

Grooming is a fundamental component of animal husbandry that provides numerous health benefits. In addition to maintaining hoof and integument health, grooming encourages a thorough daily inspection of animals. During this time minor problems can be identified before they adversely affect the animal.

Purpose

- Remove dirt and debris from coats
- Ensure hooves are free of any foreign objects

Complications

- Personnel injury
- Inadvertent disease transmission

Equipment

- Rubber curry comb
- Metal curry comb
- Stiff brush (Dandy)
- Soft brush
- Clean rag or towel
- Mane and tail comb
- Hoof pick
- Hoof polish and moisturizer
- Fly spray with mitt
- Llama equipment: slicker brush, comb, towel

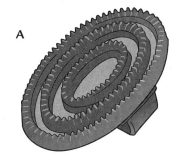
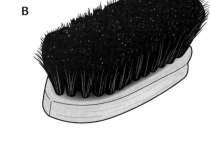
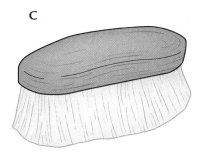
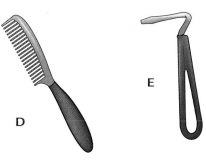

FIG. 3-1 Tools used in daily grooming of horses. (A) Rubber curry comb loosens and removes mud and deep dirt. (B) Stiff (Dandy) brush brings loose dirt to the surface. (C) Soft bristle brush removes surface dirt and shines the coat. (D) Comb untangles the mane and tail without pulling out hairs. (E) Hoof pick is used to clean debris from the bottom of the foot.

Procedure for Grooming the Horse

TECHNICAL ACTION	RATIONALE/AMPLIFICATION
1. Secure horse using halter or lead rope.	1a. Many horses are accustomed to being placed on cross-ties for grooming. Always confirm with owners that the horse is accustomed to the cross-ties.
2. Beginning at the neck of the left side, use the rubber curry comb in a circular motion to remove mud and debris. Curry both sides of the horse (Fig. 3–1).	2a. Horses are accustomed to being handled from the left side. Beginning the grooming process on this side will permit evaluation of the horse's temperament.
	2b. Curry combs should not be used over bony prominences on the legs distal to the knee or hock.
3. Use a stiff brush to remove deep dirt. Be certain to brush distal limbs. Do not use this brush on face.	3a. Use a flicking wrist motion when brushing. This helps to lift the dirt out of the coat.
	3b. Brush in the direction of hair growth.
	3c. Every fifth or sixth stroke, run the brush over the metal curry comb. This keeps the brush clean. Never use the metal curry comb on the horse.
4. Use the soft brush over entire body, including face, to remove surface dirt and bring out the shine in the coat.	4a. Every fifth or sixth stroke, run the soft brush over the metal curry comb. This keeps the brush clean. Never use the metal curry comb on the horse.
5. Wipe entire body using clean towel.	5a. Toweling brings out the shine in the coat.
6. Untangle and remove shavings from mane and tail using comb.	6a. Do not pull out hair. If mane or tail is very tangled, use your fingers instead of comb to prevent hair loss.
	6b. Always stand to the side of the horse when brushing tail.
7. Clean eye and nostril area using clean, damp towel.	—
8. Pick out the hooves. Begin with left forefoot. a. Facing the tail, run your left hand along the leg to the level of the fetlock and sub-	8a. Many techniques can be used to encourage a horse to lift its foot. This includes leaning against the horse with your shoulder to discourage the horse from placing weight on the hoof, pinching the

TECHNICAL ACTION	RATIONALE/AMPLIFICATION
sequently apply pressure to the fetlock. **b.** Once the horse has lifted foot, hold the hoof in left hand and use hoof pick to remove debris from ventral aspect of hoof. **c.** Place hoof on ground. **d.** Continue picking hooves in the following order: left hind, right hind, right forefoot. **e.** Apply moisturizer to external hoof wall. **9.** Apply fly repellent if warranted. **a.** Set bottle sprayer to mist. **b.** Begin at the level of neck and apply fine mist over entire body. **c.** Wet hand mitt or small towel and gently apply repellent to the face and ears.	chestnuts, or pulling on the feathers (hair on the caudal aspect of the fetlock). **8b.** The hoof pick should be grasped such that the tip, or point, of the pick points toward the caudal aspect of the horse. **8c.** Fig. 3–2. **9a.** Never use the spray bottle to apply repellent to the head. **9b.** Horses are often very sensitive about their ears. Apply with caution. **9c.** Some horses strongly object to fly sprayers. If needed, repellent can be applied to the entire body using a mitt.

FIG. 3–2 How to safely hold a back foot up to clean or examine it.

Procedure for Grooming the Cow

TECHNICAL ACTION	RATIONALE/AMPLIFICATION
1. Secure cow using stock, head catch, or lead rope.	1a. Cattle are rarely groomed. Show or 4-H animals have been trained to accept this procedure. Attempts to groom any other type of cattle will most likely result in severe injury to the handler and should be avoided.
	1b. The grooming procedure described here would be used in a clinical setting for hospitalized show animals. These techniques are not synonymous with the fitting for show or sale grooming techniques used by professional cattle handlers.
2. Beginning at the neck on the left side, use a stiff brush to remove mud and debris or shavings. Brush both sides of the cow.	2a. If the animal is contained in stocks, care should be taken to keep arms and hands from being pinned between animal and chute.
	2b. The amount of attention given to the legs depends on the temperament of the cow. Unless the cow is very gentle and accustomed to handling, do not groom legs.
	2c. Use a flicking wrist motion when brushing. This helps to lift the dirt out of the hair.
	2d. Brush in the direction of hair growth.
	2e. Every fifth or sixth stroke, run the brush over the metal curry comb. This keeps the brush clean. Never use the metal curry comb on the cow.
3. Use the soft brush over entire body, including face, to remove surface dirt and bring out shine in coat.	3a. Every fifth or sixth stroke, run the soft brush over the metal curry comb. This keeps the brush clean. Never use the metal curry comb on the cow.
4. Wipe entire body using clean towel.	4a. Toweling brings out the shine in the coat.
5. Remove shavings from tail end using fingers.	5a. Use caution to avoid being kicked when working with the tail.

TECHNICAL ACTION	RATIONALE/AMPLIFICATION
6. Wipe eye and nostril area using clean damp towel.	6a. To avoid a blow to the face, handlers should keep their faces back and avoid leaning over the cows' head.
7. Do not lift limbs or clean hooves as is done with horses.	—

8. Apply fly repellent if warranted
   a. Set bottle sprayer to mist.
   b. Begin at the level of the neck and apply fine mist over entire body.
   c. Wet hand mitt or small towel and gently apply repellent to the face and ears.
	8a. Never use the spray bottle to apply repellent to the face.

Procedure for Grooming the Llama

TECHNICAL ACTION	RATIONALE/AMPLIFICATION
1. Halter llama and secure to post.	—
2. Remove large debris or straw from wool using your hands.	2a. Shavings are extremely difficult to remove from wool and therefore should not be used for bedding.
3. Use a short slicker or bristle brush and begin brushing on lateral aspect of neck. Brush the entire trunk.	3a. Use firm brush strokes, but do not pull on wool. This is very painful to llamas.
4. For animals with long wool, part the wool and brush down away from part. Continue working in this manner until entire coat has been groomed.	—
5. Wipe face using damp cloth.	—
6. Llama toenails are not cleaned during routine grooming.	6b. Legs can be brushed using a soft bristle brush if needed. Llamas are often very sensitive about their legs being touched.

BLANKETS AND FLY MASKS

Blankets are used to protect horses from cold, inclement weather, ultraviolet light, or insects, with selection according to the blanket's purpose. Proper application of both blankets and **fly masks** ensures that animals receive the intended benefit of the apparel while remaining free of complications such as rubs and abrasions.

Purpose

- Provide additional warmth to critically ill animals
- Provide warmth to animals that have received a full body clip
- Protect hair coat and skin from soiled bedding
- Assist in cooling after exercise

Complications

- Abrasions or rubbing caused by improper blanket application
- Entrapment of animal within ill-fitting blanket

Equipment

- Grooming materials
- Blanket
- Halter and lead rope

Blanketing Procedure

TECHNICAL ACTION	RATIONALE/AMPLIFICATION
1. Apply halter and secure horse.	1a. Do not place in stocks. This will interfere with blanket application
2. Brush to remove dirt and bedding from coat.	2a. Debris under blanket will result in pressure sores.
3. Working from the left or near side, fold blanket in half and place on withers	—
4. Secure chest buckle. Slide blanket caudally and secure ventral surcingles.	4a. Surcingles are straps that go underneath a horse.
	4b. Ventral surcingles should be secure. A maximum of 6 inches between ventral abdomen and strap is advisable.
	4c. Some blankets have additional surcingles that pass medial to the hind leg. These prevent the blanket from slipping from side to side.
	4d. Blankets should always be applied in a caudal direction to move in the direction of hair growth.

TYPES OF BLANKETS

Various types of blankets are used depending on the circumstance. Table 3–1 lists blanket types commonly used in the equine industry.

TABLE 3–1 EQUINE BLANKETS

NAME	COMPOSITION	PURPOSE
Sheet	Lightweight, often nylon	Modest coat protection from ultraviolet light and mud, provides minimal warmth
Rug	Heavy quilted natural fibers	Multipurpose, provides significant warmth, also used as cool-out blanket
Turn-out blanket	Often lined, Thinsulate, durable, waterproof outer layers	Provides warmth in severe weather conditions
Stable blanket	Lined, heavy, not weatherproof	Provides warmth, indoor use only, most common type used in veterinary clinic

FLY MASKS

Purpose

- Decrease ocular, otic irritation caused by insects

Complications

- Rub marks from ill-fitting mask
- Trauma or abrasion caused by catching mask on fencing or other object

Equipment

- Fly mask
- Halter and lead rope

Fly Mask Application Procedure

TECHNICAL ACTION	RATIONALE/AMPLIFICATION
1. Place lead rope around neck and remove halter.	—
2. Place mask over eyes and secure Velcro under mandible.	2a. Masks that cover both ears and eyes are also available.
3. Replace halter over mask for moving horse.	3a. Always remove halter when a horse is turned loose in stall or pasture.

GENERAL HUSBANDRY PROCEDURES

Cleanliness is essential to every veterinary facility. Efforts by all members of the veterinary team should be made to ensure a clean, appropriately disinfected environment. Adherence to strict sanitation protocols will minimize **nosocomial infections** and contribute to the health of hospitalized animals.

Purpose

- Maintain a clean environment to promote general health
- Provide appropriate feed and water to promote healing and health
- Decrease consumption of dirt or foreign material while eating

Complications

- Nosocomial infections due to improper sanitation techniques
- Entanglement of limbs in improperly secured **hay nets**
- Anorexia, colic, or bloating due to inappropriate type or delivery of feeds

Equipment

- Wheelbarrow
- Shovel
- Manure fork
- Rake
- Bedding
- Scrub brush
- Disinfectant
- Broom
- Buckets
- Feed manger or tub
- Hay net

Daily Stall Cleaning

TECHNICAL ACTION	RATIONALE/AMPLIFICATION
1. Remove animal from stall.	1a. If the horse cannot be removed from stall, apply a halter and secure the animal. Never bring the wheelbarrow into stall with the horse.
	1b. Stalls adapted for cattle often have crowding panels and head catches.
	1c. Goats and llamas can remain in the stall, but be certain to secure the animal.

TECHNICAL ACTION	RATIONALE/AMPLIFICATION
2. Using pitchfork or shovel, remove large piles of manure.	2a. Selection of equipment depends upon the size of the fecal material (horse vs. goat) (Fig. 3–3).
3. Use pitchfork to sift through bedding to remove remaining manure.	—
4. Remove wet bedding using shovel and rake.	4a. Mares and cows tend to urinate in perimeter of stall, while stallions and bulls will urinate in center.
	4b. Failure to remove wet bedding can result in severe skin scalding and ocular irritation from the ammonia content of urine.
5. Replace bedding.	5a. Shavings should be maintained at a depth of 6 inches.
	5b. Do not use shavings for llama bedding. This material will stick in the wool. Use straw.
	5c. Straw is also used as bedding. Watch to ensure that animals do not eat an excessive amount of straw.
6. Clean and refill water buckets.	—
7. Replace animal in stall.	7a. Halter should always be removed when animal is turned loose in stall.
	7b. The collar can remain on the goat.

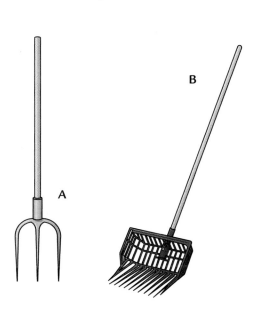

FIG. 3–3 (A) A pitch fork generally has three heavy tines and is used for carrying flakes of hay or other heavy loads. (B) A manure fork has many thin tines, a basket shape, and is used for sifting dirty bedding.

Disinfecting and Stripping Stalls

TECHNICAL ACTION	RATIONALE/AMPLIFICATION
1. Remove all bedding and waste from stall.	1a. **Stripping stalls** means removal of all bedding. This is done for sanitation purposes. 1b. Stalls should always be disinfected between patients.
2. Further procedures will depend on stall surfaces.	2a. See procedures listed for appropriate stall surface type.

Dirt Floor Stalls

TECHNICAL ACTION	RATIONALE/AMPLIFICATION
1. Rake stall.	—
2. Apply lime powder. Lime must remain in contact for 48 hours.	2a. Obviously disinfection of dirt is impossible, but this method eliminates much of the contamination.
3. Remove lime.	3a. Failure to properly remove lime can cause severe chemical burns.
4. Use hand sprayer to apply disinfectant to walls.	4a. Length of application will depend on the disinfectant selected.
5. Disinfect manger and buckets.	—
6. Bed stall with shavings or straw.	6a. Do not use shavings with llamas. 6b. Bedding should be 6 inches deep.

Cement Floors with Rubber Mats

TECHNICAL ACTION	RATIONALE/AMPLIFICATION
1. Remove mats from stall and sweep.	1a. Rubber mats provide a degree of cushioning and support. Animals should never stand directly on cement.
2. Wash floor, walls, and mats with appropriate disinfectant.	2a. A pressure sprayer will facilitate this process. 2b. Follow manufacturer's recommendation for contact time. Adequate contact time is essential for disinfection. In general, a disinfectant should remain in contact for at least 15 minutes.

TECHNICAL ACTION	RATIONALE/AMPLIFICATION
3. Rinse all surfaces. Allow to air dry.	—
4. Disinfect manger and buckets.	—
5. Bed with shavings or straw.	5a. Do not use shavings with llamas.
	5b. Bedding should be 6 inches deep.

Llama-specific Stall Care and Dung Piles

TECHNICAL ACTION	RATIONALE/AMPLIFICATION
1. Llamas defecate almost exclusively on dung piles and will not defecate randomly throughout the stall.	1a. Advise owners to bring some fresh feces if llama is to be hospitalized so as to encourage the llama to defecate in an existing dung pile.
2. If animal is physically able, lead llama to a grass or dirt area and allow to defecate before confining in stall.	—
3. Bed stall using straw. Do not use shavings or sawdust.	3a. Watch animal after entering stall to ensure that llama does not consume large amounts of straw.
	3b. Shavings entangle in wool and are extremely difficult to remove.

WATER BUCKETS, FEED TUBS, SALT BLOCKS, AND HAY NETS

Daily Care of Water Buckets, Tubs, and Salt Blocks

TECHNICAL ACTION	RATIONALE/AMPLIFICATION
1. Remove water bucket and feed tub from stall a minimum of 1 time per day for cleaning.	1a. Water and feed consumption are critical monitoring parameters of the hospitalized animal. The amount of feed and water consumed should be noted in the patient's chart at least 2 times per day.
	1b. It is advisable to hang two 5-gallon buckets in each stall. This ensures that animals will have constant access to water.

TECHNICAL ACTION	RATIONALE/AMPLIFICATION
2. Rinse buckets and feed tubs, and scrub using stiff brush.	—
3. Fill and replace buckets. Secure feed tub.	3a. If the feed tub is permanently secured to wall, manually remove any debris.
4. Note condition and any consumption of salt block.	—

Care of Water Buckets and Salt Blocks Between Animals

TECHNICAL ACTION	RATIONALE/AMPLIFICATION
1. Remove buckets and feed tubs from stall.	—
2. Thoroughly scrub inside and outside using appropriate disinfecting solution and stiff brush.	—
3. Rinse with clean water and allow to air dry.	3a. Thorough rinsing helps to remove unpleasant odors thus encouraging water consumption.
4. If grain tub is permanently secured, manually remove any debris, then disinfect. Hand dry with towel.	—
5. Place new salt block in stall.	5a. Blocks typically are secured to the wall or placed in feed tubs.
	5b. Salt blocks should never be reused in a medical facility.

Hay Nets

TECHNICAL ACTION	RATIONALE/AMPLIFICATION
1. Open the hay net by loosening the purse-string cord at end of net. Place appropriate amount of hay inside net.	1a. Hay nets provide a means of feeding hay off the ground without a rack or manger. They are much easier to sanitize between patients. Nylon materials are preferable to natural-fiber ropes (Fig. 3–4).
	1b. Some medical conditions (heaves) warrant feeding from the ground. Hay nets should not be used if medically contraindicated, with foals, or with cattle.

TECHNICAL ACTION	RATIONALE/AMPLIFICATION
2. Secure hay net in stall to beam or eye nut such that the bottom of the net is a minimum of 4 feet from the ground.	2a. Lower hay nets can be a source of limb entanglement.
3. Check net periodically throughout the day to ensure appropriate height.	—
4. Remove net when empty.	—

FIG. 3-4 A hay net filled with hay.

IDENTIFYING FEED

Nutrition plays an important role in maintaining or improving animal health. Individuals responsible for feeding must select quality, nutritionally appropriate feedstuff. To this end, recognizing forage types and quality is a critical skill that all technicians must acquire (Figs. 3–5, 3–6).

HAY TYPES

There are two basic types of hay; these are grass and **legume** hays. Comparatively, legume hays are higher in nitrogen, calcium, phosphorus, and many vitamins. Legume hays also have a greater association with colic and laminitis if fed by the free choice method (Table 3–2).

TABLE 3-2　HAY TYPES

TYPE	EXAMPLES	COMMENTS
Grass hay	Timothy, Bermuda grass, bromegrass, johnsongrass, orchard grass, prairie grass	A different type of grass hay typically grows in each area of the country.
Legume hay	Alfalfa, lespedeza, clover	Alfalfa is by far the most common type of legume. Legumes have much higher protein and calcium levels than grass hays.

FIG. 3-5　Photograph of grass hay.

FIG. 3-6　Photograph of alfalfa hay.

HAY QUALITY

Good-quality hay has the following characteristics: sweet smelling, green color, soft pliable stems, no mold. It is not uncommon for the outside portion of hay bales to become brown or discolored. This does not significantly affect quality. It is very important to remember that moldy hay should never be fed to hospitalized animals. Mold can be grossly visible as damp black areas or it can appear as a fine dust that billows out when the bale is opened. A distinct odor can also be detected with many molds.

HAY QUANTITY

The amount of hay each animal receives is dependent on many variables. These can include age, sex, body condition, pregnancy status, and disease state. As a general rule, it is acceptable to feed hospitalized horses and cattle 2–3 lb of hay per 100 lb of body weight.

In actuality most clinics feed by a per-flake measurement, with an average-sized horse receiving two flakes 2 times per day. A flake is a section of the hay bale that breaks apart naturally. Each **hay flake** is typically 4–5 inches in width. Relatively speaking, a flake of alfalfa (a legume) would weigh much more than a flake of grass hay. Llamas should receive one-half of a flake of grass hay 2 times per day. Do not feed alfalfa to llamas.

GRAINS

Most veterinary clinics use a commercial mixed-grain feed. Table 3–3 lists common components found in the rations. To determine the content of any given feed, examine the ingredient label of the grain sack. Always confirm that the grain mixture is appropriate for the species before administering it. Feed additives that are beneficial in one species can be highly toxic to another.

Clinics specializing in equine patients often purchase mixed-grain feeds with added molasses, vitamins, and minerals. Also known as **sweet feeds**, the grain mixes with added molasses are much more palatable to horses.

TABLE 3–3 GRAIN TYPES

GRAIN	FORMS	COMMENTS
Oat	Whole, rolled, crimped	Most common grain used for horses
Corn	Whole, cracked, crushed	Second most common grain, much more energy-dense feed than oats
Barley	Rolled, crushed	Less palatable than oats
Wheat	Bran, middlings	Both forms are byproducts of milling, bran commonly used for bran mashes

REVIEW QUESTIONS

1. List equipment used to groom a horse and a llama.
2. State the reason for initiating grooming from the left side of the horse.
3. Identify four reasons for blanketing a horse.

4. List three complications associated with poor husbandry techniques.
5. Identify locations in the stall where wet bedding would most likely be found.
6. Name the compound that is used to sanitize dirt floors.
7. Describe the difference between red and white salt blocks. Which one is preferable?
8. State a medical condition in which hay net use is contraindicated.
9. Identify two basic types of hay. Give examples of each type.
10. Describe the qualities of good hay.

BIBLIOGRAPHY

Gillespie, J. (1998). *Animal science* (1st ed). Clifton Park, NY: Thomson Delmar Learning.

Research staff of Equine Research, Inc. (1992). *Feeding to win II*. Grand Prairie, Texas.

SECTION TWO

Physical Examination

Chapter 4
Physical Examination

CHAPTER 4

Physical Examination

Medicine, to produce health, has to examine disease; and music, to create harmony, must investigate discord.

PLUTARCH

KEY TERMS

ataxia
auscultate
ballottement
borborygmi
diagnosis
emaciation
fetus

icteric
obese
palpate
parturition
prognosis
tympany

OBJECTIVES

- Describe information that should be obtained when taking a routine history.
- Describe the steps taken to complete a basic physical examination on the horse, cow, llama, and pig.
- Describe a routine reproductive examination for the horse, cow, small ruminant, and pig.
- Briefly describe the examination required for health certificates, insurance, and before purchase.

BASIC HISTORY TAKING

Invaluable to the busy practitioner is an individual who can gather a complete history from the owner before the patient is examined. The information can prove invaluable in diagnosing and treating either the herd or the individual. A complete history also helps us decide if the individual that was brought for examination has a disease or condition that may be affecting an entire herd or group of animals. This is very important in production animal medicine.

Purpose

- Gather patient data, including age, sex, and species.
- Ascertain the owner's primary complaint or concern about the individual patient, flock, or herd.
- Obtain information on the care the animals are currently receiving.
- Determine what disease prevention measures have been undertaken.
- Determine the degree to which the herd, flock, or group is affected.

Complications

- Poor communication between owner and history taker leading to misdiagnosis

Equipment

- Medical record form
- Pen

PROCEDURES FOR HISTORY TAKING

General History Questions

TECHNICAL ACTION	RATIONALE/AMPLIFICATION
1. What is the primary complaint regarding this patient or herd?	1a. Sometimes what the owner perceives the problem to be is different from what the veterinarian sees as the problem. It is important to address the owner's concerns first.
2. How long have these signs been present?	2a. Tells us if the condition is acute or chronic.
3. Have you treated it (them) with anything?	3a. Residual effects of some medications may affect clinical signs and blood work results.

TECHNICAL ACTION	RATIONALE/AMPLIFICATION
	3b. Injection site reaction may be present and causing current disorder.
4. Has your therapy shown any success?	**4a.** Success or failure of previous treatments may affect diagnosis and future treatment.
5. Have any animals died? If so, how many and how long ago?	**5a.** Indicative of severity of problem.
	5b. Access to deceased animals will aide in diagnosis and **prognosis**.
6. Is more than one animal showing the same signs?	**6a.** Lets us know if this is a herd or individual problem. This can affect diagnosis, treatment, and prognosis.

Vaccination History Questions

TECHNICAL ACTION	RATIONALE/AMPLIFICATION
1. When?	**1a.** How long ago was the vaccine given and was it given at an age appropriate to the animal and the disease?
2. With what?	**2a.** Elicit specific name and type of vaccine.
	2b. If possible, get a lot number from the bottle used.
3. How?	**3a.** By what route was it given?
	3b. Was the route appropriate for the vaccine?
	3c. Where on the animal(s) was it given?
	3d. Was vaccine handled according to manufacturer's recommendations?
4. How many?	**4a.** Was the whole herd or group vaccinated at once or just selected individuals?
5. Reactions?	**5a.** Note any vaccine reactions reported by owner, such as swelling, abscesses, painful muscles, allergy.

Special Note

A wide variety of vaccinations are available for horses and food animal species. Some, like brucellosis vaccine, are state and federally regulated, so use is governed by law. The vast majority of vaccines, however, are used in accordance with local disease prevalence and upon recommendation of the veterinarian caring for those animals. Some of the vaccines available for each species are listed in the appendix.

Dairy Animal Production Questions

TECHNICAL ACTION	RATIONALE/AMPLIFICATION
1. Concerning the individual cow or goat: Is there a sudden drop in milk production?	1a. Indicative of systemic disease such as left-displaced abomasum (LDA), mastitis, or organic disease.
2. Is there a recent change in the rolling herd average (milk production of the herd over time)?	2a. Drop in herd average may be indicative of nutritional problems, dry-off problems, or change in genetics.
3. Has there been an increase in the individual's or herd's somatic cell count (SCC)?	3a. May indicate hidden mastitis problems or management issues.
4. Is there a change, or do you desire a change, in the butter fat-to-protein ratio?	4a. These are affected mostly by genetics and nutrition.
5. Other: Have you noticed any changes in the milk parlor?	5a. Problems in the milk parlor or barn such as: 1. Squawks from the milking apparatus 2. Stray voltage from machinery causing restless, poorly milking cows, afraid to enter the parlor. 3. Inconsistent power to the machinery 4. Use of cow milker on dairy goats

Hoof Care and Health Questions

TECHNICAL ACTION	RATIONALE/AMPLIFICATION
1. What is the condition of the animal's or herd's feet?	1a. Splayed? Dished? Over grown? Appropriate for species?
2. Have you noticed any of the following foot health problems in the patient or herd?	2a. Inspect for cracks, flaking, dry, bad odor, painful, bounding digital pulse, bulging coronary band.
3. What kind of surface does the animal or herd stand and walk on?	3a. Animals made to stand on concrete all day will suffer from wear and tear on their feet and legs.
	3b. Ground that is always wet and boggy tends to cause foot rot and other hoof or leg infections.

TECHNICAL ACTION	RATIONALE/AMPLIFICATION

4. How often are the feet trimmed in the individual or herd?	4a. Regular trim routine:
	  1. Horse 6–8 wks
	  2. Dairy cows: quarterly
	  3. Sheep and goats: semi-annually
	  4. Beef cows: as needed, on an individual basis

5. Do you use a foot bath in your herd? If so, how often are they walked through it?	5a. Foot baths are used to prevent hairy wart disease in dairy cattle and foot rot in sheep and goats.
6. Shoeing: How often?	6a. Only horses are shod. Shoes should be in good wear, appropriate for the horse's work. Nails should not be too high or too close to heel.

Horse-specific Questions

TECHNICAL ACTION	RATIONALE/AMPLIFICATION
1. How long has the current problem existed?	1a. Differentiate between a chronic (longstanding) or an acute (sudden onset) condition.
2. How much exercise can the horse do before it tires?	2a. Indicative of a chronic (ongoing) problem.
3. How often is the horse transported and how far?	3a. Respiratory and digestive disorders are associated with frequent, prolonged travel.
4. Does the horse chew or crib?	4a. This can lead to dental disease, colic, and malnutrition.
5. Has there been a change in attitude?	5a. Indicative of pain or discomfort.
6. A change in appetite?	6a. Indicative of feed problem, dental disease, or chronic illness or pain.
7. A change in feed?	7a. Sudden feed changes can lead to digestive upset, ulcers, colic, or laminitis (founder). 7b. Either the type of feed or the source of the feed may have changed.
8. What is the animal fed? Pasture, hay, grain?	8a. Diet should be appropriate to level of work and age of horse. 8b. Hays and pastures may contain toxic plants or insects.

TECHNICAL ACTION	RATIONALE/AMPLIFICATION
9. Where and how is it housed? Is it alone or with other horses? Is it housed with other species (donkey, goat, cattle)?	9a. Pastured animals often experience different problems than animals that live in stalls or pens.
	9b. Other species may carry diseases or parasites not common to horses.
10. How is water supplied? Automatic waterer or bucket? Is a heater used in the winter?	10a. Automatic watering systems may malfunction, causing stray voltage problems or inadequate water flow.
	10b. Frozen water or near-freezing water can reduce water intake and cause constipation colic.
11. Have any medications been given? If so, when, where on the animal, and by whom?	11a. Failure of previous therapy may aid in current diagnosis.
	11b. Residual effects of some medications may affect clinical signs and blood work results.
	11c. Injection site reaction may be present and causing current disorder.
12. What is your vaccination and deworming program?	12a. Refer to Vaccination History Questions for vaccination questions.
	12b. Horses should be dewormed every 8–10 weeks, alternating chemical compounds.
	12c. Horses may also be dewormed continuously with a feed-through deworming agent.
13. If the horse is a young foal, ask the following questions:	
a. How many days did the mare carry this foal?	13a. Newborn foals have unique problems.
	13b. A foal must be carried a minimum of 335 days to be mature enough to survive unaided.
b. How long did it take the foal to stand up after being born?	13c. A foal should stand and attempt to nurse within 1 hour of birth.
c. Have you seen the foal nursing?	13d. The foal should nurse vigorously within 3 hours of birth in order to obtain antibody-containing colostrum.
d. Does the mare have milk in her udder?	13e. Inadequate ingestion of colostrum within 12 hours of birth results in failure of

TECHNICAL ACTION	RATIONALE/AMPLIFICATION
	passive transfer, meaning that the foal will not have received enough antibodies to protect it from infection during the first few months of life.
e. Did the foal defecate within 12 hours of birth?	13f. Meconium (feces produced while the foal was in utero) sometimes causes life-threatening constipation.
f. Is the foal nursing every couple of hours?	13g. A healthy foal bonded to a healthy dam will nurse often and vigorously.

Beef Cattle-specific Questions

TECHNICAL ACTION	RATIONALE/AMPLIFICATION
1. What is the age of the affected population?	1a. Some diseases are age specific.
2. Has this happened before?	2a. A recurrent problem may point toward management problems such as nutrition.
3. Have any medications been given?	3a. Success or failure of previous treatments may affect diagnosis and future treatment.
4. How many are affected?	4a. Indicates severity of the problem and potential economic loss.
5. Have any died? If so, how many and when?	5a. Indicates severity of the problem.
	5b. Access to deceased animals will aid in diagnosis and prognosis
6. What is your vaccination program?	6a. Refer to Vaccination History Questions.
7. Have you made recent purchases or has the animal been transported lately?	7a. Bringing new animals into the herd may cause an outbreak of contagious diseases.
	7b. Transportation stress can induce latent disease.
8. What are they fed?	8a. Is the diet appropriate to the age of the animals and production expectations?
9. How are they fed?	9a. Inadequately mixed feed may cause digestive problems.
	9b. Molds can lead to respiratory and digestive problems, and sudden death.

TECHNICAL ACTION	RATIONALE/AMPLIFICATION
	9c. Crowding at the bunks will cause some to be injured, some to starve, and some to overeat.
10. Do the animals have free-choice access to minerals and salt?	**10a.** Salt and minerals must be accessible for all cattle of all ages at all times.
	10b. What minerals are in the mix and how is it fed (loose salt, block, or lick tank)?
	10c. Can the baby calves reach the mineral? They need minerals, too.
11. Do they have access to water?	**11a.** Water in a tank may be contaminated with toxic algae.
	11b. Water source may freeze in winter or run dry in summer.
12. Are they out on range, on a pasture, or in a feedlot?	**12a.** Pastured animals are exposed to more parasites.
	12b. Range cattle may be exposed to toxic plants and dangerous animals, and are more apt to be injured.
	12c. Feedlot animals may suffer effects from of overcrowding and overfeeding.

Dairy-specific Questions

TECHNICAL ACTION	RATIONALE/AMPLIFICATION
1. Where is she in her lactation cycle?	**1a.** Lactation cycle can be broken down into four parts: 　**1.** Early lactation 　**2.** Midlactation 　**3.** Late lactation 　**4.** Dry period
	1b. Some conditions (such as left-displaced abomasum, mastitis, and metritis) are specific to timing during the lactation cycle.
2. What is her annual and current daily production?	**2a.** A sudden drop in milk production is associated with displaced abomasum.
	2b. Long-term milk drop may be nutritional in origin.
3. How many are affected?	**3a.** Determines degree of problem and potential economic loss.

TECHNICAL ACTION	RATIONALE/AMPLIFICATION
4. Do you use bovine somatotropin (BST)?	4a. BST is a synthetic growth hormone administered to dairy cows to increase milk production.
	4b. BST affects many aspects of health and nutrition.
5. Were there any problems during the last cycle?	5a. Cycle refers to the calving interval (how many months between calves), as well as breed-back problems (how many days between parturition and the start of the next pregnancy).
	5b. Helps us identify how closely the lactation and reproductive cycles of the cows are managed.
6. What do you feed? Is it a total mixed ration?	6a. May bring up nutrition problems. Doctor will want to know what is in the ration and how it is mixed.
7. Are the animals penned and fed according to production, place in lactation cycle, or all grouped together?	7a. It is important to feed dairy cows according to their production needs.
8. How and when are they fed?	8a. Timing of feeding can affect digestive function and incidence of mastitis.
9. Is she eating?	9a. This question helps identify how well the cows are watched for appetite.
	9b. Some diseases, such as Johnes disease, will cause the cow to become emaciated despite an excellent appetite.
10. How often are they milked?	10a. Milking frequency affects annual production as well as stress to cow.
11. Has she been treated with anything? When and how often?	11a. Affects diagnosis, prognosis, and treatment protocol.
	11b. Some medications may contribute to the disease process.
12. Has the milker noted any problems?	12a. This speaks to management. Does management pay attention to what goes on in the milking parlor?
13. Have new cows been introduced to the herd or pen?	13a. Some disease may be carried by a cow and passed on to other cows without the carrier ever showing signs of the disease.

TECHNICAL ACTION	RATIONALE/AMPLIFICATION
	13b. New cows may either injure the established herd animals or be injured by the established herd.
14. What is your vaccination program?	**14a.** Refer to Vaccination History Questions.
15. Do you practice fly control?	**15a.** This question addresses communicable disease and mastitis.
16. Do you use 24-hour lighting?	**16a.** Lighting may contribute to the stress the cows are under if they are not permitted a dark or low light time period.

Dairy Calf-specific Questions

TECHNICAL ACTION	RATIONALE/AMPLIFICATION
1. How old is the affected calf?	**1a.** Neonates are handled differently and have different problems than calves over 3 months old.
2. How long has the calf been sick?	**2a.** Duration suggests the severity of a problem, whether it is chronic or acute.
3. How many are showing the same signs?	**3a.** Number suggests the severity of the problem, communicable disease, nutritional problem, or management.
4. Have any died? How many and when?	**4a.** Deaths influence the severity of the problem, economic loss, and potential for necropsy and diagnostic sampling.
5. Did this calf receive colostrum at birth?	**5a.** Colostrum is vital for immune function.
	5b. Timing is critical. Colostrum given after 24 hours is not effective.
6. What do you feed?	**6a.** Be specific about milk replacer, grain, hay, and other feed.
	6b. If possible, get a feed tag listing the ingredients.
7. How often are they fed?	**7a.** Frequency of feeding affects digestive function.
	7b. If calves are still on milk, find out if they are bucket fed, bottle fed, or nipple tank fed.
8. What treatments has this calf received?	**8a.** Treatments affect diagnosis, prognosis, and therapy.
9. What is the vaccination history?	**9a.** Refer to Vaccination History Questions.

Llama-specific Questions

TECHNICAL ACTION	RATIONALE/AMPLIFICATION
1. What type of housing is (are) your llama(s) kept in?	1a. This lets us know if they have adequate room for a dung pile, exercise, and related factors.
2. Do you own more than one llama?	2a. Single llamas are prone to mental disorders such as berserk male syndrome.
3. What do you feed your llamas?	3a. Diets for llamas are often too rich and can cause digestive disorders and foot problems.
4. Are they on a vaccination and deworming program?	4a. Refer to Vaccination History Questions.
	4b. What is the deworming program: What product and how often?
5. If llama is a castrated male, at what age was he castrated?	5a. Castration before 2 years of age can cause skeletal defects and weaknesses.
6. What do you use your llamas for?	6a. Some llamas are pasture ornaments, some carry loads, and some guard sheep or goats.

Pig-specific Questions

TECHNICAL ACTION	RATIONALE/AMPLIFICATION
1. What type of housing do you keep them in?	1a. Pigs housed in total confinement suffer from different diseases than those housed outside.
2. If housing is indoors, is it temperature controlled?	2a. Hypothermia is a significant risk to young pigs, hyperthermia a severe problem for mature pigs.
3. Is there the smell of ammonia in or near the facility?	3a. High ammonia levels cause chemical pneumonia and are a management issue.
4. What is done with manure and urine?	4a. Answer identifies exposure to ammonia, parasites, and pathogens.
5. What vaccinations do you give and when?	5a. Refer to Vaccination History Questions.
6. What is done for parasite control?	6a. Internal parasites cause large economic losses, yet they often go untreated.

TECHNICAL ACTION	RATIONALE/AMPLIFICATION
7. What are the pigs fed?	7a. Nutrition affects production, litter size, and sow longevity.
8. How many litters does each sow average per year?	8a. Number lets us know what sort of breeding stress a sow may be under.
9. At what age are the litters weaned and what do the piglets weigh?	9a. Early weaning increases mortality, and poor weaning weights suggest parasitism.
10. How many are sick or dead?	10a. Number demonstrates severity of problem and economic loss.
11. What is average weaned litter size?	11a. A sow's production is determined by the number of pigs weaned.

PATIENT OBSERVATION

Observation of the patient (individual or herd) in its home environment is ideal. Often, however, the patient is observed after being transported to the veterinary facility. Taking the time to observe the patient moving freely about in a pen, pasture, or other holding facility will give us clues as to how the patient will need to be restrained and areas that may require special attention during the physical examination.

Purpose

- Identify postural, behavioral, or gait abnormalities before physical examination.
- Ascertain mental status of patient before handling.

Complications

- Personal injury
- Patient injury

Equipment

- Enclosed pen or corral

Procedure

- Observe patient in pen or home environment with no halter or other restraint.
- Have assistant or handler then move animal around pen or home environment and observe for ataxia, gait abnormalities, or vision impairment.
- Animals trained to lead should then be led by the handler at a walk and then a trot on a firm surface.

FIG. 4-1 Healthy llamas in a pen.

PHYSICAL EXAMINATION

Vital to any veterinary workup, the physical examination encompasses both history-taking techniques and the physical attributes of the animal or herd. In food animal medicine, often a herd physical examination is just as important as examination of individual members of that group.

Purpose

- Obtain objective findings such as temperature, pulse, and respiration.
- Identify any findings outside generally accepted norms for the species.
- Collect and record data for comparison on subsequent examinations.

Complications

- Injury to examiner due to inadequate restraint

Equipment

- Stethoscope
- Watch with a second hand
- Thermometer
- Paper and pen
- Medical record

PROCEDURES USED IN PHYSICAL EXAMINATIONS

Physical Examination of the Horse

TECHNICAL ACTION	RATIONALE/AMPLIFICATION
1. Place a halter and lead rope on the horse.	1a. Usually the client will present the horse with halter and lead rope already in place.
2. Observe the horse being led at a walk.	2a. Watch for irregular gait, staggering, dragging toes, or other abnormality.
3. Restrain horse in stock, crossties, or have someone hold it.	3a. Depends on facility and initial complaint.
4. Examine hair, coat, and skin.	4a. Note condition of the hair coat: Shiny, dull, sparse, thick, shaggy.
	4b. Note any hair loss, describing character (patchy, generalized, symmetrical) and location.
5. Note body condition.	5a. Degree of muscling depends on use of the horse and breed.
6. Examine the head.	6a. Is the face symmetrical? Do jaws align correctly?
	6b. Eyes: Note any discharge, squinting or discoloration.
	6c. Nostrils: Note any discharge or asymmetry.
	6d. Lips: Note any sores, flaccidity, drooling.
	6e. Cheeks: Symmetry, note any swelling.
	6f. Ears: Note any discharge, drooping, head shaking.
	6g. Lymph nodes: **Palpate** submandibular and parotid lymph nodes for swelling, heat, or pain.
	6h. Pulse: Palpate ventral edge of mandible.
7. Examine inside of mouth.	7a. May require extra assistance in restraint. Note: A complete dental examination requires sedation and full mouth speculum and is not covered here.
	7b. Breath: Should smell like feed.
	7c. Mucous membranes: Should be pink or pale pink; check capillary refill time.

TECHNICAL ACTION	RATIONALE/AMPLIFICATION
	7d. Gingiva: Should be pink or pale pink, with no sores, vesicles, swelling, or redness.
	7e. Teeth: Incisors should meet without an overshot or undershot jaw. Canines should be intact; note presence of wolf teeth. Note any sharp projections from cheek teeth.
8. Examine neck.	**8a.** Observe symmetry of musculature.
	8b. Observe symmetry of mobility, turning neck right and left.
	8c. Check for presence of a jugular pulse wave. A normal horse will have one if the head is dropped below the level of the heart.
9. Examine body.	**9a.** Shoulders: Should be symmetrically muscled.
	9b. Saddle area: Look for white hairs or muscle wasting on either side of the withers, indicating poorly fitting saddle.
	9c. Ribcage: Ribs should be palpable but covered with fat and muscle. Examine girth area for sores or lumps.
10. Examine hindquarters.	**10a.** Symmetry: Viewed from behind, both sides should be same shape and size.
	10b. Muscling: Equal bilaterally; hips and sacrum covered with tissue.
	10c. Tail: Moves in all directions, resists your efforts to move it, horse can swat flies.
11. Assess temperature.	**11a.** Obtain a rectal temperature, and stand to the side when inserting the thermometer so as not to get kicked. Normal temperature in the adult horse is 37.2°C to 38°C (99°F to 100.5°F) (Fig. 4–2A).
	11b. Check vulva for swelling or discharge.
	11c. Note color and consistency of feces.

TECHNICAL ACTION	RATIONALE/AMPLIFICATION
12. **Auscultate** heart and record heart rate.	**12a.** Listen to both sides of chest, just behind the elbow. Normal heart rate in an adult horse is 30–45 bpm. A young foal may have a heart rate of 70–80 bpm (Fig. 4–2B).
	12b. Heart should beat in a regular rhythm, with no variation with respirations.
	12c. Note any murmurs or abnormal sounds.
13. Auscultate lungs and observe respirations.	**13a.** Listen to both sides of the chest, breaking the ribcage down into four quadrants. It is not unusual to hear **borborygmi** through the chest.
	13b. Airflow is sometimes difficult to hear in the horse. To make the horse breathe more deeply, it may be exercised or a large plastic bag may be placed over the nostrils. Do not do this without veterinary supervision.
	13c. Watch nostrils to see if they flare with inspiration or expiration.
	13d. Listen for an expiratory grunt or coughing/heave line.
	13e. Palpate the trachea and note if this causes the horse to cough.
14. Auscultate abdomen.	**14a.** Place the stethoscope in the paralumbar fossa to listen.
	14b. Listen dorsally and ventrally for borborygmi on both sides of the horse. They sound like low rolls of thunder.
	14c. Horses with ileus have either no borborygmi or just the occasional sound of bubbles bursting.
	14d. If cecal **tympany** is suspected, flick your finger against the hide in the right dorsocranial paralumbar fossa while listening with the stethoscope. Listen for a sound similar to flicking your finger on a rubber ball (Fig. 4–2C).

TECHNICAL ACTION	RATIONALE/AMPLIFICATION
15. Examine all four legs.	**15a.** Look for old or new scars, lumps, bumps, or swellings.
	15b. Front legs: Run your hands down from elbow to hoof to feel for abnormalities.
	15c. Front legs: Stand in front and to the side of the horse to look for conformational abnormalities, such as knock-kneed or pigeon-toed conformation.
	15d. Hind legs: Keep one hand on or near hip while you run the other hand all the way down the leg to the hoof, feeling for abnormalities.
	15e. Stand behind and to the side of the horse to look for conformational abnormalities such as cow-hocked or bow-legged conformation.
	15f. Palpate digital pulses at the caudal aspect of the pastern. Pulses that can be seen or are very easily palpated are described as bounding and are indicative of laminitis.
	15g. Examine hooves after cleaning them out with a hoof pick. Note any foul odor, large cracks, loose shoes, or sole injuries.

 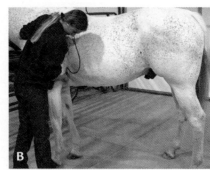 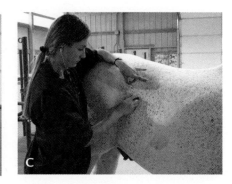

FIG. 4–2 (A) Safely taking the temperature of a horse not restrained in a stock. (B) Auscultation of the heart. Best done in a quiet place. (C) Auscultation of the right paralumbar fossa while flicking the skin with a finger to detect cecal gas.

Physical Examination of the Dairy Cow

TECHNICAL ACTION	RATIONALE/AMPLIFICATION
1. Observe cow walking in pen or alleyway.	**1a.** Note symmetry of gait and awareness of surroundings.
	1b. Note any staggering, weakness.
	1c. Note attitude of cow: Is she lethargic, agitated, belligerent?
2. Place cow in stock or head catch.	**2a.** Dairy cows are not used to chutes and will stand more quietly with just the head caught.
3. Score body condition (1–5).	**3a.** Score cow as follows: 1. **Emaciated:** Deep cavity either side of tail head. Bones of pelvis, spine, and ribs painfully visible, with deep depression of paralumbar fossa. Cow is weak. 2. **Thin:** Slight fat covering of tail head, pelvis, and ribs. All still readily visible, including spine. Paralumbar fossa depression still visible. Cow is not weak. 3. **Good:** No cavity around tail head, with fat easily palpated over entire area. Pelvis can be palpated easily, but ribs and spine are well covered with fat. Slight depression of paralumbar fossa. 4. **Fat:** Lumps of fat tissue over and around tail head and covering pelvic bones. Pelvis can be palpated with firm pressure, but ribs and spine can no longer be felt. No depression of paralumbar fossa. 5. **Obese:** Tail head, pelvis, spine, and ribs all are buried in a thick layer of fat and are no longer palpable. This condition is rare in dairy cows.

TECHNICAL ACTION	RATIONALE/AMPLIFICATION
4. Observe posture and respirations.	4a. Cows with abdominal pain tend to stand humped up.
	4b. Cows with dyspnea tend to stand with elbows away from the body.
5. Measure rectal temperature with 5-inch ring-top thermometer.	5a. Normal adult cow temperature is 37.8°C to 39.2°C (100°F to 102.5°F) but may be higher on very hot days or when the cow is highly agitated.
	5b. While checking temperature, observe vulva for swelling or discharge.
	5c. Note the odor and consistency of the fecal material around the anus or on the thermometer.
6. Auscultate heart and record number of beats per minute.	6a. Auscultate both sides of the chest.
	6b. The heart is heard best by placing your stethoscope on the ribs just behind the elbow.
	6c. Normal heart rate for an adult cow is 50–80 bpm.
	6d. Note any murmur or abnormal rhythm.
7. Auscultate lungs and record number of respirations per minute.	7a. Listen to both sides of the chest.
	7b. The breath sounds are heard best over the midthorax over ribs 8 to 10. Normal respiratory rate in the adult cow is 10–30 breaths per minute.
	7c. Note any squeaks (friction rubs), gurgles, (moisture), or rough sounds (rales).
8. Auscultate rumen.	8a. Listen to and feel for rumen motility. Count the number of contractions per minute. Normal rate is 1–3 contractions per minute.
	8b. Using your fist, use **ballottement** (rhythmic pushing) to assess the rumen, to feel the character of the rumen content. Excess fluid or excessive dryness may be detected this way.

TECHNICAL ACTION	RATIONALE/AMPLIFICATION
	8c. Ping rumen by flicking finger on skin just caudal to last rib and overlying last two ribs while listening with your stethoscope (Fig. 4-3).
	8d. A ping suggests the presence of a left-displaced abomasum.
9. Auscultate right side of cow's abdomen.	**9a.** Listen for progressive intestinal contractions (borborygmi).
	9b. Ping the abdomen in the right paralumbar fossa to detect cecal gas or tympany.
10. Palpate udder and express milk from each teat.	**10a.** Note the color and consistency of discharge from each teat. Colostrum is creamy yellow and sticky. Milk is white and watery.
	10b. Heat in or tightness of the udder is suggestive of mastitis.
11. Examine all four feet and legs.	**11a.** Look at the feet for cracks, swellings, or sores. Note if the claws are very long, misshapen, or unevenly worn.

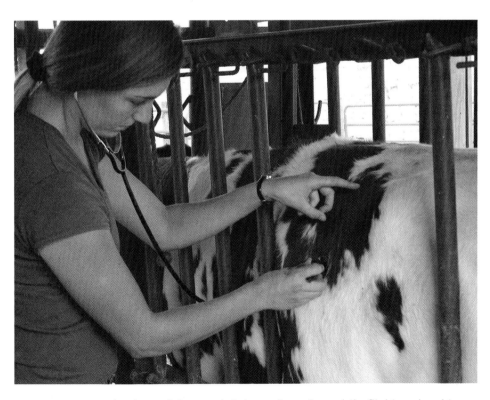

FIG. 4-3 Auscultation of the caudal thoracic region while flicking the skin with a finger to detect a left-displaced abomasum.

TECHNICAL ACTION	RATIONALE/AMPLIFICATION
	11b. Foot problems in a dairy cow can lead to mastitis, weight loss, and poor production.
	11c. Look at the legs for signs of injury.
12. Examine hide for swelling, hair loss, sores.	**12a.** Note any abnormalities.
13. Examine head and neck.	**13a.** Try to palpate prescapular lymph nodes; they are normally too small to find. They are located just rostral to the shoulder.
	13b. Check brisket (chest) for edema or swelling; if swelling is present, note the nature or consistency of the swelling.
	13c. Observe jugular groove for pulse wave. A pulse wave is not normal in a cow.
	13d. Palpate jaws and intermandibular space. Lymph nodes here are normally the size of a walnut.
	13e. Observe nares for presence and character of nasal discharge. A healthy cow usually has a clean, moist nose.
	13f. Examine gingiva and tongue, looking for ulcers or erosions. The tongue should be symmetrical and mobile.
	13g. Note any breath odor; It should smell like feed.
	13h. Observe mandibular incisors and maxillary plate, looking for broken or missing teeth, sores, blisters, or tumors.
	13i. Eyes should be clear and bright, with no discharge.
	13j. Ears should be symmetrical. There should be an orange metal Bang's tag and a tattoo in the right ear.

Physical Examination of Beef Cattle

TECHNICAL ACTION	RATIONALE/AMPLIFICATION
1. Observe the patient in pen or alleyway.	1a. Procedure is the same as for a dairy cow.
2. Place the patient in a squeeze chute.	2a. Beef cattle (unless show animals) are not used to handling and are best examined in a squeeze chute.
3. Score body condition (1–9).	3a. Score as follows: 1. Emaciated: Tail head, spine, ribs, and hips painfully visible, pronounced muscle atrophy, physically weak animal. 2. Very thin: Ribs, hips, spine, and tail head easily visible; muscle atrophy is present but animal is not weak. 3. Thin: Ribs, hips, spine visible, slight muscle atrophy, slight fat cover is visible. 4. Borderline: Three to five ribs are visible, hips visible but slightly rounded, no muscle atrophy. 5. Good: One to two ribs are visible, hips visible but rounded, spine not visible. 6. Very good: No ribs or spine are visible, but they are easily palpated. Hips are rounded. Some fat present in brisket and flanks (Fig. 4–4). 7. Fleshy: Fat in brisket, slight fat deposits are visible at tail head and udder. 8. Fat: Bone structure is not visible. Moderate fat deposits are present over ribs, hips, tail head, and brisket. 9. Obese: Large fat deposits over animal. It is rare to see a beef cow this fat.
4. Observe posture and respirations.	4a. Refer to Step 4 of Physical Examination of the Dairy Cow.

FIG. 4-4 (A) Healthy young adult cow with a body condition score of 6.

TECHNICAL ACTION	RATIONALE/AMPLIFICATION
	4b. Note if the patient is hyperresponsive or overly reactive. She will have the ears up, head up, eyes wide open, and may even be snorting. Cows exhibiting this behavior may be deficient in magnesium (or they just may be wild).
5. Measure rectal temperature with 5-inch ring-top thermometer.	**5a.** Refer to Step 5 of Physical Examination of the Dairy Cow.
	5b. Check for protrusion of rectal or vaginal tissue.
6. Auscultate heart and record number of beats per minute.	**6a.** Refer to Step 6 of Physical Examination of the Dairy Cow. Young calves (less than 6 months old) have a heart rate of 100–120 bpm.
	6b. Cattle that have been living on high-altitude (greater than 6,000 feet) pastures may have a murmur and congestive heart failure.
7. Auscultate lungs and record number of respirations per minute.	**7a.** Refer to Step 7 of Physical Examination of the Dairy Cow.
	7b. Listen carefully for crackles (moisture), rales (rough sounds), or squeaks (friction rubs) in feedlot cattle, because pneumonia is common in these animals.

TECHNICAL ACTION	RATIONALE/AMPLIFICATION
8. Auscultate rumen.	8a. Refer to Step 8 of Physical Examination of the Dairy Cow.
	8b. Displaced abomasum is a rare occurrence in beef cows. Bloat (rumen tympany) is much more common in beef cattle, especially feedlot cattle.
	8c. Feedlot cattle and those pastured on beet or potato fields should be observed closely for signs of bloat (tympany). They will have a large gas distention bulging from the left paralumbar fossa (Fig. 4–4B).
	8d. Cattle that are bloated need immediate treatment, because they often are unable to breathe adequately. They may be difficult to handle and may not fit in the squeeze chute.
	8e. In general, adult beef cows with bloat are more likely to have an esophageal obstruction, whereas feedlot cattle are more likely to have a rumen imbalance secondary to a high concentrate diet.

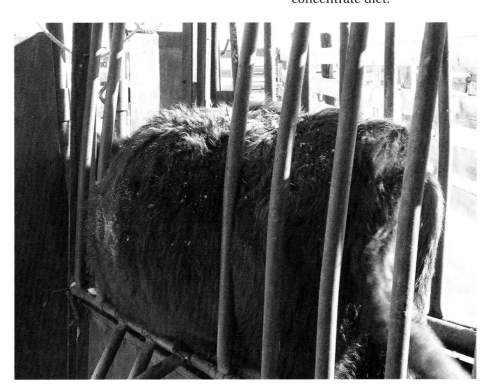

FIG. 4–4 (B) Bloated thin steer in chute. Note how left side of abdomen bulges past the bars of the chute.

TECHNICAL ACTION	RATIONALE/AMPLIFICATION
9. Auscultate the right side of the animal's abdomen.	9a. Refer to Step 9 of Physical Examination of the Dairy Cow.
10. Examine all four feet and legs.	10a. Refer to Step 11 of Physical Examination of the Dairy Cow.
	10b. Foot injuries are more likely to occur in range cattle.
11. Examine hide for swelling, hair loss, sores.	11a. Refer to Step 12 of Physical Examination of the Dairy Cow.
	11b. Lice and mites often are located at the base of the ears, tail, and other warm, moist areas.
	11c. Ringworm (dermatophytosis) appears as grey crusty lesions devoid of hair. They are more common in young cattle (less than 18 months) on dry feed.
12. Examine head and neck.	12a. Refer to Step 13 of Physical Examination of the Dairy Cow.
	12b. Pay close attention to the eyes of adult beef cattle, looking for lesions on the third eyelid and/or the surface of the cornea. Squamous cell carcinoma and malignant melanoma are cancers that occur on or near the eyes of older beef cattle. Signs include:
	1. White lesion on the cornea or sclera
	2. Bumpy or erosive lesion on the third eyelid or conjunctiva
	3. Blepharospasm and excessive tearing
	12c. Calves less than 18 months old are more susceptible to pink eye, or infection of the cornea. Signs include:
	1. Excessive tearing
	2. Blepharospasm
	3. Blue cornea
	4. Central white or red corneal lesion that may protrude from the surface of the eye.

TECHNICAL ACTION	RATIONALE/AMPLIFICATION
	12d. Examine mouth and note missing or worn incisors. Observe the flexibility, strength, and symmetry of the tongue. Note any sores, erosions, or vesicles.
13. Observe prepuce or udder.	**13a.** Feedlot steers may develop urethral blockage, in which case you may see urine dribbling from the prepuce and an enlarged abdomen.
	13b. Beef cows can develop mastitis or blocked teats, which usually become apparent at parturition.

Physical Examination of the Llama

TECHNICAL ACTION	RATIONALE/AMPLIFICATION
1. Place halter and lead rope on llama if possible.	**1a.** Not all llamas are halter broken (Fig. 4–5).
2. Observe the llama in a walk (being led or free in a pen).	**2a.** Watch for irregular gait, staggering, dragging toes.
3. Examine hair coat and skin.	**3a.** Note if llama has been sheared recently.

FIG. 4–5 Healthy young llama in good body condition.

TECHNICAL ACTION	RATIONALE/AMPLIFICATION
	3b. Note condition of the hair coat and skin. Feel for any lumps, bumps, or sores.
4. Assess body condition (obese, thin, emaciated).	**4a.** You must palpate ribs and hips through the hair coat to get an accurate assessment of body condition.
	4b. At a normal weight, ribs and hips are palpable but are covered with a layer of fat.
	4c. A thin llama has a thin layer of tissue covering hips, ribs, and vertebrae. Hips are not visible through hair.
	4d. An emaciated llama has hips that are prominent and visible. Ribs, sacrum, and neck vertebrae are palpated easily, with no tissue cover. Loss of medial thigh musculature is apparent when viewed from behind.
	4e. An obese llama has fat thick enough that the ribs and hips cannot be palpated without deep pressure.
5. Obtain temperature measurement.	**5a.** Use a rectal thermometer; stand off to one side to avoid getting kicked. Normal temperature in the adult llama is 37.2°C to 38.7°C (99°F to 101.5°F).
	5b. Check vulva for swelling or discharge.
	5c. Note if feces are stuck to perineal wool or if there is soft stool on the thermometer.
6. Auscultate heart.	**6a.** Listen to both sides of chest, placing the stethoscope head on the hairless area just behind elbows. Normal heart rate in the adult llama ranges 60–80 bpm.
	6b. Regular rhythm that does not change with respirations.
7. Auscultate lungs.	**7a.** This is difficult to do in heavily wooled animals. Normal respiratory rate in the adult llama is 10–30 rpm.

TECHNICAL ACTION	RATIONALE/AMPLIFICATION
	7b. Regular pattern, nostrils not flared, elbows relaxed (not held away from body).
8. Auscultate abdomen.	**8a.** On the left side: Listen to and palpate the rumen. Use your fist (ballottement) to feel the consistency of rumen contents.
	8b. On the right side: Ping for cecal gas and listen for regular pattern of borborygmi.
9. Examine head and neck.	**9a.** All facial structures should be symmetrical.
	9b. Nostrils should be clean and dry (no discharge).
	9c. Upper lip should be freely mobile.
	9d. Mouth may contain regurgitated feed. In adult males, fighting teeth (canines) may need to be trimmed.
	9e. Eyes are large and should be bright and clear with no discharge.
10. Examine legs.	**10a.** Look for old or new scars, lumps, bumps, or swellings.
	10b. Legs should be straight, strong, and aligned.
11. Examine feet.	**11a.** Most halter-broken llamas will allow their feet to be picked up in the same way you do with a horse.
	11b. Llamas with less training or experience may lie down when the foot is picked up.
	11c. Llamas have spongy soles. There should not be any foul odor, sores, or cracks. The wall of the hoof should not roll over the sole.
	11d. Llamas pastured on lush grass or soft soils may need their hooves trimmed several times a year.

Physical Examination of the Pig

TECHNICAL ACTION	RATIONALE/AMPLIFICATION
1. Observe pig in pen, trailer. or alleyway (Fig. 4–6).	1a. Note character and symmetry of gait. Watch for goose-stepping, lameness, or foot dragging.
	1b. Note attitude: Is it lethargic, belligerent, or aggressive?
	1c. Note exercise tolerance. Does it lie down after only a few steps?
	1d. Note respiratory depth and pattern. Does it breathe hard, grunt to exhale, wheeze, or exhibit open-mouth breathing?
	1e. Note ability to negotiate in environment. Can it negotiate obstacles; does it run into walls or other objects?
	1f. Does it act as if it itches? Does it rub up against any hard, rough surface or sit down and scoot across the ground?

FIG. 4–6 Variety of pigs housed in a large pen. Note that all are alert to the presence of an observer and have similar body condition scoring.

TECHNICAL ACTION	RATIONALE/AMPLIFICATION
2. Score the body condition of the pig.	**2a.** Score body condition as follows: 　**1. Emaciated:** Hips, ribs and backbone are readily visible and project sharply. Body is bony appearing. 　**2. Thin:** Hips, ribs, and backbone are noticeable but rounded and easily felt. Body is slab-sided. 　**3. Normal:** Hips, ribs, and backbone are palpable with firm pressure. Body is cylindrical. 　**4. Fat:** Hips, ribs, and backbone cannot be palpated even with firm pressure. The body has a smooth but round appearance. 　**5. Obese:** Hips, ribs, and backbone cannot be palpated, and the pig has bulging areas of fat deposits. **2b.** Back fat depth is best assessed with ultrasonography and is not done in a routine physical examination.
3. Having answered all of those questions, it is now time to restrain the pig.	**3a.** Chapter 2. **3b.** Usually, any restraint technique employed will cause the pig to start screaming, making the rest of the physical examination difficult.
4. Examine skin and hair coat.	**4a.** Pay particular attention to areas of redness (erythema) or excessive dander and scale (hyperkeratosis). **4b.** Lice and mites often are located at the base of the ears and other warm, moist areas. **4c.** Skin that is yellowed (**icteric**) should be noted and correlated with mucous membrane color. This may indicate liver disease. **4d.** Pig hair, although not soft, should not be brittle.
5. Auscultate heart or palpate pulse rate.	**5a.** The heart may be difficult to auscultate when the pig is screaming.

TECHNICAL ACTION	RATIONALE/AMPLIFICATION
	5b. The femoral pulse is easiest to palpate in a young pig. In adult pigs, the caudal auricular vein may be used.
	5c. Normal pulse rate in the piglet is 100–130 bpm, while in the adult it is 60–90 bpm. The pulse will of course be elevated if the pig is fighting restraint.
6. Auscultate lungs.	**6a.** This is difficult unless the pig is calm or very sick.
	6b. Listen for moist rales (rough breath sounds) or crackles (moisture).
7. Examine head and neck.	**7a.** Make sure face is symmetrical, including jowl musculature.
	7b. Nostrils should be clean and moist, with no discharge.
	7c. Carefully inspect mouth for fractured teeth, abscesses, ulcers, or sores.
	7d. Eyes are normally small, and there may be some clear discharge. They should be otherwise bright and clear.
	7e. Neck should flow smoothly from jowls to shoulder. Palpate the larynx for any swelling.
8. Examine mouth.	**8a.** Check for broken teeth, sores, vesicles, or erosion. Be careful, their teeth are sharp.
	8b. Normal mucous membrane color is pink to pale pink.
	8c. Capillary refill time (CRT) may be checked by pressing on the gums with your finger until the gum tissue blanches. Release the pressure and count the number of seconds until the color reappears. Normal capillary refill time is 1–2 seconds.
9. Examine feet and legs.	**9a.** Hooves should be short and evenly worn. Make sure there are no sores on the coronary bands or between the toes.

TECHNICAL ACTION	RATIONALE/AMPLIFICATION
	9b. Feel the legs for lumps, bumps, or other injuries. Note any crookedness or asymmetry.
10. Palpate ventral abdomen.	**10a.** Sows: Palpate and visually inspect mammary glands for heat, swelling, or discoloration.
	10b. Boars: Palpate and visually inspect prepuce and scrotum.
	10c. All pigs: Palpate and visually inspect ventral abdomen for skin lesions and swellings. Umbilical hernias are fairly common in pigs and appear as a lump on the underside of the pig.

FEMALE REPRODUCTIVE EXAMINATION

In most states, reproductive examinations are to be done by a veterinarian. Done by either rectal palpation or ultrasonographic visualization, examination of the female requires hands-on experience which cannot be obtained from reading a book.

Purpose

- Ascertain the animal's ability to reproduce effectively
- Determine whether the patient is pregnant and with how many fetuses
- Determine the stage of pregnancy and predicted date of **parturition**
- Aid in the selection of replacement breed stock

Complications

- Injury to the patient because of restraint or examination
- Injury to the examiner
- Incorrect **diagnosis** of pregnancy

Equipment

- Obstetrical sleeves
- Lubricant
- Ultrasound machine and rectal or transabdominal probe
- Pelvimeter (optional)
- Medical record and pen

PROCEDURES USED IN REPRODUCTIVE EXAMINATION OF THE FEMALE

Mare Reproductive Examination

TECHNICAL ACTION	RATIONALE/AMPLIFICATION
1. Restrain mare.	1a. Mares are best restrained in stocks designed for palpation. They have an adjustable butt bar or half gate to reduce the chance of injury to the examiner.
	1b. Recalcitrant mares may require sedation or a lip twitch to keep them still.
2. Prepare mare for examination.	2a. Wrap the tail to keep hair out of the way (Fig. 4–7).
	2b. Wash perineum and vulva with dilute antiseptic solution.
	2c. External conformation of the perineum is examined. A mare with a sunken anus and tilted vulva may be more likely to suffer from fecal contamination and reduced fertility.

FIG. 4–7 Mare in a stock, readied for fertility examination. Ultrasound machine is off to the left, placed where the operator can see it easily. The bucket hanging on the stock contains a vaginal speculum in warm disinfectant solution.

TECHNICAL ACTION	RATIONALE/AMPLIFICATION
3. Set up ultrasound machine if applicable.	3a. Machine should be placed by the hind end of the horse, angled such that the palpator will be able to see the screen.
	3b. Attach the probe to the machine, taking care not to drop it or bang it into anything.
4. Examiner dons obstetrical sleeve.	4a. Some wear a sleeve on both arms.
	4b. Some tear the fingers off the obstetrical sleeves and use examination or surgical gloves to cover their hands.
	4c. Apply liberal amounts of lubricant to sleeve and hand.
5. Mare is palpated rectally.	5a. Examiner will first remove any feces from the rectum so as to reduce interference.
	5b. Once the rectum is cleared, the examiner palpates the cervix, uterus, and both ovaries, describing the structures as they are palpated.
	5c. If the mare is pregnant, gestational age is estimated either by determining the size of the vesicle in early pregnancies or the size of the foal's crown in late pregnancy.
	5d. At this point, the examiner may request the well-lubricated ultrasound probe.
6. Rectal ultrasonography is performed.	6a. The probe is passed carefully per rectum in a well-restrained horse.
	6b. The uterus is checked for fluid, lining abnormalities, or presence of a **fetus**.
	6c. If a fetus is present, it appears as a fluid-filled vesicle at 10–15 days, after which the embryo may be visualized. By 24 days the heart beat can be seen and the rate counted. The foal's sex is best identified between days 59 and 68.

TECHNICAL ACTION	RATIONALE/AMPLIFICATION
	6d. During the first trimester, the veterinarian checks for twins. If twins are present, one will be eliminated to protect the health of the mare and remaining fetus.
	6e. Later in pregnancy the placenta and fluids may be monitored to ascertain the health of the fetus.
	6f. The ovaries are examined identifying follicles, corpora lutea, and any ovulation sites. All structures are measured and recorded.
	6g. Often the examination is recorded either on video cassette or as pictures on special paper.
7. Vaginal examination.	**7a.** Generally done before breeding. May be done immediately postparturition as well.
	7b. The perineum and vulva are washed, and then a clean, well-lubricated vaginal speculum is inserted.
	7c. A light is shown into the vaginal vault and the walls are examined for tumors, lacerations, or urine pooling.
	7d. The cervix is inspected for injuries, defects, or mucopurulent discharge.
	7e. Specimens for uterine culture and biopsy may be acquired at this time.

Special Note

The rectum of a horse is easily torn. It is vital that the mare be well restrained and prevented from straining during rectal palpation. The danger is particularly high when an ultrasonographic probe is being used. Rectal tears in horses are life-threatening conditions and must be treated on an emergent basis.

Cow Reproductive Examination

TECHNICAL ACTION	RATIONALE/AMPLIFICATION
1. Restraint.	**1a.** Beef cows are placed in a squeeze chute.
	1b. Dairy cows may be placed in a squeeze chute, lined up in a narrow alleyway, or locked in head locks at the feed bunk.
2. Preparation of cow.	**2a.** No preparation is necessary.
3. Set-up for palpation.	**3a.** If the veterinarian will be using ultrasonography, the machine needs to be set up on the left side of the chute near the back gate, angled so that the veterinarian can see the screen.
	3b. Hook up the probe and (if available) the belt clip controls for the machine.
	3c. Most of the time, reproductive examinations are done on herds of cows, as many as 600 a day. In this case, the veterinarian will wear a sleeve protector on the palpation arm and shoulder.
	3d. Many veterinarians use a disposable obstetrical sleeve, while some wear a nylon or rubber obstetrical sleeve.
	3e. Set lubricant out near the back of the chute where the veterinarian can reach it easily.
4. Reproductive examination.	**4a.** The veterinarian inserts an arm (usually the left) into the rectum of the cow and scoops out some of the feces.
	4b. When palpating open (cows that are not pregnant) cows, the veterinarian gathers the uterus in the hand, identifying the cervix, uterine body, and uterine horns. The uterus is examined for fluid or irregularities and the cervix for degree of turgidity or flaccidity.

TECHNICAL ACTION	RATIONALE/AMPLIFICATION
	4c. The veterinarian then palpates both ovaries, noting the size and number of gravid follicles and presence of corpora lutea.
	4d. Pregnant cows are palpated to identify the stage of pregnancy. This is accomplished by palpating for cotyledons (placental attachment sites), size of uterine arteries, and if in late pregnancy, the size of the calf's head and feet. The number of months pregnant is often marked with a paint stick on the cow.
	4e. If twins are palpated early in pregnancy (first trimester), one twin is eliminated by pinching the vesicle. This is for the safety of the cow as well as of the other twin.
	4f. Some veterinarians use ultrasonography to determine the sex of the fetus.
5. Pelvic measurement.	**5a.** Most often this is done on heifers before breeding as a means of deciding which ones to keep in the herd or sell as breeding animals versus which ones to send to the feedlot.
	5b. The veterinarian introduces a caliper-style, hydraulic-style, or digital-style pelvimeter into the rectum (Fig. 4–8).
	5c. The width of the pelvis is measured at its widest point between the shafts of the ilia.
	5d. Then the distance from the pubic tubercle to the sacrum is measured.
	5e. The two numbers (in centimeters) are then multiplied to get pelvic area.

FIG. 4-8 An example of hydraulic-style pelvimeter used to measure the pelvic diameter of cattle.

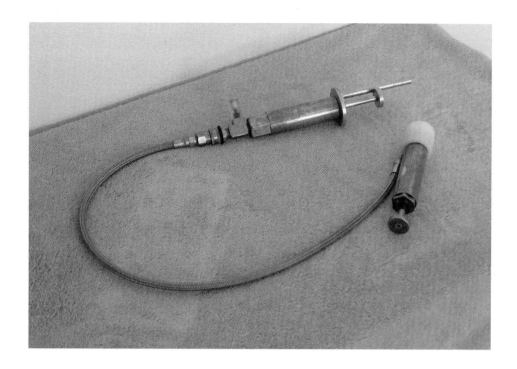

Small Ruminant Reproductive Examination

TECHNICAL ACTION	RATIONALE/AMPLIFICATION
1. Restraint.	1a. Goats may be restrained manually. Dairy goats may be induced to stand on a table stanchion.
	1b. Sheep may be restrained in a small chute, a narrow elevated alleyway, or held in a sitting position by an assistant.
	1c. Llamas may be placed in stocks or held by an assistant. If palpating the rectum or doing an ultrasonographic examination on the llama, give it a lidocaine caudal epidural to reduce the chance of rectal tear.
2. Pregnancy examination by ultrasonography.	2a. The easiest way to check a small ruminant for pregnancy is to examine her with ultrasonography (Fig. 4-9).
	2b. The ultrasound machine is set up in a place where the examiner can see the screen easily.
	2c. Ultrasonography gel is applied to the ventral caudal abdomen (inguinal area). If hair or wool is

FIG. 4-9 This ewe was in heavy pregnancy. The wool may need to be shaved just ahead of the udder. The ewe may be examined standing or sitting.

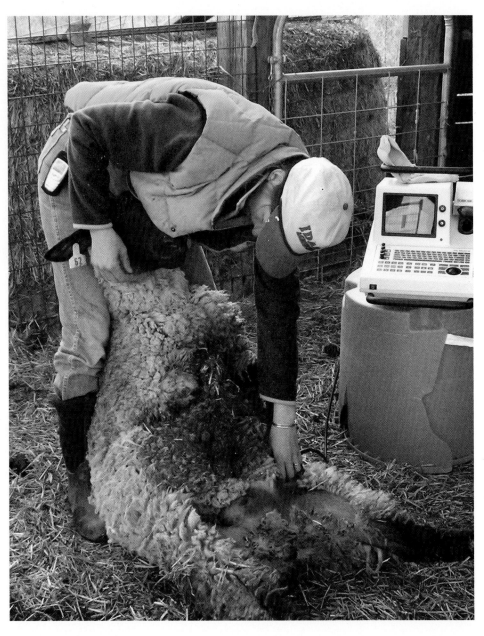

TECHNICAL ACTION	RATIONALE/AMPLIFICATION
	present, a 4-inch by 5-inch patch should be clipped away. The probe must come in direct contact with the skin to reduce the level of interference.
	2d. Dairy goats must be checked lateral to the udder.
	2e. Pregnancy may be diagnosed as early as 25 days in the sheep and goat; however, for litter size and near 100% accuracy, one should wait until 40–120 days after breeding.

Sow Reproductive Examination

TECHNICAL ACTION	RATIONALE/AMPLIFICATION
1. Restraint.	**1a.** Sows in large units are run through an elevated, narrow alleyway with a lower rail that can be removed to make it safer for the examiner to reach through.
2. Pregnancy examination.	**2a.** In some large pig farms, the sow may be examined via transabdominal ultrasonography to determine if a pregnancy exists and litter size.
	2b. Apply liberal amounts of ultrasonography gel to the sow's ventral abdomen.
	2c. The probe often has a gel cover to protect it from the stiff hairs on the sow.
	2d. The ultrasonographic device may be a small handheld unit or one of the larger machines used in the rest of veterinary practice.

REPRODUCTIVE EXAMINATION OF THE MALE

The male of the species should also be examined for reproductive capacity. A reproductive examination may be a part of routine herd management, may be done for insurance purposes, or may be done for sale purposes. The procedure is the same regardless of purpose.

Purpose

- Establish the likelihood that the male will be physically able to perform intromission
- Establish the presence of viable sperm
- Eliminate or decrease the spread of venereal disease
- Remove nonproductive animals from the herd

Equipment

- Obstetrical sleeves
- Lubricant
- Sterile, nonspermicidal lubricant (stallion)
- Artificial vagina and collection bottle (stallion)

- Water bucket and mild antiseptic soap
- Mounting dummy or jump mare (stallion)
- Electroejaculator and rectal probe (bull, ram, buck)
- Disposable or reusable collection sleeves
- Collection cup
- Test tube
- Scrotal calipers (stallion, llama, boar, bull)
- Scrotal tape (bull, ram, goat)
- Microscope and slides
- Live-dead stain
- Pasteur pipettes
- Spectrophotometer or hemocytometer (horse)
- Collection pipette and syringe (llama)

Complications

- Injury to handler: All personnel involved in collecting semen from a stallion should wear safety helmets.
- Injury to stallion from trying to mount the mare or dummy
- Injury to bull from falling in the chute
- Injury to patient from overzealous use of the electroejaculator
- Poor sample quality because of improper sampling techniques

PROCEDURES FOR REPRODUCTIVE EXAMINATION OF THE MALE

Stallion Breeding Soundness Evaluation

TECHNICAL ACTION	RATIONALE/AMPLIFICATION
1. Restraint.	1a. Halter, lead rope, and chain.
2. General physical examination.	2a. Refer to Physical Examination of the Horse.
3. Examination of external genitalia.	3a. The stallion's prepuce and penis are washed with warm water and dilute detergent. Be sure to rinse well (Fig. 4–10A).
	3b. The washing procedure may need to be done in the presence of a mare in heat in order to stimulate the stallion to extend his penis.

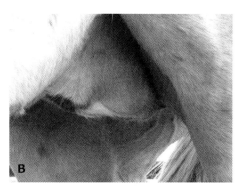

FIG. 4-10 (A) Stallion being washed in preparation for breeding. Handler is careful to stay out of reach of the hind legs. (B) Close-up of the prepuce and scrotum of a stallion.

TECHNICAL ACTION	RATIONALE/AMPLIFICATION
	3c. At this point the prepuce and penis may be examined, taking care not to get kicked.
	3d. The end of the penis (urethral process, fossa, and surrounding tissues) is examined for lesions and foreign material.
	3e. A small amount of greasy-feeling smegma may be present in the prepuce or even on the shaft of the penis.
	3f. The shaft of the penis should be free of scars, tumors, or other lesions.
	3g. The scrotum and testes may be examined immediately after semen collection, when the stallion is more docile.
	3h. Scrotal skin should be soft and pliable, with the testes moving freely within. Normal testes are oriented horizontally and are held relatively close to the body (Fig. 4-10B).
	3i. Scrotal width is measured with special calipers placed at the widest point.
4. Examination of accessory sex glands.	**4a.** The stallion is palpated rectally either before or after ejaculation, depending on the examiner's preference and cooperation of the stallion.

TECHNICAL ACTION	RATIONALE/AMPLIFICATION
	4b. Horses have bulbourethral glands, vesicular glands, and a prostate gland. They all supply a portion of the seminal fluid.
	4c. During rectal palpation, the veterinarian will also check the inguinal rings to see that there is no herniation of abdominal contents through the rings.
5. Semen collection process.	**5a.** A trained mare or a breeding dummy is required for the stallion to mount. Use of a breeding dummy is safer for both stallion and handlers.
	5b. The entire ejaculate is collected in a specially designed artificial vagina held by a helmeted assistant.
	5c. The artificial vagina is prepared by putting warm water (45°C to 48°C or 113°F to 115°F) in the water jacket that surrounds the artificial vagina. The amount of water is dependent on the stallion's preference for amount of pressure.
	5d. Sterile, nonspermicidal lubricant is added to the inner lining of the artificial vagina. Again, the amount is dependent on stallion preference.
	5e. The collection bottle is warmed to and kept warm at 38°C (100.4°F).
	5f. The stallion is exposed to a mare in heat, stimulated to reach a full erection, and then allowed to mount the trained mare or breeding dummy.
	5g. The person holding the artificial vagina gently guides the stallion's penis into the artificial vagina and braces against his thrusts.
	5h. When ejaculation occurs, the stallion raises and waves his tail (flagging).

TECHNICAL ACTION	RATIONALE/AMPLIFICATION
	5i. The semen is placed immediately in an incubator heated to 37°C (98.6°F) until evaluation.
6. Semen evaluation.	**6a.** Semen evaluation is based on four criteria. All materials the semen comes in contact with must be warmed to 37°C (98.6°F). **1.** Volume is the total amount of gel-free ejaculate and is measured in milliliters. **2.** Concentration is determined with a hemocytometer or spectrophotometer. **3.** Sperm motility is determined as soon after collection as possible, by placing a drop of semen on a warm slide and observing it with a microscope under high power. The technician will estimate the percentage of progressively motile spermatozoa. **4.** Sperm morphology is evaluated by microscopic examination of the individual sperm cells using specialized stain. One hundred sperm cells are counted, identifying those with abnormalities such as bent tails, missing tails, and double heads. At least 70% of the sperm cells on the slide should be normal.

Bull Breeding Soundness Evaluation

TECHNICAL ACTION	RATIONALE/AMPLIFICATION
1. Restraint.	**1a.** The bull is restrained in a squeeze chute. **1b.** Mature bulls have necks thicker than the width of their heads. For safety's sake, place a pole behind the bull through the chute's bars to keep him from backing out when the gate is open.

TECHNICAL ACTION	RATIONALE/AMPLIFICATION
2. General physical examination.	2a. Refer to Physical Examination of Beef Cattle.
	2b. Particular attention is paid to the feet and legs. A bull that can't walk well cannot catch and mount a cow well, either.
	2c. Eyesight is also important, because the bull uses visual cues to identify a cow in heat.
	2d. The body condition score should lie somewhere between 2 and 3.5. An obese bull will not have the stamina to pursue the cows and could even injure smaller cows with his weight. Very thin bulls may be more intent on eating than on breeding or may not have the reserves to last the season.
3. Scrotal circumference.	3a. Scrotal circumference in ruminants is correlated directly with volume of sperm produced.
	3b. The scrotum is first palpated. The testes should be freely moveable within the scrotal sac and oriented vertically.
	3c. The testes and epididymis are palpated for consistency, size, shape, and symmetry.
	3d. The testes should be of equal size. They should have the consistency of a flexed biceps muscle with no irregularities.
	3e. The epididymis should be palpable and symmetrical with the contralateral epididymis.
	3f. Scrotal circumference is measured by pushing the testes fully into the scrotal sac and pulling the scrotal tape snugly around the widest portion of the sac. The measurement is taken in centimeters and is read where the bar crosses the tape (Fig. 4–11A).

TECHNICAL ACTION	RATIONALE/AMPLIFICATION
4. Internal examination.	4a. The accessory sex glands (seminal vesicles, prostate, and ampullae) are palpated for size, shape, consistency, and symmetry.
	4b. Most often if there is a problem, it will be in the seminal vesicles.
	4c. Often the veterinarian will massage these organs and the urethra to provide stimulation prior to electroejaculation.
	4d. Inguinal rings are checked for herniation.
5. Semen collection.	5a. Semen collection in ruminants (except the llama) most often is done by electroejaculation.
	5b. A well-lubricated rectal probe is placed in the rectum, with the electrodes in a ventral position. An assistant should hold the probe in place throughout the procedure (Fig. 4–11B).
	5c. The probe is hooked up to a box that supplies the low-amperage, low-voltage electricity to the probe.
	5d. Starting on the lowest setting, the collector sends electricity to the probe in a pulsatile manner. This may be done manually by dialing a rheostat, or the machine may be programmed to do it automatically.
	5e. Gradually the amount of electricity going to the bull is increased, until the bull extends his penis and ejaculates.
	5f. The ejaculate is caught in a warmed test tube attached to a collection cup and sleeve (Fig. 4–11C).
	5g. While the penis is extended, the examiner checks for any lesions or scarring and checks to see that it is freely mobile within the prepuce.
	5h. The semen is evaluated immediately under a microscope.

SECTION 2 Physical Examination

FIG. 4-11 (A) Scrotal tape around the scrotum of a bull. *Arrow* points to the measurement marker on the tape. (B) Assistant placing the rectal probe in the anus of the bull. *Arrow* points to electrode strips. (C) Collecting the semen sample. Some veterinarians have the collection cup attached to a stick. Semen is visible in the bottom of the test tube.

TECHNICAL ACTION	RATIONALE/AMPLIFICATION
6. Semen evaluation.	**6a.** A drop of semen is placed on a warm slide and observed under the microscope for degree of progressive motility.
	6b. A slide is then made with special stain to evaluate the individual sperm morphology. One hundred sperm cells are counted, identifying those with abnormalities such as bent tails, missing tails, and double heads. At least 70% of the sperm cells on the slide must be normal for the bull to pass.
7. Breeding soundness evaluation form.	**7a.** In North America, veterinarians use a standardized system for evaluating bulls and other ruminants for breeding soundness.
	7b. The Society for Theriogenology has set out criteria the animal must meet to be considered a satisfactory breeder. The animal must pass all categories.
	7c. Categories include: physical examination, scrotal circumference, sperm motility, and sperm morphology. Fig. 4-12 shows the form used by this author.

BREEDING SOUNDNESS EXAM RESULTS

Animal Id._____

Sperm Motility:	Very Good	Good	Fair	Poor
	(Rapid Linear)	(Mod. Linear)	(Slow Linear)	(Erratic)

Sperm Morphology:
%Primary abnormals:_____ %Secondary abnormals:_____
%Normal sperm:_____ Must be minimum of 70% to pass

Scrotal Circ/Age:	Very Good	Good	Unsatisfactory
<15 months	34+	30-34	<30
15-18mos	36+	31-36	<31
19-21mos	38+	32-38	<32
22-24mos	39	33-38	<33
Over 31 mos	39+	34-39	<34

Physical Exam:

General Condition		Norm/Abnorm			Norm/Abnorm	
Obese ☐	Vesic Glands	☐	☐	Penis	☐	☐
Good ☐	Epididymes	☐	☐	Eyes	☐	☐
Fair ☐	Testes	☐	☐	Feet	☐	☐
Poor ☐	Prepuce	☐	☐	Other_____		

Final Result
☐ Satisfactory Potential Breeder
  Bull must pass all categories to pass a breeding soundness evaluation.
☐ Unsatisfactory Potential Breeder
  Failed following category(ies)_____
☐ Questionable Breeder: Re-evaluation suggested in_____days.

Comments:

Trichomoniasis Tested: Date:_____ Result:_____

Signed:_____ Date:_____

FIG. 4-12 Report form used by author's clinic, based on recommendations of the Society of Theriogenology.

Special Note

Some bulls may lie down when semen is being collected via electroejaculation. If it is not possible to collect the sample, these bulls can be stimulated manually per rectum.

Small Ruminant Breeding Soundness Evaluation

TECHNICAL ACTION	RATIONALE/AMPLIFICATION
1. Sheep (ram).	1a. Semen is not tested often except for ram sale or unless a fertility problem exists in the flock. This depends on your local area.
	1b. Rams are evaluated for conformational defects and overall general health first.
	1c. In large groups, the rams are run into a narrow alleyway. The testes and epididymis are palpated, with special attention paid to the head of the epididymis where infection tends to lie.
	1d. A full breeding soundness examination may be done on an individual using the same procedure as for the bull.
2. Goats (buck).	2a. Because of smaller herd size (than sheep) goats are more likely to undergo a full breeding soundness examination.
	2b. The only differences from examination of the bull are that a rectal probe of appropriate size is needed, goats are often held by the handler, and goats make a lot more noise when electroejaculation is done.
	2c. A trained goat may be induced to mount a dummy in the presence of a doe in heat, and then ejaculate into a collection sleeve and cup.
3. Llamas (male).	3a. Llama breeding soundness examinations are more similar to those of horses than other ruminants.
	3b. The penis in the llama is directed caudally in a relaxed state and the scrotum is tight to

TECHNICAL ACTION	RATIONALE/AMPLIFICATION
	the ventral perineum, much like that of a boar (Fig. 4–13).
	3c. The most effective way to obtain a semen sample is to allow the llama to breed a female (done with the female in sternal recumbency) for a minimum of 12 minutes. Seminal fluid is then collected from the vagina via pipette.
	3d. Electroejaculation can be done on lamas, but it requires general anesthesia.

FIG. 4–13   Close-up of the scrotum of a llama.

EXAMINATION FOR LEGAL PURPOSES

Examinations done for legal purposes include those done for a health certificate, insurance purposes, and for prepurchase evaluation. Legally, the veterinarian must do the entire examination of the patient or patients. The completeness of the examination and the diagnostic tests that will be performed depend upon the purpose for the examination.

HEALTH CERTIFICATE

A health certificate is required for interstate transport of all livestock (Fig. 4–14). For groups of animals such as cattle or sheep, each animal may need to be inspected and assigned an identity tag, or the entire group may be inspected as one. Diagnostic testing depends on the state regulations concerning that particular species, age, and sex. The owner, the location from which they are leaving, and the location to which they are going must all be filled out legibly on the health certificate. Even the name of the transporter or transport company must be given. Often the state to which the animals are traveling must be called in advance to get up-to-date regulations and a permit number. If any line is missed or filled out improperly, the animals may be denied entry into the state, and the veterinarian or owner, or both, may be fined. In general, health certificates are valid for only 30 days from the date of examination.

International transport regulations are even more complicated. Each country has its own import rules and regulations. A special health certificate issued by that country must be filled out and signed by the examining veterinarian. Then it must be presented to and signed by a federal veterinarian who usually is

IDAHO DEPARTMENT OF AGRICULTURE
BUREAU OF ANIMAL HEALTH, BOISE, IDAHO

CERTIFICATE OF VETERINARY INSPECTION **82-** 110288

Name and Address of Consignor	Name and Address of Consignee	Brand Inspection No.	**PERMIT NO.**
Origin Address *(if Different From Above)*	Destination Address *(if Different From Above)*	Carrier Name and Address	

Species: ☐ Cattle ☐ Horses ☐ Sheep ☐ Swine ☐ Poultry ☐ Other________
Number of Animals Shipped
ORIGIN OF SHIPMENT: County; Market
Area Status: ☐ TB Free ☐ TB Mod. Accred. ☐ Brucellosis Free ☐ Brucellosis A ☐ Brucellosis B ☐ Brucellosis C ☐ Other __________
Herd or Flock Status: ☐ Accredited Herd No. ______ ☐ Certified Herd No. ______ ☐ Validated Herd No. ______ ☐ Other__________
QUAL. HERD TEST DATES: 1. 2. 3.
VACCINATION and/or TREATMENT: For; Date and Product

INDIVIDUAL ANIMAL IDENTIFICATION AND TEST DATA

	OFFICIAL PERMANENT INDIVIDUAL IDENTIFICATION AND DESCRIPTION	**AGE**	**BREED**	**SEX**	**TUBERCULIN TEST (INTRADERMAL)** INJ Date / Hour; OBS Date / Hour	**BRUCELLOSIS** Date; Lab (Name and Address); TEST: CARD	PLATE	RIV	CF	**TEST RESULTS**	**BRUC. VACC. TATTOO SYMBOL**	**OTHER TESTS** Test For; Lab and Date; **RESULTS**	Test For; Lab and Date; **RESULTS**
1													
2													
3													
4													
5													
6													
7													
8													
9													
10													
11													
12													
13													
14													
15													
16													
17													
18													
19													
20													
21													
22													
23													
24													

VETERINARY CERTIFICATION
I certify, as an accredited veterinarian, that the above-described animals have been inspected by me and that they are not showing signs of infectious, contagious, and/or communicable disease (except where noted). The vaccinations and results of tests are as indicated on the certificate. To the best of my knowledge, the animals listed on this certificate meet the state of destination and federal interstate requirements. No further warranty is made or implied.
Date | Accredited Veterinarian
Address

OWNER/AGENT STATEMENT (Where applicable)
"The animals in this shipment are those certified to and listed on this certificate."
Date | Owner/Agent
Address

State Verification

FIG. 4–14 Example of a state health certificate.

located in the state's capitol. This means there must be time to get the health certificate to and from the federal veterinarian's office. Often, many diagnostic blood tests are required at very specific times before transport. Food animals headed to Canada may need to be transported in a vehicle that has been sealed by either the veterinarian or livestock inspector. All access doors on the livestock trailer will have a special numbered band secured to the latch, making it impossible to open the doors without breaking the seal. Keeping track of all the paperwork is vital to the veterinarian, the client, and the buyer. Without correct paperwork, the animals will not be allowed to cross international borders.

Horses require an up-to-date negative test result for equine infectious anemia (Coggins test), and each horse must be identified and described individually (Fig. 4-15). Coggins tests must be done at a certified laboratory, which may necessitate mailing blood samples off and waiting for results to come back. The Coggins test can be ordered only by a veterinarian or certified agent of the government. Again, the state to which the horse is going should be called to verify transport regulations and to get any permit numbers that may be needed. Some states work cooperatively and have developed a 3-month passport system so the owner can compete across state lines without having to get a health certificate every 30 days. Most states have a brand inspection program and require that an inspector verify that the horse being transported is the same horse listed on the transport papers.

International horse travel is far more complicated and may involve housing the animal in a certified quarantine facility either before leaving this country, on arrival at the destination country, or when and if the horse returns to this country. Regulations change often, so a call to the federal veterinarian's office is necessary. International show horses travel with a special passport that identifies the horse and verifies its health.

INSURANCE EXAMINATIONS

Livestock can be insured for a variety of reasons, and coverage ranges from mortality to loss of production. Bulls commonly are insured at the time of purchase against loss of production. In other words, if something happens to the bull so that he can no longer mount and breed a cow, the insurance will pay the owner a portion of the purchase price. For the animal to be insured, he must have a complete breeding soundness evaluation done by a veterinarian. More valuable bulls and cows will need a complete physical examination, complete with diagnostic tests, done as well. All paperwork must be filled out accurately and promptly.

Horses may be insured against mortality or illness or loss of use. Some owners will insure the horse against mortality just for the time it is being transported. The value of the horse is based on the purchase price, any monies it has earned, and its show or breeding record (if applicable). Again, stallions must have a complete breeding soundness evaluation performed by a veterinarian in order to qualify, and mares must be certified as fertile. Any horse will need a complete physical examination, vaccination, and deworming history submitted for insurance. Paperwork varies with the insurance company, and it is vital that it be filled out correctly.

PREPURCHASE EXAMINATION OF HORSES

Most of the time, a prepurchase examination is done on a performance horse to verify that it is physically capable of doing the work for which it has been sold. Every blemish, mark, and injury (old or new) is carefully recorded. Often the horses undergo many diagnostic tests in addition to the general physical

USE TYPEWRITER OR PRINT CLEARLY-PRESS HARD-YOU ARE MAKING 5 COPIES

See reverse for more OMB information. FORM APPROVED - OMB NUMBER 0579 - 0127

U.S. DEPARTMENT OF AGRICULTURE ANIMAL AND PLANT HEALTH INSPECTION SERVICE **EQUINE INFECTIOUS ANEMIA LABORATORY TEST** (VS Memorandum 555.8)	SERIAL NO. D 155273	1. ACCESSION NUMBER	2. DATE BLOOD DRAWN

Forms Without Adequate Descriptions Of The Horse and Complete Addresses Including Zip Codes, Counties, and Telephone Numbers Will Not Be Processed.

3. REASON FOR TESTING ☐ Show ☐ First Test ☐ Market ☐ Change of Ownership ☐ Retest ☐ Export

7. NAME AND ADDRESS OR STABLE/MARKET *(Please print or type)* Zip Code Tel No. County

4. GEOGRAPHIC INFORMATION SYSTEMS (GIS) LAT: LONG:

5. VETERINARY LICENSE OR ACCREDITATION NO.

6. TEST TYPE ☐ ELISA ☐ AGID

8. NAME AND ADDRESS OF OWNER *(Please print or type)* Zip Code Tel No. County

8. NAME AND ADDRESS OF VETERINARIAN *(Please print or type)* Zip Code Tel No. County

CERTIFICATION OF FEDERALLY ACCREDITED VETERINARIAN

I certify the specimen submitted with this Form was drawn by me from the horse described below on the date indicated above.

10. SIGNATURE OF FEDERALLY ACCREDITED VETERINARIAN	11. TYPE OR PRINT SIGNATURE NAME	12. SIGNATURE DATE

CERTIFICATION OF OWNER OR OWNER'S AGENT

I certify that I have examined this form and, to the best of my knowledge and belief, this form is true, correct and complete.

13. SIGNATURE OF OWNER OR OWNER'S AGENT	14. TYPE OR PRINT SIGNATURE NAME	15. SIGNATURE DATE

16. Tube No.	17. Official Tag. No.	18. Tattoo/Brand	19. Name of Horse	20. Color	21. Breed	22. Electronic I.D. No.	23. Age or DOB	24. Sex	M - Male F - Female G - Gelding N - Neuter

SHOW ALL SIGNIFICANT MARKINGS, WHORLS, BRANDS, AND SCARS

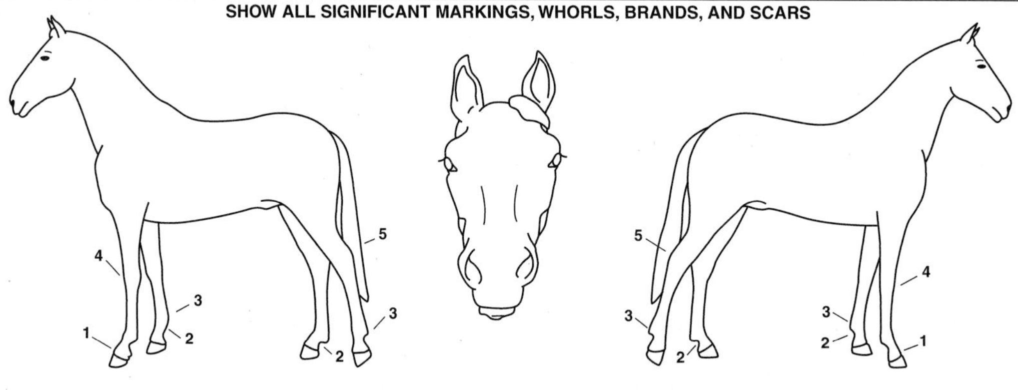

1 - Coronet, 2 - Pastern, 3 - Fetlock, 4 - Knee, 5 - Hock

NARRATIVE DESCRIPTION AND REMARKS

25. HEAD	26. OTHER MARKS AND BRANDS
27. LEFT FORELIMB	28. RIGHT FORELIMB
29. LEFT HINDLIMB	30. RIGHT HINDLIMB

FOR LABORATORY USE ONLY

31. LABORATORY NAME/CITY/STATE	32. DATE RECEIVED	33. DATE REPORTED OUT	34. TEST RESULTS ☐ Negative ☐ Positive ☐ AGID ☐ ELISA
	36. SIGNATURE OF TECHNICIAN		35. REMARKS

Falsification of this form or knowingly using a falsified form is a criminal offense and may result in a fine of not more than $10,000 or imprisonment for not more than 5 years or both (U.S.C. Section 1001).

FIG. 4–15 Example of an equine infectious anemia (Coggins) test form, which must be filled out for sample submission.

examination. Radiographs may be taken of the feet and legs, looking for hidden damage. In-depth lameness examinations often are done to test every joint for soundness. Sometimes even nerve blocks are done. The horse may have endoscopy done on its upper airway to look for breathing difficulties. Blood may be taken to test for inapparent infections, or kidney or liver damage. A drug test may be run as well, or some of the serum kept in the freezer for future testing should the purchaser request it. All horses should also have a negative Coggins test result within 6 months of sale. The extent to which the purchaser wants the horse examined is often dependent upon the purchase price. All results are the property of the veterinarian and the purchaser; the owner of the horse must not be communicated with unless the purchaser agrees to it in writing. Complete, accurate records safely stored away are vital to the veterinarian should a complaint be filed in the future. All veterinary records are legal documents and subject to review.

REVIEW QUESTIONS

1. Why is taking a good history important to the diagnosis and treatment of the patient?
2. What five questions are important to ask when inquiring about vaccinations?
3. List five questions important to ask a dairy producer.
4. How do the questions asked of a dairy producer differ from the questions asked of a beef producer?
5. Why is it important to know if any treatments or medications were given to the patient prior to the visit?
6. List three questions one should ask a llama owner.
7. List three questions asked of a pig producer.
8. Why are dairy cattle usually examined in stocks or a head lock, while beef cattle are examined in a squeeze chute?
9. What is the importance of body condition scoring in the physical examination?
10. Why is it important to observe the patient before restraining it for a physical examination?
11. How does the breeding soundness evaluation of a stallion differ from that of a bull?
12. Transabdominal ultrasonography can be used to diagnose pregnancy in what species?
13. Why is it important to have a mare or stallion well restrained for rectal palpation?
14. All items used in semen evaluation should be warmed to _____ degrees Celsius.
15. How does a prepurchase or insurance examination differ from a routine physical examination?

BIBLIOGRAPHY

Ball, L., Ott, R. S., Mortimer, R. G., Simmons, J. C. (1983). Manual for breeding soundness evaluation of bulls. *J Soc Theriogenology, XII,* 1–65.

Connor, J. F., & Tubbs, R. C. (1992). Management of gestating sows. *Compend Contin Educ Pract Vet, 14*(10), 1395–1419.

Dascanion, J. J., Parker, N. A., Purswell, B. J., Digrassie, W. A., Bailey, T. L., Ley, W. B., et al. (1997). Diagnostic procedures in mare reproduction: Basic evaluation. *Compend Contin Educ Pract Vet, 19*(8), 980–985.

Fowler, M. (1998). *Medicine and surgery of South American camelids* (2nd ed.). Ames: Iowa State Press.

Kelly, W. R. (1967). *Veterinary clinical diagnosis.* Baltimore: Williams & Wilkins.

Kennedy, S. P., Spitzer, J. C., Hopkins, F. M., Higdon, H. L., Bridges, W. C. Jr. (2002). Breeding soundness evaluations of 3,648 yearling beef bulls using the 1993 Society for Theriogenology guidelines. *Theriogenology, 58*(5), 947–961.

Levis, D. G. (1997). Managing postpubertal boars for optimum fertility. *Compend Contin Educ Pract Vet, 19*(Suppl 1), S17–S23.

Lowman, B. G., Scott, N. A., & Sumerville, S. H. (1976). *Condition scoring of cattle.* East of Scotland College of Agriculture Bulletin 6.

Ott, R. S., & Memon, M. A. (1980). *Breeding soundness examinations of rams and bucks* (38–43). Sheep and goat manual from Society for Theriogenology vol.X, Hastings NE.

Ott, R. S., & Memon, M. A. (1980). *Pregnancy diagnosis* (34–37). Sheep and goat manual from Society for Theriogenology vol.X, Hastings NE.

Ruegg, P. L. (1991). Body condition scoring dairy cows: Relationships with production, reproduction, nutrition, and health. *Compend Contin Educ Pract Vet, 13*(8), 1309–1312.

Smith, B. P. (1990). *Large animal internal medicine.* Philadelphia: C. V. Mosby.

Spitzer, J. C. (2000). Bull breeding soundness evaluation: Current status. In P. S. Chenoweth (Ed.). *Topics in bull fertility.* Ithaca NY: IVIS. Retrieved January 16, 2006, from *http://www.ivis.org* Search term: A0501.1000.

Stroud, B. K. (1994). Clinical applications of bovine reproductive ultrasonography. *Compend Contin Educ Pract Vet, 16*(8), 1085–1097.

Tibary, A., & Anouassi, A. (2000). Reproductive disorders in the female camelid. In L. Skidmore & G. P. Adams (Eds.). *Recent advances in camelid reproduction.* Ithaca NY: IVIS. Retrieved January 16, 2006, from *http://www.ivis.org* Search term: A1007.1100.

Westendorf, M., Absher, C. W., Burris, R. W., Gay, N., Johns, J. T., Miksh, J. D. (1988). *Scoring beef cattle condition.* Kentucky Extension Service, ASC–110. Lexington.

Wolverton, D. J., Perkins, N. R., & Hoffis, G. F. (1991). Veterinary application of pelvimetry in beef cattle. *Compend Contin Educ Pract Vet, 13*(8), 1315–1320.

SECTION THREE

Sample Collection and Clinical Procedures

Chapter 5
Sample Collection

Chapter 6
Clinical Procedures

Chapter 7
Neonatal Clinical Procedures

CHAPTER 5

Sample Collection

> *As for butter verses margarine, I trust cows more than chemists.*
> — JOAN GUSSOW

KEY TERMS

- abdominocentesis
- aspiration
- biopsy
- California mastitis test
- carotid
- catheterization
- coagulation studies
- coccygeal
- dermatophyte
- impression smear
- jugular
- necropsy
- rumenocentesis
- thoracocentesis
- transtracheal wash
- venipuncture

OBJECTIVES

- Identify purposes for routine sample collection procedures in the horse, cow, llama, goat, and pig.
- List equipment and techniques required to perform common sample collections.
- Discuss complications associated with sample collection in the various animal species.
- Compare and contrast sample collection procedures used in large animal practice.

BLOOD COLLECTION

Blood collection is one of the most frequently performed procedures in clinical practice. Accurate complete blood counts, chemistry profiles, blood gas analyses, and serology all hinge upon obtaining a quality sample. Adherence to appropriate technique minimizes animal discomfort and ensures that results are of diagnostic value.

VENIPUNCTURE

Purpose

- Obtain blood samples for diagnostic clinical tests including complete blood cell counts, antibody titers, or serum chemistry analysis
- Obtain blood samples to evaluate response to therapy
- Administer various substances including medication, fluids, or diagnostic chemicals

Complications

- Iatrogenic infection
- Hematoma formation
- Inadvertent esophageal puncture (left side)
- Inadvertent puncture of internal carotid artery (jugular collection)
- Ocular trauma (facial collection)
- Thrombophlebitis
- Minor hemorrhage

Equipment

- 20g × 1.5-inch needle (horse, llama, large pig jugular vein)
- 22g × 1-inch needle (small pig jugular vein)
- 18g × 1.5-inch needle (cow jugular vein)
- 20g × 1-inch needle (cow coccygeal vein)
- 18g × 1-inch needle (goat jugular vein)
- 20–23g × 1-inch needle (pig marginal ear vein)
- 12-cc syringe
- Double-ended collection needle with sheath (Vacutainer system, alternative to needle and syringe)
- Isopropyl alcohol 70%
- Cotton

- Microhematocrit tube, 22g × 1-inch needle (facial sinus, marginal ear vein)
- Collection tubes: Type depends on types of diagnostic tests (Table 5-1 and Fig. 5-1)

TABLE 5-1 COLLECTION TUBES

TUBE TOP COLOR	CONTENT	FUNCTION
Purple top	EDTA	Complete blood cell counts
Red top	No additive	Serum studies
Blue top	Sodium citrate	Coagulation studies
Green top	Heparin	Blood smears

EDTA, ethylamine diamine tetraacetic acid.

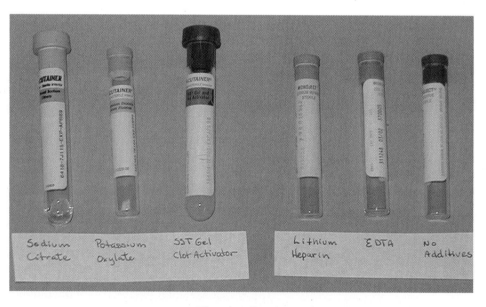

FIG. 5-1 Blood collection tubes.

Procedure for Jugular Venipuncture of the Horse

TECHNICAL ACTION	RATIONALE/AMPLIFICATION
1. Halter horse and place in stock. Alternatively, an assistant standing on the same side as the collector can restrain the horse.	1a. Never stand directly in front of horse or under neck.
2. Identify jugular furrow.	2a. Fig. 5-2.

TECHNICAL ACTION	RATIONALE/AMPLIFICATION
3. Wet furrow using alcohol-soaked cotton swab.	—
4. Occlude jugular vein.	4a. Occluding both veins provides greater filling and visualization of the jugular vein.
5. Hold needle bevel side up and place in distended jugular vein.	—
6. Using both hands attach syringe to needle.	—
7. Reocclude vein and collect desired blood volume.	—
8. Alternatively, a Vacutainer system may be used for collection. Once the double-ended needle has been inserted, a collection tube is used instead of a syringe.	8a. The vacuum within the tube allows blood to flow in freely. 8b. Vacutainer needles will not exhibit a flash (dripping of blood from the needle). 8c. Multiple tubes can be collected without withdrawing the needle.
9. After collection, apply digital pressure for 20 seconds.	9a. Pressure prevents hematoma formation.

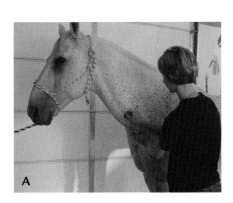
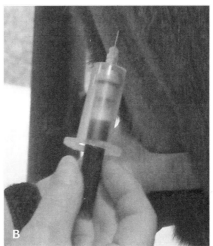
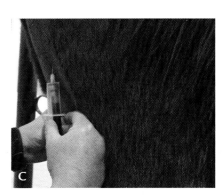

FIG. 5-2 (A) Position of person collecting blood, their hands, and the horse. (B) Close-up of the hand holding off the vein while collecting blood. (C) Close-up of Vacutainer system during blood collection.

Special Notes

Needle may be placed in a downward, toward the heart, direction. This technique will lessen the chance of inadvertent carotid artery puncture. Alternatively, needle may be placed in an upward direction, away from the heart. This technique minimizes the amount of air entering the vein.

Procedure for Facial Sinus Venipuncture of the Horse

TECHNICAL ACTION	RATIONALE/AMPLIFICATION
1. Halter and place horse in stock.	—
2. Identify point of insertion for venous sinus. Sinus is located ventral to facial crest, midway between the medial canthus of the eye and the rostral end of the facial crest.	2a. Facial collections are recommended only when the jugular vein must be preserved for catheters. 2b. This site is limited to collection of small blood volumes. 2c. Fig. 5–3.
3. Clean site using alcohol swab.	—
4. Insert 20–22g × 1-inch needle perpendicular to skin.	—
5. Allow blood to drip into microhematocrit tubes or attach 3-cc needle for collection.	5a. It is difficult to attach syringe and keep the needle in the sinus.
6. Remove needle and apply digital pressure for 20 seconds.	—

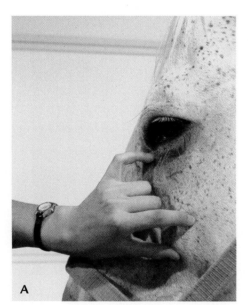 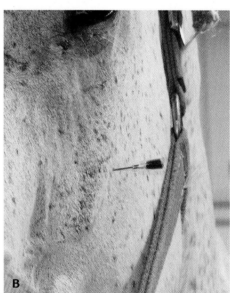

FIG. 5–3 (A) Landmark for locating the facial venous sinus. Pinkie finger is on the medial canthus of the eye. Thumb is placed on rostral end of facial crest. Forefinger is dropped midway between ventral to the facial crest. (B) Close-up of collecting blood from the facial sinus using a needle.

Special Note

This technique should be attempted only by the experienced technician. Inadvertent puncturing of the eye can occur if the horse moves suddenly.

Procedure for Jugular Venipuncture of the Cow

TECHNICAL ACTION	RATIONALE/AMPLIFICATION
1. Place cow in chute or stock.	—
2. Secure head to side of stock using halter.	2a. Head can be secured to either side.
3. Identify jugular furrow.	3a. Fig. 5–4.
4. Wet furrow using alcohol-soaked cotton swab.	4a. Alcohol assists in raising vein.
5. Occlude vein (using left hand) two-thirds of the way caudal (down) on the neck.	—
6. Grasp needle by hub with thumb and forefinger.	—
7. Thrust needle through skin at a 45–90 degree angle to the vein.	—
8. Confirm vein penetration by witnessing blood stream from hub of needle.	8a. Cows will have a stream of blood from the hub; horses will have a drip.
	8b. If needle has only penetrated skin and not entered the vein, relocate the vein and thrust needle into vein without removing from skin.
	8c. If the needle has been injected too deeply and has penetrated through the vein, pull the needle back slowly.
9. Thread the needle by laying it parallel to the skin and advancing it into the vein. Thread needle to its hub.	—
10. Using both hands, attach syringe to needle.	—
11. Reocclude vein (using left hand) and collect desired blood volume (using right hand).	—
12. After collection, apply digital pressure for 20 seconds.	12a. Pressure prevents hematoma formation.

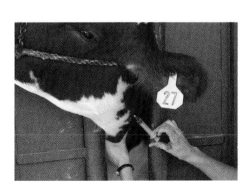

FIG. 5–4 Position of cow and handler's hands for jugular blood collection.

Special Note

Needles typically are placed in a downward direction (toward the heart) in cattle.

Procedure for Coccygeal Venipuncture of the Cow

TECHNICAL ACTION	RATIONALE/AMPLIFICATION
1. Place cow in chute or stock.	1a. Dairy cattle, which are much less fractious than beef cattle, can often be haltered and tied for this procedure.
2. Stand directly behind and tail-up the cow.	2a. Fig. 5–5. 2b. Tailing-up decreases animal movement and aids in visualization.
3. Clean manure from ventral side of tail using alcohol wipe.	—
4. Palpate vertebral prominences and the groove that runs directly along the ventral midline of tail.	—
5. Attach a 3–6 cc syringe to needle (20g × 1 inch) before starting collection.	5a. Attaching syringe after needle insertion often will result in laceration of the vessel.
6. Insert needle at a 90-degree angle (perpendicular) directly on midline groove, approximately 3 inches from base of tail. Avoid inserting directly over bony prominence.	—
7. Collect 2–5 cc of blood.	7a. Limited amounts of blood can be collected from this site.
8. Remove needle and apply digital pressure for 30 seconds.	8a. Digital pressure prevents hematoma formation. This procedure is not performed in many cattle.

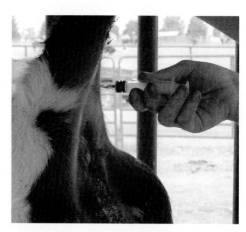

FIG. 5–5 Collecting blood from the coccygeal vein of a cow.

Special Note

Coccygeal (tail vein) collection typically requires less restraint and is less stressful for the cow. Collection at this site, however, is limited to small volumes.

Procedure for High-neck Jugular Venipuncture of the Llama

TECHNICAL ACTION	RATIONALE/AMPLIFICATION
1. Halter llama and secure in stock if available. Flex head slightly and allow neck to bow so that convex side is toward the collector.	1a. If stock is not available, stand llama next to fence or building to minimize movement.
	1b. Head position is critical to obtaining this sample.
2. With head in identified position, draw an imaginary line from ventral mandible to neck.	2a. Do not clip animal unless permission is obtained from owner. Regrowth of fleece can take up to 18 months.
	2b. Llama jugular puncture is more difficult than in other large animal species because of the vessel's protected location. It is protected by the enlarged transverse processes of the cervical vertebrae. Skin in the cranial cervical area is almost ½-inch thick.
	2c. These features were developed to protect animals from exsanguination caused by bites from fighting males.
3. Palpate sternomandibularis tendon. Draw a parallel line along this tendon. Needle insertion is the point just dorsal and caudal to the intersection of these two lines.	3a. Fig. 5–6.

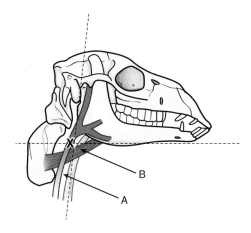

FIG. 5–6 Close-up of llama's head for high jugular venipuncture. Landmarks include mandible, the sternohyoideus tendon (**a**), and the omohyoideus muscle (**b**).

TECHNICAL ACTION	RATIONALE/AMPLIFICATION
4. Apply pressure to ventrum of vertebra to occlude vein.	4a. The distended vein typically can not be visualized.
5. Collector can attempt to confirm vessel location by strumming finger of collecting hand along the vessel and feeling for fluid wave against the occluding hand.	5a. This technique is not extremely reliable, and collector may need to attempt collection using landmark palpation only.
6. Insert 18–20g × 1.5-inch needle with attached syringe at a 45-degree angle to the skin.	6a. Sample collection is facilitated if the needle is inserted with syringe already attached.
7. Collect sample.	—

Special Note

High jugular collection technique decreases the chance of inadvertent arterial puncture, because the jugular vein is very superficial in this area. Unfortunately, the jugular vein is not readily visible in this area and the collector must rely on landmarks to obtain the sample.

Procedure for Low-neck Jugular Venipuncture of the Llama

TECHNICAL ACTION	RATIONALE/AMPLIFICATION
1. Place llama in stock or other restraint device to prevent forward motion.	—
2. Elevate head.	—
3. In lower third of neck, palpate for the enlarged transverse processe of the sixth cervical vertebra.	3a. Fig. 5–7.
	3b. The jugular vein lies just medial to this process.
	3b. The pulsating carotid artery typically can be palpated medial to the transverse process.
4. Occlude vein. Confirm jugular penetration point by occluding and releasing vein several times.	4a. Watch for jugular vein distention between the fifth and sixth cervical vertebrae.
	4b. Do not clip animal unless permission is obtained from owner. Regrowth of fleece can take up to 18 months. If permission is obtained, clipping can assist with visualization.

TECHNICAL ACTION	RATIONALE/AMPLIFICATION
5. Attach 18–20g × 1.5-inch needle to syringe and insert slightly medial to the process, aiming toward the center of the neck.	5a. Attaching the syringe before insertion aids in collection.
6. Collect sample.	—

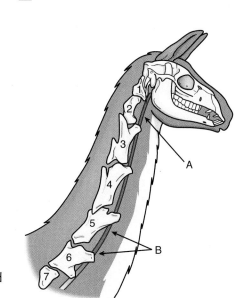

FIG. 5-7 Jugular venipuncture sites in the llama. High jugular site (**a**) and two low jugular sites (**b**).

Special Note

Thinner skin, better visualization, and fewer problems associated with animal movement are clear advantages of the low-neck collection site. Disadvantages include heavier fleece and increased likelihood of arterial puncture.

Procedure for Jugular Venipuncture of the Goat

TECHNICAL ACTION	RATIONALE/AMPLIFICATION
1. Place goat in stanchion, apply halter, and secure head to the side. Alternatively, an assistant can back goat into a corner then straddle neck.	1a. If the goat is too large to straddle, the assistant can back animal into corner or against fence.
2. Grasp mandible and elevate head while turning head slightly to the side.	2a. Fig. 5–8.
3. Identify jugular furrow and wipe with alcohol.	3a. Alcohol will remove dirt and assist with visualization, especially in goats with long or thick hair.

TECHNICAL ACTION	RATIONALE/AMPLIFICATION
4. Occlude vein at the level of the lower third of the neck or thoracic inlet.	4a. Occluding and releasing vein several times can help with visualization.
5. Attach an 18g × 1-inch needle to syringe. Insert needle bevel up at a 30-degree angle to the skin, parallel to the jugular vein.	—
6. Collect sample.	—

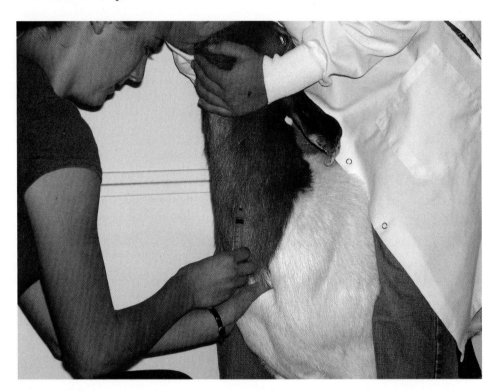

FIG. 5–8 Restraint and positioning of the goat for jugular venipuncture.

Procedure for Jugular Venipuncture of the Small Pig

TECHNICAL ACTION	RATIONALE/AMPLIFICATION
1. Place pig in dorsal recumbency on thighs of assistant who is seated. The pig's snout should be at assistant's knees and tail toward the assistant's abdomen.	1a. This technique is used on pigs weighing 30 lbs or less. 1b. Assistant is advised to sit on an object or in a chair that is stable and will not easily move.
2. Assistant will use left hand to pull the front limbs of the piglet against its abdomen and the right hand to hold snout and extend neck.	2a. Do not overextend neck, because this can cause respiratory distress.
3. Collector should attach a 22–23g × 1-inch needle to syringe.	—

TECHNICAL ACTION	RATIONALE/AMPLIFICATION
4. Identify manubrium. Insert needle 1 inch craniolateral to the manubrium at an angle 45 degrees to the skin.	4a. Fig. 5–9. 4b. Needle tip is pointing caudally toward the heart.
5. Once needle has penetrated skin, begin aspirating as the needle is pushed forward. Vessels typically are encountered within 10–20mm.	—
6. Once a flash is visible, maintain needle position and collect sample.	6a. Flash is the presence of blood in the needle hub. 6b. Placing the collecting hand against the pig during collection will help stabilize the needle.

FIG. 5–9 Positioning and collection of blood from a pig weighing less than 30 pounds.

Procedure for Jugular Venipuncture of the Large Pig

TECHNICAL ACTION	RATIONALE/AMPLIFICATION
1. Secure pig using a snout snare and stretch neck upward.	1a. This technique should be used on all pigs weighing 75 lb or more. 1b. The pig must be bearing weight on all four limbs.

TECHNICAL ACTION	RATIONALE/AMPLIFICATION
	1c. Holding technique for pigs 31–74 lbs is at the discretion of the assistant who is holding.
2. Identify the deepest point in the jugular groove in the lower third of the neck.	**2a.** Fig. 5–10.
	2b. This point is usually 4 inches from the point of the shoulder, or 7 inches from the mandible.
	2c. Right-handed collectors will find it easiest to use the right jugular vein.
3. Attach a 20g × 1.5-inch needle to syringe.	—
4. Insert needle perpendicular to skin. Aspirate as needle is advanced slowly.	—
5. Once flash is visible, maintain needle position and collect sample.	—

FIG. 5-10 Positioning of a large pig for jugular venipuncture. Animal is restrained by a hog snare. Note angle of syringe and needle.

Procedure for Venipuncture of the Marginal Ear Vein of the Pig

TECHNICAL ACTION	RATIONALE/AMPLIFICATION
1. Secure pig using snare, chute, or assistant holding against body.	1a. Restraint technique will depend on pig size.
2. Marginal ear veins usually are visible without occlusion. Three veins are present on the ear, with the lateral or central being the easiest to visualize.	2a. Fig. 5–11.
	2b. Pigs are unable to sweat, so the ears are an important component of temperature regulation.
	2c. Increasing the ambient temperature will result in vasodilation and assist with vein visualization.
3. Assistant should compress the base of the ear to occlude vein. A pumping motion can be used to help raise the vein.	3a. Alternatively, the collector can use the nondrawing hand to occlude the vein.
4. Attach a needle to the syringe. Insert needle along vessel at a very shallow angle of penetration. Collect sample.	4a. Small pigs: 25g × 1-inch; medium pigs: 22g × 1-inch; large pigs: 20g × 1-inch needle.
	4b. Needle must be attached to the syringe before insertion, because ear veins are very fragile.
	4c. When very small blood volumes are needed, such as for packed red blood cell volumes or blood smears, a needle is inserted in vessel without syringe attached. A microhematocrit tube is then used to collect sample via capillary action.

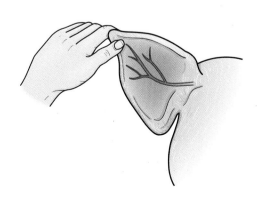

FIG. 5–11 Marginal ear veins on the pig.

Alternative Collection Sites

The following veins may be used for sample collection, but their use is not recommended.

- Equine: Cephalic vein dorsal to carpus and saphenous veins
- Complication: Extreme risk of personnel injury
- Bovine: Lateral thoracic or subcutaneous abdominal (milk) vein
- Complication: Severe subcutaneous hemorrhage

ARTERIAL PUNCTURE

Purpose

- Obtain sample for blood gas and pH analysis
- Assess pulmonary function in adult and neonatal horses

Complications

- Hematoma formation
- Iatrogenic infection
- Hemorrhage

Equipment

- 18g × 1.5-inch needle
- 6-cc syringe
- Povidone-iodine scrub and solution
- Isopropyl alcohol 70%
- Sterile cotton or gauze squares
- Clippers with #40 blade
- Cooling container (ice)

Procedure for Arterial Puncture of the Carotid Artery of the Horse

TECHNICAL ACTION	RATIONALE/AMPLIFICATION
1. Halter horse and place in stock.	—
2. Palpate carotid artery deep to the jugular vein in lower third of neck.	2a. The femoral and brachial arteries can also be used in neonates.

TECHNICAL ACTION	RATIONALE/AMPLIFICATION
3. Clip 4 × 4-inch area and perform a three-step surgical preparation.	—
4. Palpate carotid artery pulsations using two fingers of the left hand, and insert an 18g × 1.5-inch needle using right hand.	4a. Use a 22–25g needle in neonates. 4b. Bright pulsing red blood confirms placement.
5. Attach syringe and collect sample.	—
6. Apply compression for 5 minutes after withdrawing needle.	6a. A large hematoma and severe hemorrhage can occur without sufficient compression.

Procedure for Arterial Puncture of the Facial, Transverse Facial, and Greater Metatarsal Arteries

TECHNICAL ACTION	RATIONALE/AMPLIFICATION
1. Facial or greater metatarsal artery puncture should be attempted only on anesthetized horses.	—
2. Clip 4 × 4-inch area and perform a three-step surgical preparation.	—
3. Palpate facial, transverse facial, or greater metatarsal artery pulse using left hand, and insert a 20g × 1.5-inch needle using right hand.	3a. Fig. 5–12.
4. Attach syringe and collect sample.	—
5. Apply compression for 5 minutes after withdrawing needle.	5a. A large hematoma and severe hemorrhage can occur without sufficient compression.

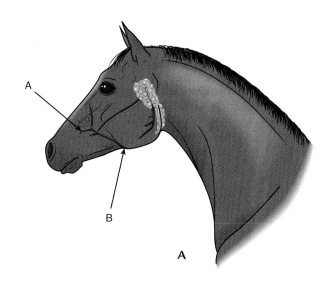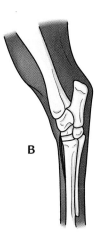

FIG. 5–12 (A) Drawing of locations of lateral facial artery (**a**) and the facial artery (**b**). (B) Schematic of the location of the greater metatarsal artery relative to palpable landmarks.

Procedure for Arterial Puncture of the Coccygeal Artery of the Cow

TECHNICAL ACTION	RATIONALE/AMPLIFICATION
1. Use in cattle only.	1a. Do not attempt coccygeal puncture in the horse, goat, pig, or llama.
2. Place in stock and tail-up the animal.	—
3. Palpate coccygeal artery on ventral midline of tail.	—
4. Wipe with alcohol.	—
5. Insert an 18g × 1-inch needle with syringe attached. Collect sample.	5a. Apply digital pressure for 5 minutes after removing needle to prevent hematoma formation.

Procedure for Arterial Puncture of the Median Auricular Artery

TECHNICAL ACTION	RATIONALE/AMPLIFICATION
1. Place cow in stock and secure head using halter.	1a. Do not use this technique on horses unless they are under general anesthesia.
2. Clip 4 × 4-inch area over dorsal aspect of ear pinna and perform three-step preparation.	—
3. Locate artery cranial to auricular vein.	3a. Fig. 5–13.
4. Insert needle and draw blood into heparinized syringe.	4a. Needle sizes vary with animal weight. Calf: 22g × 1-inch; cow: 18–20g × 1-inch needle.
5. Withdraw needle and apply pressure for 5 minutes.	5a. Pressure prevents hematoma formation.
6. Place needle and syringe on ice and process sample immediately.	6a. Arterial blood gas analysis must be completed within 1 hour of collection.

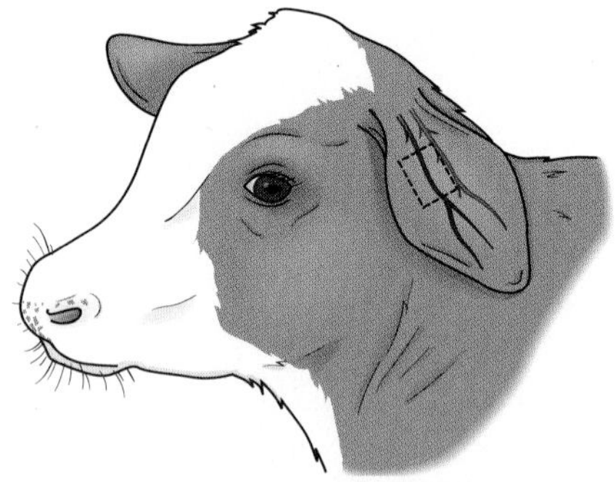

FIG. 5–13 Location of the median auricular artery (*red*) next to the auricular veins (*blue*). *Dashed lines* depict area to shave.

COAGULATION STUDY BLOOD COLLECTION

Purpose

Various diagnostic tests are performed to evaluate clotting mechanisms. Common tests will include one-stage prothrombin time, activated partial thromboplastin time, and activated clotting times.

Complications

- Invalid sample due to poor or traumatic sampling technique
- All complications listed under routine venipuncture

Equipment

- 20g × 1.5-inch (horse) or 18g × 1.5-inch (cow) needle
- 6-cc and 12-cc syringes (several)
- Double-ended collection needle with sheath (Vacutainer system, preferred alternative to needle and syringe)
- Isopropyl alcohol 70%
- Cotton
- Blue top (sodium citrate) and purple top (ethylenediamine tetra-acetic acid, EDTA) tubes.

Procedure for Coagulation Study Blood Collection

TECHNICAL ACTION	RATIONALE/AMPLIFICATION
1. Follow initial procedure outlined in jugular venipuncture.	—
2. Once syringe is attached, aspirate 1 cc into syringe.	—
3. Leave needle in vein; detach and discard syringe.	—
4. Attach second syringe and aspirate 5–6 cc of blood.	4a. The two-syringe technique minimizes the likelihood of blood collection trauma, which would prematurely activate the clotting factors.
5. After collection, apply digital pressure to collection site for 3 minutes.	5a. Additional time for compression is recommended if animal has blood clotting problems.
6. Place samples on ice.	—
7. Process immediately.	—
8. Check animal in 10–15 minutes to ensure that hemorrhage from collection site has not occurred.	—

Special Notes

Whenever possible, blood samples for coagulation study should be collected using the Vacutainer system.

Control samples from normal animals should be run concurrently with those of the test subject.

FECAL SAMPLE COLLECTION

Despite its lack of glamour, fecal material is a valuable diagnostic substance. Obtaining fresh feces permits accurate culturing and assessment of sand or parasite loads. Feces can be collected from the ground, although this is less than ideal and collection directly from the rectum is advised.

Purpose

- Perform parasitologic examination
- Perform microscopic examinations, chemical determinations, and cultures
- Evaluate feces for presence of sand

Complications

- Rectal tears
- Trauma to personnel (kicked)

Contraindications

- Suspected rectal tear

Equipment

- Obstetrical sleeve
- Lubrication gel
- Tail wrap
- Fecal sample container

Procedure for Fecal Sample Collection from the Horse

TECHNICAL ACTION	RATIONALE/AMPLIFICATION
1. Place horse in stock.	—
2. Apply tail wrap.	2a. Fig. 6–26.
3. Put on an obstetrical sleeve and use lots of lubrication on sleeve.	3a. Remember to remove all jewelry and keep fingernails short to minimize chance of rectal lacerations.

TECHNICAL ACTION	RATIONALE/AMPLIFICATION
4. Standing to the side of horse, remove handful of feces directly from rectum.	—
5. If sample collection is for purposes of sand evaluation, recover a minimum of 3 lbs of manure.	5a. Alternatively, manure for sand evaluation can be recovered fresh from the trailer or cement surface.

Procedure for Fecal Sample Collection from the Cow

TECHNICAL ACTION	RATIONALE/AMPLIFICATION
1. See equine procedure above.	1a. Cow tails do not require tail wraps.
	1b. Cow manure is not evaluated routinely for sand.

Procedure for Fecal Sample Collection from the Llama

TECHNICAL ACTION	RATIONALE/AMPLIFICATION
1. Halter and bring llama to dung pile.	1a. Llamas typically will defecate only on a dung pile. If a fresh sample is required, it is advisable to ask owners to bring a small bag of feces from the dung pile at home. The feces are placed in a specified area or corner of a large stall. New feces can then be collected from the area.
2. Rectal stimulation can be attempted by lubricating a finger and applying gentle stimulation.	2a. This technique has limited value with llamas but is very effective with goats.

Special Note

Samples to be evaluated for *Salmonella* can be placed in enrichment media such as trypticase soy agar (TSA), tetrathionate, or selenite. A sample should be obtained daily for 3 consecutive days to confirm absence of *Salmonella*.

URINE COLLECTION

Urinalysis is a valuable diagnostic tool for evaluation of infection, metabolic disturbances, or drug residues. Technicians usually are responsible for collection of these samples. Both free-catch and catheterization collection methods are performed, with technique selection reflecting patient species and sex. Cystocentesis as a means of collecting urine is not performed in the large animal patient.

Purpose

- Evaluate urine pH and presence of ketones, glucose
- Urine culture and cellular analysis
- Detection of drug residues

Complications

- Urinary tract infection
- Mucosal irritation
- Transient stranguria
- Personnel injury
- Bacterial contamination of sample

Equipment

- Free-catch collection container
- Test strips
- Povidone-iodine solution and scrub
- Sterile gloves
- Sterile Foley catheter or red rubber feeding tube
- Stallion: 6–7 mm outside-diameter tube
- Llama: 5-French tube
- Cotton
- Sterile lubricant
- Catheter-tip syringe
- Collection container

Procedure for Urinary Catheterization of the Horse

TECHNICAL ACTION	RATIONALE/AMPLIFICATION
1. Place horse in stock.	—
2. Sedate males.	2a. Sedation facilitates extension of the penis.

TECHNICAL ACTION	RATIONALE/AMPLIFICATION
	2b. Anesthetic agents can affect urine volume and some chemistry values.
	2c. Females generally do not require sedation unless very fractious.
3. Thoroughly clean distal penis and prepuce (including urethral fossa) or vulva area using iodine scrub and sterile water.	—
4. Apply sterile gloves.	—
5. Select appropriate size and type of catheter.	**5a.** Foals and fillies: 12-French Foley. Colts: 12-French red rubber feeding tube catheter. Males: use a stallion catheter, which includes a stylet.
	5b. Use the smallest diameter possible to minimize urethral trauma.
6. Lubricate catheter.	—
7. Males: Place catheter tip in urethral orifice and advance through urethra into bladder.	—
8. Females: Digitally palpate urethral orifice along ventral aspect of vaginal vault. Advance catheter along finger into urethral orifice.	—
9. Attach syringe and aspirate sample.	**9a.** Alternatively, free-flowing urine can be caught in a sterile container.
	9b. Excessive aspiration pressure can alter cellular content of sample and cause minor hemorrhage.

Procedure for Free-catch Urine Collection in the Horse

TECHNICAL ACTION	RATIONALE/AMPLIFICATION
1. Place horse in stock or secure with halter and lead rope.	—
2. Thoroughly clean distal penis and prepuce (including urethral fossa) or vulva area using iodine scrub and sterile water.	**2a.** Cleansing of external genital area is not warranted in urine drug test collection.

TECHNICAL ACTION	RATIONALE/AMPLIFICATION
3. Wait by horse with urinary container.	3a. In an attempt to encourage urination, the following strategies can be used: Bring horse to a freshly bedded stall, run water on cement, tickle prepuce with straw.

Special Notes

Recumbent foals frequently learn to urinate when held in standing position. Free-catch urine samples can be obtained by placing a container under the foal after lifting.

Catheterization results in significantly less bacterial contamination of urine. Typical catheterization values of 500 colony forming units (CFU) versus 20,000 CFU associated with free-catch samples are noted. Samples collected from females exhibit much greater contamination than those from males.

Some types of chemical restraint can alter urine values. Document any use of chemicals to ensure proper interpretation of laboratory values.

Procedure for Urine Collection Using the Digital Technique in Cows

TECHNICAL ACTION	RATIONALE/AMPLIFICATION
1. Place cow in chute or stock or secure head in stanchion.	—
2. Lift tail using left hand.	—
3. Rub gently up and down just below the vulvar lips.	3a. Small volumes of urine are collected in this manner.
	3b. Technique is used to monitor urine pH and ketones.
4. Collect sample or place urine dipstick directly in stream.	4b. Bacterial contamination is significant.

Special Notes

Urethral catheterization is not performed in the male bovine.

Urethral catheterization of the cow is rare but would be performed using techniques used for mares.

Procedure for Llama Urine Free-catch Technique

TECHNICAL ACTION	RATIONALE/AMPLIFICATION
1. Halter llama and lead it to dung pile.	1a. Llamas urinate and defecate almost exclusively on dung piles.
	1b. Sample collection is most successful if attempted early in the morning.
2. Hold cup or a cup attached to a 4-foot pole. Stand caudal and lateral to animal.	2a. Both males and females will squat and eject urine caudally.
3. Collect sample and process immediately.	3a. Complete urination normally occurs over 30–60 seconds.

Procedure for Llama Urinary Catheterization

Special Notes

Urethral catheterization is not performed in the male llama. Retrograde catheterization into the bladder is prevented by a membranous flap at the ischial arch.

TECHNICAL ACTION	RATIONALE/AMPLIFICATION
1. Halter llama and place in stock.	1a. Male llamas can not be catheterized. A dorsal recess at the level of the ischial arch makes catheterization virtually impossible.
2. Clean vulvar lips using chlorhexidine and water solution. Dry area.	2a. A dilute iodine solution may also be used.
3. Apply sterile glove and small amount of sterile lubricant.	—
4. Insert finger into vulva and palpate ventrally for external urethral orifice on floor of vulva.	4a. Fig. 5–14.
5. Once orifice has been palpated, withdraw finger slightly, about ¾ inch, and insert a 5-French red rubber or polypropylene catheter along dorsal aspect of index finger into orifice.	5a. Insertion along the dorsal aspect of the finger prevents inadvertent insertion into the diverticulum located just ventral and caudal to the urethral orifice.

TECHNICAL ACTION	RATIONALE/AMPLIFICATION
6. Advance catheter slowly. Bladder should be reached within 25 cm of vulvar lips.	—
7. Attach catheter-tip syringe and collect sample.	7a. If urine is flowing freely, it can be deposited directly in container.

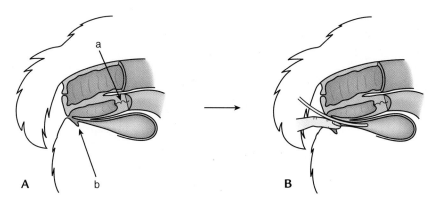

FIG. 5-14 (A) Schematic of anatomy of the female llama: (**a**) is the cervix and (**b**) is the suburethral diverticulum. (B) Fingertip covering the opening to the urethral diverticulum, allowing the catheter to pass into the urethra.

CENTESIS

Centesis involves placing a needle within a body cavity. Rumen, thoracic, and abdominal cavities are of particular interest in large animal medicine. Samples from these sites contribute to accurate diagnosis of a myriad of respiratory and gastrointestinal diseases.

ABDOMINOCENTESIS

Purpose

- Obtain a sample from the abdominal cavity for chemical, microscopic, and cellular diagnostic evaluation

Complications

- Peritonitis
- Bowel laceration (especially foals)
- Personnel injury
- Accidental fetal injury (pregnant mares)
- Minor cutaneous hemorrhage
- Subcutaneous hematoma
- Accidental puncture of abdominal organs
- Minor cutaneous hemorrhage

Contraindications

- Late gestation in pregnant animals
- Severe tympanitis
- Sand impactions
- Extreme uncontrolled abdominal pain with kicking, rolling

Equipment

Needle Method

- 18g × 1.5-inch needle
- 6-cc syringe
- Purple top tube (EDTA)
- Povidone-iodine scrub and solution
- Isopropyl alcohol 70%
- Sterile cotton or gauze squares
- Clippers with #40 blade
- Sterile gloves
- Antibiotic (suitable for intraabdominal placement)

Teat Cannula Method

- Equipment listed above
- Number 15 scalpel blade (horse, cow) or #12 blade (llama)
- Blunt-tipped metal cannula such as teat cannula (14g × 3-inch) or female dog urinary catheter
- Lidocaine 2%
- 3-cc syringes

Preparation for Abdominocentesis of the Horse: For Both Needle and Teat Cannula Method

TECHNICAL ACTION	RATIONALE/AMPLIFICATION
1. Place horse in stock.	—
2. Sedate, administer analgesics if necessary.	2a. Analgesics may be necessary if horse is in extreme pain.
3. Clip a 4 × 4-inch area on the ventral-most portion of the abdomen caudal to the xiphoid process	3a. Avoid linea alba and superficial veins.
4. Prepare site aseptically using standard three-step preparation.	—

Procedure for Abdominocentesis of the Horse: Needle Method Continued

TECHNICAL ACTION	RATIONALE/AMPLIFICATION
1. Apply sterile gloves.	—
2. Thrust an 18g × 1.5-inch needle quickly through skin, muscle, subcutaneous tissue in the center of 4 × 4 inch clipped area.	—
3. Slowly reposition needle tip in a fluid space.	3a. This is accomplished by slowly pulling needle back or pushing forward. Rotating the needle is also helpful.
4. If no fluid is obtained, insert a second needle a few inches cranial or caudal to the first needle.	4a. Do *not* remove the first needle.
5. If sample is still not obtained, use syringe to inject 5–10 cc of air.	5a. Injecting air breaks the peritoneal vacuum.
6. Collect free-flowing fluid in purple top tube.	6a. Fluid typically drips rapidly.
7. Remove needle and spray area with iodine solution.	—
8. Evaluate sample immediately.	8a. See appendices for normal values.

Procedure for Abdominocentesis of the Horse: Teat Cannula Method Continued

TECHNICAL ACTION	RATIONALE/AMPLIFICATION
1. Inject a 3-cc lidocaine bleb in subcutaneous tissue in the center of 4 × 4 inch clipped area.	—
2. Apply sterile gloves.	—
3. Use a #15 blade and make a stab incision. Do not enter abdominal cavity using scalpel.	3a. Scalpel is used to penetrate skin only.
4. Insert teat cannula in incision and thrust into abdominal cavity.	—
5. Reposition cannula tip as needed to enter fluid space.	5a. This is accomplished by slowly pulling cannula back or pushing forward. Rotating the cannula is also helpful.
6. Collect free-flowing fluid in purple top tube	6a. Fluid typically drips from the cannula.
7. Remove cannula and spray with iodine solution.	—
8. Evaluate sample immediately.	8a. See appendices for normal values.

Special Notes: Method Comparison

Needle Method

- Advantages: Quick, minimal equipment
- Disadvantage: Increased risk of laceration

Teat Cannula Method

- Advantage: Less risk of perforation
- Disadvantages: Increased time, equipment

Procedure for Abdominocentesis of the Cow

TECHNICAL ACTION	RATIONALE/AMPLIFICATION
1. Place cow in stock.	—
2. Clip 4 × 4-inch square in appropriate quadrant.	2a. Because of the rumen size and the location of bovine abdominal disorders, samples should be obtained from all four quadrants. Order of collection is not critical. Most veterinarians prefer to begin with the abnormal quadrant.
3. Quadrant location. • Left cranial: 2 inches caudal to xiphoid process, left of midline • Left caudal: Directly anterior to left anterior attachment of mammary gland to body wall • Right cranial: 8–12 inches cranial to right anterior attachment of mammary gland to body wall • Right caudal: Directly anterior to right anterior attachment of mammary gland to body wall	—
4. Perform standard three-step surgical preparation.	—
5. Tail-up the cow.	—
6. Follow steps 1–8 of equine procedure in Procedure for Abdominocentesis of the Horse: Needle Method Continued.	6a. The teat cannula method of collection can also be used in cattle. Due to the lower risk of laceration, this technique is seldom used in cattle.

Procedure for Abdominocentesis of the Llama

TECHNICAL ACTION	RATIONALE/AMPLIFICATION
1. Halter and place llama in stock.	—
2. Clip a 4 × 4-inch area on ventral midline just caudal to umbilicus.	2a. Entering caudal to the umbilicus decreases the likelihood of entangling the omentum while collecting the sample.
	2b. It is extremely important to stay directly on the linea alba (midline). Llamas have significant retroperitoneal fat pads on either side of the linea. These 3-inch-deep fat deposits will prevent sample collection if cannula insertion is off midline.
3. Clean area thoroughly using iodine scrub. Rinse with water or alcohol.	—
4. Inject a 2–5 cc lidocaine bleb subcutaneously at the intended insertion site.	—
5. Perform a standard three-step preparation using iodine and alcohol.	—
6. Use a #12 scalpel blade to make a stab incision through the skin. Do not enter peritoneal cavity.	—
7. Insert a 14g × 3-inch teat cannula into peritoneum using a quick, thrusting motion.	—
8. If sample is flowing freely, collect at this time.	—

9. If sample does not flow, attempt the following:
  a. Reposition cannula tip by moving cranially, caudally, up, down.
  b. Attach syringe and apply slight negative pressure while repositioning cannula tip.
  c. Inject 10 cc air into abdominal cavity, then reposition cannula tip.
	9a. Sample acquisition is dependent on the presence of a fluid pocket within the abdomen.

Complication Management

- Bowel entered: Sample will be greenish in color. Inject antibiotic *before* removing needle or cannula. Initiate parenteral antibiotic administration.
- Failure to obtain sample is not uncommon. This is most likely caused by absence of a fluid pocket at the collection site. Should this occur in horses or cattle, reevaluate collection site to determine if it is the most ventral aspect of the abdomen.

THORACOCENTESIS

Purpose

- Obtain sample for diagnostic analysis. Common examinations include microscopic, cytologic, and chemical analysis.
- Therapeutic removal of pleural effusions
- Treatment of tension pneumothorax

Complications

- Iatrogenic infection
- Pneumothorax
- Dyspnea

Equipment

- Clippers with #40 blade
- Povidone-iodine scrub, solution, and 70% isopropyl alcohol
- Sterile gauze squares
- 18g × 1.5-inch and 20g × 1-inch needles
- 5-cc and 12-cc syringes
- Sterile gloves
- Red top (no additive) and purple top (EDTA) tubes
- Microscope slides
- 2-inch to 3-inch teat cannula or bitch catheter (horse, cow)
- 16g × 2-inch needle (llama)
- Masking tape (llamas)
- Lidocaine 2%

Horse and cows only:

- Number 15 scalpel blade

- Catheter extension set
- 3-way stopcock valve
- Ultrasound machine (optional)
- Suture material (size 0, nonabsorbable)
- Needle holders
- Suture scissors

Procedure for Thoracocentesis of the Horse and Cow

TECHNICAL ACTION	RATIONALE/AMPLIFICATION
1. Place horse or cow in stock.	—
2. Sedation and analgesics generally are not required.	—
3. Clip from olecranon to tenth intercostal space. Clip should extend from point of shoulder dorsally to 2 inches below the olecranon ventrally.	3a. Fig. 5–15.
4. Aseptically prepare site using a three-step surgical preparation.	—
5. Spray with iodine solution.	—

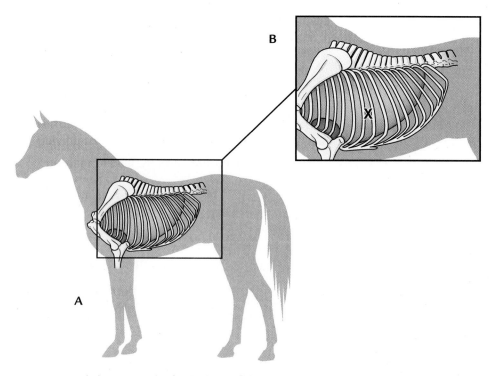

FIG. 5–15 (A) Anatomic depiction of the pulmonary system of the horse. (B) Close-up marking the exact spot to enter the chest for a **thoracocentesis**.

TECHNICAL ACTION	RATIONALE/AMPLIFICATION
6. Apply sterile gloves.	—
7. Inject a lidocaine bleb the size of a quarter on the cranial aspect of seventh rib 4 inches dorsal to the olecranon. Inject an additional 2–3 cc into the intercostal space, deep to the bleb.	7a. Incision will be made cranial to the rib. This will avoid vessels and nerves running along caudal aspect. 7b. If ultrasonography is available, identify fluid pocket and change the aspiration site accordingly.
8. Use a #15 blade and make stab incision.	—
9. Introduce teat cannula with attached extension tubing and 3-way stopcock. Bluntly advance through parietal pleura.	9a. You will know you are in the thoracic cavity when sudden loss of resistance is felt.
10. Attach 6-cc syringe and aspirate fluid.	10a. Normally one will not retrieve more than a few cc's of straw-colored fluid.
11. If fluid is not obtained, redirect cannula tip and reattempt aspiration.	—
12. Sew purse-string suture around stab incision.	—
13. Withdraw cannula as purse-string suture is tightened.	—
14. Immediately place part of the sample in purple and red top tubes, and with the remainder prepare a slide for microscopic examination.	—

Procedure for Thoracocentesis of the Llama

TECHNICAL ACTION	RATIONALE/AMPLIFICATION
1. Halter and place llama in stock.	—
2. Clip a 4 × 4-inch area at the level of the sixth to seventh intercostal space, 1–2 inches dorsal to the costochondral junction.	2a. Llamas have 12 ribs. Count back from the last rib to identify the sixth intercostal space. 2b. The costochondral junction is located about 4–6 inches from the sternum. 2c. Whether the left or right side is used varies and is determined by sample need.

TECHNICAL ACTION	RATIONALE/AMPLIFICATION
3. Apply masking tape to fleece borders and perform a standard three-step surgical preparation.	3a. Masking tape keeps long hairs from contaminating the site.
4. Insert a 16g × 2-inch needle with syringe attached along the cranial boarder of the seventh rib. Advance needle 1.0–1.5 inches into pleural cavity.	4a. Intercostal vessels are located on the caudal border of each rib. These are avoided by entering on cranial border. 4b. A 12-cc syringe is best.
5. Aspirate sample.	5a. Process immediately.
6. Withdraw needle and spray area with iodine. Remove tape.	6a. A small bleb of antibiotic ointment may be spread over the collection site.

RUMENOCENTESIS

Purpose

- Collect sample of rumen content for diagnostic evaluation. Rumen acidosis (pH) measurement is most common.

Complications

- Peritonitis
- Abscess in body wall at needle insertion site
- Needle lumen occlusion
- Personnel injury from kicking

Equipment

- 16g × 5-inch needle
- 12-cc syringe
- Clippers with #40 blade
- Povidone-iodine scrub, solution, and 70% isopropyl alcohol
- Gauze or cotton
- pH paper or meter

Procedure for Rumenocentesis

TECHNICAL ACTION	RATIONALE/AMPLIFICATION
1. Place cow in stock or hobble.	—
2. Sedate if necessary.	2a. Sedation generally is not required.
3. Clip a 4 × 4-inch area 6–8 inches caudoventral to the costochondral junction of the last rib.	3a. This location allows for needle insertion into the fluid layer of the ventral rumen sac. 3b. Fig. 5–16.

TECHNICAL ACTION	RATIONALE/AMPLIFICATION
4. Perform a three-step surgical preparation.	—
5. Tail-up the cow.	—
6. Insert needle through skin only.	6a. Cattle generally object most to penetration of the highly innervated skin.
7. Once cow is standing quietly, insert needle to hub by thrusting smoothly.	—
8. Attach syringe and aspirate fluid.	8a. When needle lumen is occluded by ingesta, clear lumen by injecting a small volume of air.
	8b. A maximum volume of 8 cc usually is collected.
9. Measure pH immediately using pH paper or meter.	—

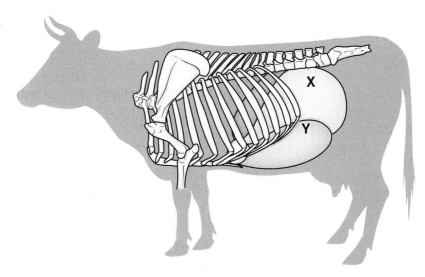

FIG. 5-16 Anatomic depiction of the bovine rumen. *X* marks the site for rumen gas collection and *Y* marks the site for rumen fluid collection.

Special Note

Collect sample 2–5 hours postprandial for most accurate measurement of rumen pH.

RESPIRATORY SAMPLING

Sampling of the upper airway facilitates diagnosis of many respiratory tract diseases. Horses and pigs are the species most likely to undergo sampling. The **transtracheal wash** often is limited to horses and focuses on diagnosing both infectious and noninfectious diseases. In contrast, nasal swabs typically are used to diagnosis transmissible diseases in the pig.

NASAL SWABS

Purpose

- Identify bacteria involved in infectious respiratory diseases. This is performed in pigs as a component of atrophic rhinitis monitoring programs.

Complications

- Contamination of sample

Equipment

- Ice packs
- Cotton
- Sterile swab

Procedure for Nasal Swabs

TECHNICAL ACTION	RATIONALE/AMPLIFICATION
1. Have assistant restrain piglet.	1a. Most sampling protocols call for sampling at 4 weeks, 8 weeks, or 3 months of age. Adults are tested when needed. 1b. A minimum of eight head are typically sampled.
2. Clean nostril area with cotton.	—
3. Grasp jaw and snout and insert swab to the back of the nose. Rotate swab several times.	3a. The swab must reach far enough into the nasal cavity. This distance is several inches (Fig. 5–17).
4. Return swab to transport tube. Squeeze moisture pad in tube to wet the swab tip.	—
5. Label samples. Place on ice and ship overnight.	5a. Samples should be cooled during transport and processed rapidly.

FIG. 5–17 (A) Measuring before inserting a nasal swab in a piglet. (B) Insertion of the swab into the nasal sinus of a pig.

TRANSTRACHEAL WASH

Definition

- Diagnostic technique used for collecting bronchial exudate samples

Purpose

- Collect sputum samples without pharyngeal contamination for histologic and microbiologic examination
- Therapeutic lavage
- Indications include persistent undiagnosed cough, or abnormal radiographic or ultrasonographic findings

Complications

- Peritracheal subcutaneous emphysema
- Cellulitis
- Pulmonary foreign body due to presence of catheter in airway
- Acute dyspnea
- Tracheal laceration
- Minor subcutaneous hemorrhage
- Iatrogenic infection

Equipment

- Povidone-iodine scrub and solution
- Isopropyl alcohol 70%
- Gauze squares
- Clippers with #40 blade
- Number 15 scalpel blade with handle
- Lidocaine 2%
- 6-cc, 30-cc, and 60-cc syringes
- Red top (no additive) and purple top (EDTA) tubes
- Microscope slide
- Trochar needle or angiocatheter
- Polyethylene tubing or angiocatheter
- Saline
- Antibiotic ointment, bandaging material

Procedure for Transtracheal Wash

TECHNICAL ACTION	RATIONALE/AMPLIFICATION
1. Place horse or cow in stock.	—
2. Mild sedation is recommended.	2a. Do not sedate heavily or cough reflex will be disrupted.
3. Palpate trachea and clip a 4 × 4-inch square over trachea in middle to lower third of neck.	3a. Fig. 5–18.
4. Aseptically prepare site using three-step surgical preparation.	—
5. Spray with iodine solution.	—
6. Apply sterile gloves.	—
7. Insert a lidocaine bleb the size of a quarter over trachea.	—
8. Make small stab incision through skin.	—
9. Introduce trochar on midline and puncture through ventral tracheal wall between cartilaginous rings.	—

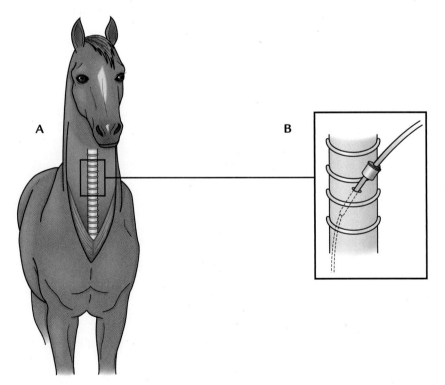

FIG. 5–18 (A) Anatomic location of site for performance of a transtracheal wash. Rectangle depicts area to be clipped and prepped. (B) Close-up of trochar (or needle) and catheter within the lumen of the trachea.

TECHNICAL ACTION	RATIONALE/AMPLIFICATION
10. Holding trochar in place, introduce approximately 12 inches of polypropylene tubing into trachea.	10a. Directing trochar tip downward will assist introduction of tubing down trachea.
11. Withdraw trocar and hold for duration of procedure.	11a. Withdrawing prevents possible laceration and loss of tubing.
	11b. Blunt tipped cannula need not be withdrawn at this point.
12. Instill 30 cc of sterile saline.	12a. Do not use commercial sterile saline with antibacterial additives.
	12b. Instilling saline often triggers a cough reflex. This will assist with sample collection.
13. Intermittently aspirate while gradually withdrawing tubing.	13a. An additional 30 cc can be instilled if no sample is obtained.
14. Place some of sample in red and purple top tubes and with the rest, immediately prepare a slide for Gram stain.	—
15. Apply antibiotic ointment and wrap neck for 24 hours.	15a. Wrap will prevent crepitus and infection.

BIOPSY, ASPIRATION, SCRAPES, AND SMEARS

Aspiration, biopsy, scrape, and impression smear techniques provide insight into various disease processes at the cellular level. Examination of tissue parenchyma or cellular composition elicits both diagnostic and prognostic information. Although the procedures are performed more frequently in companion animals, they are a valuable component of large animal medicine.

UTERINE BIOPSY

Purpose

- Determine uterine grade or prognosis for ability of uterus to sustain pregnancy
- Determine cause of decreased fertility
- Evaluate endometritis

Complications

- Iatrogenic infection
- Uterine tear
- Rectal tear

Equipment

- Roll gauze or veterinary wrap
- Sterile palpation sleeves
- Sterile surgical gloves
- Biopsy punch
- Sterile lubricant
- Bucket
- Chlorhexidine solution
- Rolled cotton
- Paper towels
- Bouin's solution or 10% formalin
- Sample vial (red top tube)
- 18g × 1-inch needle

Procedure for Uterine Biopsy of the Horse

TECHNICAL ACTION	RATIONALE/AMPLIFICATION
1. Place mare in stock.	—
2. Sedate if necessary.	2a. Sedation generally is not required.
3. Apply tail wrap.	—
4. Apply nonsterile palpation sleeve and remove feces from rectum.	—
5. Wash and dry perineum and vulvar area.	—
6. Apply two sterile palpation gloves on one hand.	6a. The two-glove technique prevents vaginal bacterial contamination of the uterus.
7. Place the biopsy instrument between the two gloves.	—
8. Apply sterile lubricant to outer glove.	8a. Do not use lubricant containing antibacterial agents.
9. Enter vagina and place index finger in cervix.	—

TECHNICAL ACTION	RATIONALE/AMPLIFICATION
10. Push biopsy instrument through outer glove and slide along index finger into uterus.	—
11. Hold biopsy instrument in place and remove hand from vagina.	—
12. Insert hand into rectum and locate (palpate) tip of biopsy instrument.	—
13. Direct biopsy tip toward left or right horn.	—
14. Open biopsy jaws, push tissue into jaw, close jaw. Pull in a quick sharp motion to remove sample.	—
15. Remove biopsy instrument.	—
16. Using a hypodermic needle, remove sample from biopsy jaws and place in Bouin's solution or 10% formalin.	16a. Bouin's solution produces a firmer specimen with fewer artifacts. After 2–24 hours in Bouin's solution, the sample should be placed in 10% formalin and sent to the laboratory.
17. Remove tail wrap.	17a. Watch mare for any signs of discomfort.

Special Note

Uterine cultures should always be obtained before biopsy is performed.

LIVER BIOPSY

Purpose

- Evaluate liver function
- Assess degree of hepatic pathology in cases of hepatic lipidosis, parasite migration, toxicity
- Mineral analysis (copper)

Complications

- Iatrogenic infection
- Hemorrhage
- Trauma to personnel

Equipment

- Sterile surgical gloves
- Biopsy punch, long alligator type
- 14g × 6-inch true-cut biopsy needle (llama)
- Masking tape (llama)
- Bucket
- Povidone-iodine scrub and solution
- Isopropyl alcohol 70%
- Clippers with #40 blade
- Lidocaine
- 12-cc syringe with 18g × 1.5-inch needle
- Lancet or #10 scalpel blade

Procedure for Liver Biopsy of the Cow

TECHNICAL ACTION	RATIONALE/AMPLIFICATION
1. Place cow in chute.	—
2. Clip a 4 × 4-inch area on the right side at the juncture of the upper and middle third of the 11th intercostal space.	2a. Draw a horizontal line from the middle of the right paralumbar fossa to the 11th intercostal space. 2b. The cow liver is located on the right side of the abdomen. 2c. Fig. 5–19.
3. Clean area using iodine scrub. Wipe with alcohol.	3a. The area should be cleaned before injecting local anesthetic.
4. Inject a 3-cc lidocaine bleb.	4a. Local anesthetic reduces discomfort during the procedure.

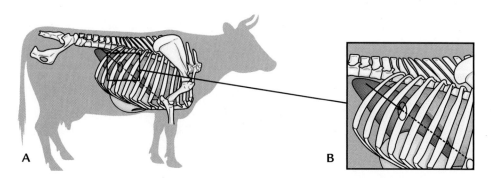

FIG. 5–19 (A) Location of the bovine liver relative to anatomic landmarks. Biopsy site is within the rectangle and marked with a *circle*. *Arrow* and *dotted line* show direction of biopsy needle. (B) Close-up of the biopsy site. *Arrow* and *dotted line* depict direction of the biopsy needle.

TECHNICAL ACTION	RATIONALE/AMPLIFICATION
5. Perform a three-step surgical preparation.	—
6. Use scalpel or lancet to make a small opening in the skin.	6a. Incision should be just large enough to accommodate biopsy instrument. 6b. Bovine skin is extremely thick, and biopsy forceps are unable to penetrate dermis.
7. Insert biopsy instrument in the direction of the left elbow.	—
8. Advance biopsy forceps until a sense of give is experienced.	8a. The sense of give signifies that the forceps tips are adjacent to the liver.
9. Quickly plunge the forceps forward ½ inch and obtain biopsy specimen.	—
10. Place specimen in appropriate container.	10a. Samples typically are preserved in 10% formalin or are submitted fresh or frozen for culture.
11. Spray area with iodine solution.	11a. The site usually is not sutured.

Procedure for Liver Biopsy of the Horse

TECHNICAL ACTION	RATIONALE/AMPLIFICATION
1. Place horse in stock.	—
2. Sedate and provide analgesia.	2a. Chapter 10.
3. Identify area of collection on right side by forming a triangle that connects tuber coxae, point of olecranon, and scapulohumeral joint. Count from the last rib and identify the 12th to 14th intercostal spaces.	3a. Fig. 5–20.
4. Clip a 8 × 8-inch square in the target area.	4a. Use #40 clipper blades.
5. Use a 3.5-MHz ultrasonography probe to identify liver parenchyma.	5a. Choose the area with greatest depth of liver parenchyma free of hepatic vessels. 5b. Ultrasonography is used to prevent inadvertent puncture of the intestines. 5c. Equine liver biopsy without the use of ultrasonographic guidance is not recommended.

TECHNICAL ACTION	RATIONALE/AMPLIFICATION
6. Inject a 2–4 cc dermal lidocaine bleb at the identified site.	—
7. Perform a three-step sterile preparation.	—
8. Inject 5–8 cc lidocaine in the underlying subcutaneous and intercostal area.	**8a.** Additional analgesia minimizes horse movement during the procedure.
9. Make a 5-mm stab incision using a #21 scalpel blade.	—
10. Insert biopsy needle oriented cranioventrally toward the contralateral shoulder.	**10a.** Use a 14g × 150-mm true-cut biopsy needle (Baxter).
11. Advance into liver parenchyma and obtain biopsy specimen.	**11a.** Depending on the amount of sample required, the cannula can be left in place while the trochar is removed and tissue is

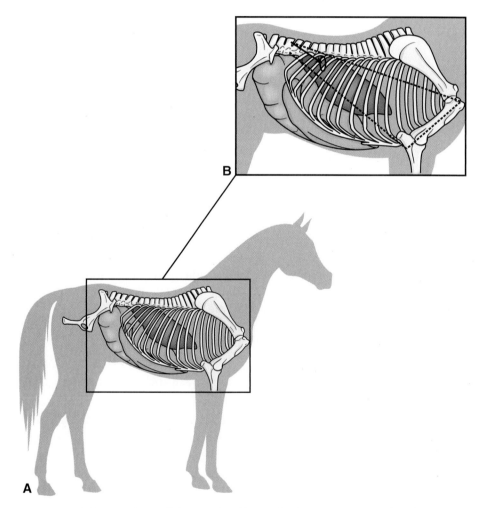

FIG. 5–20 (A) Location of the equine liver relative to anatomic landmarks. (B) Close-up of the liver. *Dotted lines* show triangulation of landmarks described in the text. The *circle* depicts location of incision for biopsy needle, and the *arrow* depicts the direction the needle should follow.

TECHNICAL ACTION	RATIONALE/AMPLIFICATION
	harvested. The trochar can then be reinserted and another specimen obtained.
	11b. Two or three samples typically yield 30–60 mg of liver.
12. Spray area with iodine solution.	**12a.** The entire procedure takes about 15–20 minutes.
13. Keep in a stall to rest, and monitor horse for 3–4 days.	**13a.** Colic is the most common complication. This typically is managed successfully with flunixin meglumine.

Procedure for Liver Biopsy of the Llama

TECHNICAL ACTION	RATIONALE/AMPLIFICATION
1. Halter llama and place in stock.	**1a.** Procedure can be attempted while animal is in sternal recumbency, but standing is much preferred.
2. On the right side, clip a 3 × 3-inch area over the ninth intercostal space, 9–10 inches from the top of the back.	**2a.** The llama liver is located entirely on the right side.
	2b. The llama has 12 ribs. Count back from the last rib to locate the ninth rib.
3. Secure wool away from site using masking tape.	—
4. Clean area using iodine scrub and alcohol.	—
5. Inject a lidocaine bleb the size of a quarter subcutaneously.	—
6. Perform a standard three-step preparation.	**6a.** If an ultrasound machine is available, exact liver biopsy site can be determined at this time.
7. Insert a 14g × 6-inch true-cut biopsy needle angling it toward the midline, caudally, and slightly ventrally.	**7a.** The diaphragm is immediately adjacent to the chest wall, which is very thin. To confirm placement, let go of needle after passing it through the chest wall. Needles correctly placed within the diaphragm will move cranially and caudally in synchrony with respiration.
	7b. Aim needle toward contralateral stifle.

TECHNICAL ACTION	RATIONALE/AMPLIFICATION
8. Advance 1 inch once chest wall has been penetrated.	—
9. Collect sample. Process as needed.	—
10. Spray area with iodine.	10a. Monitor llama for any signs of discomfort.

DERMAL BIOPSY

Purpose

- Identify bacterial, fungal, or parasitic organisms responsible for skin disease
- Diagnose immune-mediated diseases
- Diagnose and determine prognosis of neoplasia

Complications

- Iatrogenic infection
- Minor hemorrhage
- Scar formation

Equipment

- Povidone-iodine solution
- Isopropyl alcohol 70%
- Gauze
- Clippers
- Suture material (2-0 nylon)
- Scissors or punch biopsy (4–6 mm)
- Needle drivers
- Forceps
- Lidocaine 2%
- 22g × 1-inch needle with 3-cc syringe
- Container with 10% formalin

Procedure for Dermal Biopsy

TECHNICAL ACTION	RATIONALE/AMPLIFICATION
1. Identify area of biopsy and wipe with alcohol-soaked cotton.	1a. Gross contamination should be removed, but the area should not be scrubbed or shaved, which can remove epithelial tissue important for a diagnosis.

TECHNICAL ACTION	RATIONALE/AMPLIFICATION
2. Inject 0.5–1.0 cc lidocaine subcutaneously to intended biopsy site. Wait 2–5 minutes for anesthetic to take effect.	2a. Do not infiltrate dermis or use excessive lidocaine, because this will distort specimen.
3. Immobilize area of biopsy with one hand.	—
4. Place punch biopsy directly over lesion (area of biopsy); apply pressure and rotate in a continuous circular motion until subcutaneous tissue is reached.	4a. A disposable punch biopsy instrument can be sterilized and reused 2–3 times until blade becomes dull.
5. Grasp subcutaneous side of biopsy tissue using forceps. Use scissors to cut any remaining connective tissue attachments, if needed.	5a. Avoid handling dermal side to minimize artifactual changes in specimen.
6. Blot sample on gauze to remove any surface hemorrhage and place in 10% formalin.	6a. Impression smears could be made prior to placing in formalin.
7. Obtain two or three specimens.	7a. Multiple specimens improve diagnostic potential.
8. Clean biopsy site using iodine solution and place one or two simple, interrupted sutures per biopsy site.	8a. Depending on dermal condition, sites can remain unsutured to allow second-intention healing.

Special Notes

Some conditions such as ulcers are best diagnosed doing elliptical biopsies instead of using standard punches. This would allow the operator to collect a portion of normal and abnormal tissue in a single specimen.

Biopsies intended for immunofluorescence testing should be placed in Michel's fixative.

Any steroid administration should be discontinued 3 weeks before biopsy is performed.

FINE-NEEDLE ASPIRATION

Purpose

- To differentiate various causes of tissue swelling or enlargement including neoplasia, inflammation, and hyperplasia
- To differentiate benign from malignant tumors for purposes of developing therapeutic protocols

Complications

- Iatrogenic infection
- Minor hemorrhage
- Tissue damage

Equipment

- 22g × 1.5-inch needle
- Microscope slides
- 6-cc or 12-cc syringe
- Isopropyl alcohol 70%
- Cotton or gauze squares
- 84-cm flexible intravenous extension set (technique two only)

Procedure for Fine-needle Aspiration Using Technique One

TECHNICAL ACTION	RATIONALE/AMPLIFICATION
1. Identify area of aspiration and wipe with alcohol-soaked cotton.	1a. Gross contamination should be removed, but the area need not be surgically prepared.
2. Immobilize area of biopsy or aspiration with one hand.	—
3. Introduce needle with syringe attached into the lesion or mass.	—
4. Apply negative pressure to aspirate by withdrawing syringe plunger.	4a. Do this several times in rapid sequence.
5. Partially withdraw the needle and redirect into lesion a second time.	5a. Do not remove the needle completely. Be careful that needle tip does not leave desired biopsy area.
6. Aspirate again.	—
7. Release negative pressure.	7a. Let go of the syringe plunger. This helps retain the sample within the needle lumen.
8. Withdraw needle.	—
9. Detach needle from syringe.	—
10. Pull 5 cc of air into empty syringe and reattach to needle.	10a. Always detach the syringe when filling with air. Failure to detach will result in loss of the aspiration sample.

TECHNICAL ACTION	RATIONALE/AMPLIFICATION
11. Forcefully expel sample onto microscope slide.	11a. Sample is very small. The syringe can be removed and filled with air a second time, if needed, to empty needle lumen completely.
12. If needed, make a smear using the two-slide technique.	—

Procedure for Fine-needle Aspiration Using Technique Two

TECHNICAL ACTION	RATIONALE/AMPLIFICATION
1. Identify area of aspiration and wipe with alcohol-soaked cotton.	1a. Gross contamination should be removed, but the area need not be prepared using a three step surgical preparation.
2. Immobilize area of aspiration with one hand.	—
3. Connect needle to syringe via an 84-cm intravenous extension tubing.	3a. Extension tubing permits better control of the needle and more accurate placement.
4. Prefill the 12-cc syringe with 5 cc of air.	—
5. Grasp needle in a pencil-hold fashion and hang extension tubing around your neck.	5a. This keeps the syringe and extension set out of the operator's way.
6. Place needle tip rapidly into the lesion or mass.	6a. During the procedure, never allow the needle tip to exit the target tissue.
7. Rapidly move the needle tip up and down without changing the needle path.	7a. Movement must be rapid, because coagulation factors will be induced quickly, thus decreasing the sample quality.
	7b. Aspiration is not required with this technique, because rapid movement creates a cellular slurry that becomes trapped in the needle lumen.
8. Forcefully expel sample onto slide using the prefilled, attached syringe.	—
9. If needed, make a smear using standard two-slide technique.	—

Special Notes

Needle length will vary from 1 to 3 inches depending on depth of target tissue.
   Sample size is often minute or not visible.
   Three to five samples should be taken for each area of interest. This will help to ensure that a diagnostic sample was obtained.

SKIN SCRAPES

Definition

- Intentional abrasion of the skin for purposes of diagnosis

Purpose

- Detect external parasites or fungi

Equipment

- Glass slide and cover slip
- Number 10 scalpel blade
- Mineral oil

Complications

- Minor skin irritation or hemorrhage
- Accidental laceration of skin

Procedure for Skin Scrapes

TECHNICAL ACTION	RATIONALE/AMPLIFICATION
1. Halter horse and place in stock. Place cow in chute.	1a. Llamas and goats should be haltered.
2. Put a drop of mineral oil on microscope slide.	—
3. Apply drop of oil directly to area of scrape.	3a. Oil helps dermal debris to adhere to the scalpel blade and slide.
	3b. Area selected for scrape should be toward the periphery or active area of lesion.
4. Firmly pinch the scrape area.	4a. Pinching helps to force external parasites out of the hair follicle.

TECHNICAL ACTION	RATIONALE/AMPLIFICATION
5. Holding scalpel blade with thumb and forefinger, keep the cutting surface of the blade perpendicular to the skin and scrape surface.	—
6. Continue scraping until a small amount of capillary blood is seen.	6a. Parasites often are located deep within the dermis.
7. Transfer dermal material from blade to oil on slide by scraping blade along slide edge.	—
8. Also pluck two or three hairs and place on slide.	8a. Fungal elements frequently adhere to the hair follicle.
9. Place cover slip over sample.	9a. Add additional oil if necessary.
10. Examine under microscope immediately.	10a. Multiple skin scrapes often are required to assess lesion accurately.

IMPRESSION SMEARS

Purpose

- Obtain diagnostic information concerning the sampled tissue

Complications

- Excessive blood on specimen rendering impression smears nondiagnostic

Equipment

- Rat-tooth forceps
- Gauze
- Microscope slide
- Scalpel blade

Procedure for Impression Smears

TECHNICAL ACTION	RATIONALE/AMPLIFICATION
1. Biopsy specimens: Blot sample several times onto gauze.	1a. Blotting removes excess blood and fluid from the sample.
	1b. Clean surface of *in situ* mass using sterile saline and gauze. Blot surface if bleeding is induced.

TECHNICAL ACTION	RATIONALE/AMPLIFICATION
2. Press slide firmly against mass or biopsy sample.	2a. Make several imprints of mass on each slide.
	2b. Do not rub slide, because this will distort cells.
3. Air dry and fix sample with heat.	—
4. Stain as needed.	4a. Dip Quick® is probably the most commonly used stain in veterinary practice. This gives a coloration similar to a Wright-Giemsa stain.

CULTURES AND TESTING

Culturing involves the inoculation of an appropriate media with the intent of isolating bacteria, viruses, or fungi. Samples can be shipped to a diagnostic laboratory, but many veterinary clinics have in-house facilities for routine cultures.

UTERINE CULTURE

Definition

- Culture of uterine endometrial lining for purposes of diagnosis, prognosis, and response to treatment

Purpose

- Determine bacterial causes of decreased fertility
- Evaluate endometritis
- Evaluate response to treatment

Complications

- Uterine puncture
- Personnel injury

Equipment

- Roll gauze or veterinary wrap
- Sterile palpation sleeves
- Sterile lubricant
- Bucket
- Chlorhexidine solution
- Rolled cotton
- Paper towels
- Uterine guarded culture swab

Procedure for Uterine Culture of the Horse

TECHNICAL ACTION	RATIONALE/AMPLIFICATION
1. Place mare in stock.	—
2. Sedate if necessary.	2a. Sedation generally is not required.
3. Apply tail wrap.	3a. Chapter 6 Application of a tail wrap.
4. Wash and dry perineum and vulvar area.	—
5. Apply two sterile palpation gloves, one over the other.	5a. The two-glove technique prevents contamination of the uterus with vaginal bacteria.
6. Place the swab between the two gloves.	—
7. Apply sterile lubricant to outer glove.	7a. Use lubricant that does not contain antibacterial agents.
8. Enter vagina and place index finger in the cervix.	—
9. Slide the tip of the guarded swab along the finger, punching through the outer glove.	—
10. Pass swab into body of uterus.	—
11. Expose swab to endometrium for 2–3 minutes.	—
12. Withdraw swab back into its guard, then remove from mare.	12a. The guard prevents vaginal bacteria from contaminating the sample.
13. Remove tail wrap.	—
14. Process sample immediately.	—

Procedure for Uterine Culture of the Cow

TECHNICAL ACTION	RATIONALE/AMPLIFICATION
1. Place cow in stock or chute.	—
2. Clean perineum using chlorhexidine or iodine solution and water. Rinse thoroughly and dry with paper towels.	2a. Cattle often will defecate when washed. Clean area as well as circumstance allows.
3. Wrap tail using gauze.	3a. Wrapping minimizes manure contamination.
4. Apply plastic sleeve and lubricate. Insert palpation hand into rectum.	4a. To remove feces from rectum, it is usually easiest to insert three fingers into the anus and allow air to pass into rectum. This typically will stimulate cow to defecate. All feces will not be removed.

TECHNICAL ACTION	RATIONALE/AMPLIFICATION
	4b. Ensure that cow's tail is lateral to palpation arm.
5. Locate cervix on ventral floor of pelvis. This can be accomplished by sweeping hand across pelvic floor.	**5a.** Rectal constriction or tightening commonly occurs during palpation. To assist relaxation, gently massage with fingers, back and forth along constricted area.
	5b. The cervix will feel hard and rubbery.
6. Grasp cervix with thumb and first two fingers.	—
7. Spread vulvar lips by gently pressing down with palpation hand.	—
8. Insert guarded culture swab into vagina at an upward angle of 30 degrees.	**8a.** Angulation assists in avoiding the urethral orifice.
9. Push cervix forward slightly to eliminate folds in the vagina. Level culture rod and advance until the tip reaches the external opening of the cervix.	**9a.** The swab will meet with resistance when reaching the cervix.
10. Withdraw swab 1 inch, and reposition palpation hand such that thumb and fingers are holding the caudal end of the cervix.	**10a.** The palm of the palpation hand then assists with guiding the swab through the cervix.
11. Gently advance swab into and through the cervix. Keep the thumb and two fingers just ahead of the tip of the swab.	**11a.** This position assists with manipulating through the cervix.
	11b. The swab will need to be moved in an up-and-down direction to pass through the cervix.
	11c. The cow cervix, having several rings, is much more difficult to pass through than the horse cervix.
12. Once the culture tip has passed into the uterus, push the swab through the guard and collect the sample. Withdraw the culture tip back into the guard before removing the swab.	**12.** The guard helps to prevent contamination of the sample.
13. Remove tail wrap.	—
14. Process sample immediately.	—

SKIN CULTURES

Purpose

- Diagnose **dermatophyte** (fungal) or bacterial infections of dermis

Complications

- None

Equipment

- Sterile gloves
- Sterile water
- Isopropyl alcohol 70%
- 18g × 1.5-inch needle
- Number 10 scalpel blade
- Mosquito forceps
- Chlorhexidine solution
- Tooth brush
- Inoculation media and permanent marking pen or wax pencil (Special Notes)

Procedure for Bacterial Culture of the Skin

TECHNICAL ACTION	RATIONALE/AMPLIFICATION
1. Clean area around collection site.	1a. Use sterile water only.
	1b. Some clipping may be necessary. Clip carefully so as not to disrupt the affected area.
2. Gently wipe area with alcohol.	2a. Alcohol removes surface bacterial contamination.
3. Allow area to dry.	3a. Wet alcohol can burn the animal if a pustule must be incised. Additionally, the culture swab should not get wet with the alcohol, because this will interfere with the culture.
4. Puncture the pustule using a needle or scalpel blade and swab.	4a. Alternatively, the pustule contents can be aspirated using a sterile needle and syringe.
5. Inoculate media or place swab in transport media.	—
6. Gently wash collection area with mixture of chlorhexidine solution and water.	—

Procedure for Dermatophyte Culture

TECHNICAL ACTION	RATIONALE/AMPLIFICATION
1. Wipe collection area with alcohol and allow to dry.	1a. Alcohol removes bacterial and saprophytic contaminants.
2. Pluck hairs from the outer margin of the skin lesion using mosquito forceps. The scalpel blade can be used to scrape skin crusts.	2a. The outer margin will have an active infection. Central portions do not have active areas of infection and do not yield diagnostic results. 2b. The hair follicle must be removed. Do not cut the hair using scissors or scalpel blade.
3. Press hairs into the dermatophyte test media (DTM).	3a. DTM is a Sabouraud's agar that contains a pH indicator, as well as antibacterial and antifungal agents that prevent growth of contaminants. b. Pressing hair into the media ensures that the dermatophytes will contact the DTM.
4. An alternative to plucking and pressing hairs involves brushing skin firmly with toothbrush, pressing the bristles into DTM, and cutting off the bristles or brush head.	4a. This technique is used with suspected carriers or poorly defined lesions.
5. Place cap loosely on inoculated DTM.	5a. Overtightened caps will limit oxygen availability and inhibit dermatophyte growth.
6. Place in a dark area (drawer) at room temperature and examine daily for growth.	6a. A dark environment enhances fungal growth. Cultures should be examined for color changes, as well. 6b. Growth typically is seen in 5–7 days. The culture should be maintained for 30 days, however, before stating that results are negative. 6c. Culture colonies appear as white fluffy growths. Dark or brown colonies are contaminants.

TECHNICAL ACTION	RATIONALE/AMPLIFICATION
	6d. Positive cultures can exhibit color changes with minimal growth. A color change from yellow to red indicates a positive result. The alkaline waste of the fungi combined with the pH indicator of the media causes this color change.
7. A sample can be removed and stained using new methylene blue to verify spore type.	—

Special Notes

Bacterial skin cultures may be submitted in a commercially available swab and transported to a diagnostic lab or may be cultured in-house.

Identification of fungal hyphae is aided by using the following stain mixture. Combine 100 mg chlorazol black E dye in 10 cc dimethyl sulfoxide (DMSO). Add 9 cc of water and 5 g of potassium hydroxide (KOH). This stain mixture will cause fungal hyphae to appear light green against a gray background. (Ingredients can be obtained from most chemical supply companies such as Fisher Scientific or VWR.)

MILK CULTURES

Purpose

- Identify causative agent of mastitis
- Determine antibiotic sensitivity of microbial pathogens
- Determine presence of antibiotic residues within milk

Complications

- Personnel trauma during collection

Equipment

- Isopropyl alcohol 70%
- Cotton or gauze squares
- Cotton-tipped applicator sticks
- Sterile screw-top bottles or tubes
- Permanent marking pen

Procedure for Milk Culture

TECHNICAL ACTION	RATIONALE/AMPLIFICATION
1. Place cow in chute or stock.	—
2. Label tubes with owner name, date, cow identification number, and quarter of udder sampled.	2a. Samples taken within 3 days of treatment are invalid. 2b. Ideally a milk sample should be obtained upon noticing milk changes prior to initiating treatment.
3. Wipe teats clean using alcohol.	3a. Do not wash udder using soap and water. 3b. If udder is extremely dirty and needs washing, be certain to dry the entire udder surface using single-use disposable paper towels.
4. Carefully swab teat sphincter with alcohol.	—
5. Compress teat and discard the first two or three squirts of milk.	5a. Inexperienced milkers can use the following technique: Compress proximal portion (base) of teat using thumb and index finger. Without releasing pressure sequentially, tighten your third, fourth, and fifth digits around the teat. 5b. Fig. 5–21.

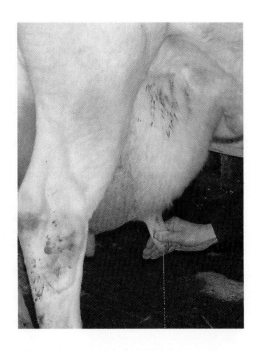

FIG. 5-21 Close-up of expressing a milk sample from a cow.

TECHNICAL ACTION	RATIONALE/AMPLIFICATION
6. Holding collection tube as horizontal as possible, put two or three squirts of milk in the tube.	6a. Maintaining the tube in a vertical upright position increases the likelihood of contamination.
7. If all four quarters must be sampled, collect milk from the closer teats first.	7a. This prevents your arm from contaminating the alcohol-prepared near teat during sampling of far teat.
8. Inoculate suitable media immediately.	8a. Samples that can not be processed immediately should be refrigerated.
	8b. Do not freeze samples.
	8c. If mycoplasma mastitis is suspected, add ampicillin to the sample at a rate of 1–10 mg/ml of milk to prevent contaminant overgrowth.

CALIFORNIA MASTITIS TEST (CMT)

Purpose

- Assess somatic cell counts to detect subclinical mastitis

Complications

- Personnel trauma during collection

Equipment

- Isopropyl alcohol 70%
- Cotton or gauze squares
- CMT reagents and paddle

Procedure for California Mastitis Test

TECHNICAL ACTION	RATIONALE/AMPLIFICATION
1. Place cow in stock.	—
2. Wipe all teats with alcohol.	—
3. The test is conducted using a paddle with four quadrants marked A, B, C, and D.	—
4. Put two or three squirts of milk from each quarter into correct quadrant of paddle.	—

TECHNICAL ACTION	RATIONALE/AMPLIFICATION
5. Mix equal amount of CMT reagent with milk samples.	—
6. Mix milk and reagent by gently rotating the paddle for 10 seconds.	—
7. Read test results immediately.	7a. Subclinical mastitis (high somatic cell count) is evidenced by a thickening of the reagent solution. This occurs when the white blood cells bind to the reagent.
	7b. The amount of thickening is proportional to the amount of infection. The test result is negative when the reagent-milk mixture appears unchanged. A thickening or gelling of the reagent-milk mixture indicates a positive test result.
8. Record results.	—

Special Notes

Antibiotic residue tests frequently are performed to ensure that no antibiotic residue remains in the milk.

The **California mastitis test** detects subclinical mastitis or elevated somatic cell counts. This test can be done in the field and typically is done before milk culture is performed.

Milk sampling is not routinely performed in species other than the cow.

NECROPSY SAMPLING

It's been said (tongue in cheek) that pathologists have all the answers, they're (the answers) just not available when you need them. Given its postmortem timing, **necropsy** is undoubtedly a bit late for the individual animal. However, its use as a diagnostic tool and its ability to facilitate the continued health of herd mates is unequaled.

Purpose

- Obtain postmortem diagnosis
- Apply information from diagnosis to other herd animals
- Comply with regulations set by owner's insurance company

Complications

- Possible zoonotic contamination
- Injury to personnel from necropsy equipment (lacerations)

Equipment

- History form and pen
- Necropsy knife
- Sharpening stone for knife
- Axe
- Saw
- Pruning shears
- Scalpel handle and blade
- Scissors
- Rat-tooth forceps
- Cutting board
- Ruler
- Permanent marking pen
- Umbilical tape
- Red top (no additive) and purple top (EDTA) tubes
- Formalin 10%
- 18g × 1.5-inch needles and 12-cc syringes
- Leak-proof specimen containers
- Gloves
- Rubber boots
- Disposable or washable coveralls
- Safety glasses or mask

Procedure for Necropsy

TECHNICAL ACTION	RATIONALE/AMPLIFICATION
1. Obtain complete clinical history.	1a. Chapter 4.
2. Determine what samples must be collected.	2a. Samples will be determined by the veterinarian based on the differential diagnosis.
	2b. Confirm required sample submission with the diagnostic laboratory.
3a. Label sample containers with permanent marker.	3a. Animal identification number, date, and tissue type should be written on all containers.
b. Cow: Place animal in left lateral recumbency.	3b. Left-side-down position prevents the rumen from obscuring other abdominal structures. Ideally a necropsy surface is skid-proof and washable.
c. Horse: Place in right lateral recumbency.	3c. Right lateral recumbency provides better access to stomach and large intestine.

TECHNICAL ACTION	RATIONALE/AMPLIFICATION
4. Perform thorough external examination.	4a. Obtain photographs for insurance if needed.
5. Incise pectoral muscles and reflect thoracic limb dorsally.	5a. Note condition of axillary lymph nodes, brachial plexus.
	5b. Fig. 5-22.
6. Incise muscles of the pelvic area and disarticulate coxofemoral joint. Reflect limb dorsally.	6a. Note condition of articular cartilage, synovia, and sciatic nerve.
7. Incise along ventral midline from pubis to chin, circumventing umbilicus and external genitalia.	—
8. Reflect skin on right side to dorsal midline. Incise muscles of hind and fore limbs several times to examine for lesions.	8a. Note condition of musculature and external lymph nodes.
9. Collect histopathologic samples of muscle and lymph node tissue. Collect samples for culture if needed.	—

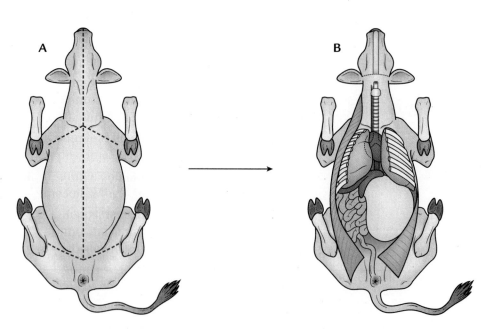

FIG. 5-22 (A) *Dotted lines* depicting the incision lines for cow or horse in a diagnostic necropsy. Animal may be rolled into dorsal recumbency after incisions are made and limbs are more mobile. (B) Depiction of cow after skin, ribs, and body wall are pulled back and the sternum has been split open.

TECHNICAL ACTION	RATIONALE/AMPLIFICATION
10. Incise stifle and hock joints. Note condition of ligaments, articular cartilage, and synovia. Collect samples for culture if needed.	10a. If these joints are abnormal, incise all joints, looking for other lesions.
11. Incise from dorsocaudal aspect of last rib to midline, and tuber coxae to midline, to open the abdominal cavity.	11a. This creates a flap that can be reflected dorsally.
12a. Examine abdominal viscera for normal anatomic size, shape, and location.	12a. Many diseases are the result of abnormal location, such as displaced abomasum and nephrosplenic entrapment.
b. Cow: Examine space between reticulum and diaphragm.	12b. Adhesions or metal objects indicate traumatic reticuloperitonitis.
13. Puncture diaphragm and listen for inflow of air.	13a. A normal thoracic cavity is under negative pressure.
14. Incise diaphragm along costal arch attachment.	—
15. Cut and remove ribs using shears or ax.	—
16. Examine thoracic structures for anatomic location, size, and shape.	—
17. Double ligate using umbilical tape.	—
a. Cow: Double ligate the proximal duodenum and rectum using umbilical tape. Cut between the ligations.	17a. Ligating helps prevent contamination of other structures and keeps work area cleaner.
b. Then double ligate the esophagus. Cut between ligations.	17b. —
c. Remove rumen, omasum, abomasum, and reticulum.	17c. The spleen is attached to the left side of the rumen.
d. Incise rumen. Examine lining and contents. Collect content sample for toxicology if needed.	17d. The rumen is the largest compartment, and the lining has a lawn-grass appearance.
e. Incise reticulum and check for metal objects. Recover magnet.	17e. The reticulum is the smallest compartment and has a distinct honeycomb-like lining. Magnets that have been placed correctly to prevent hardware should be recovered from the reticulum.
f. Incise omasum and abomasum.	17f. Note lining and contents. Collect histopathologic and culture samples if needed.
g. Horse: Rectum and esophagus.	—

TECHNICAL ACTION	RATIONALE/AMPLIFICATION
18. Remove gastrointestinal tract.	18a. Mesenteric attachment will need to be severed.
19. Open small and large intestine lengthwise using scissors. Examine for lesions.	19a. Collect histopathologic samples of colon, jejunum, ileum, duodenum, and stomach if doing a necropsy on the horse. Collect samples for culture as needed.
20. Horse: The distinctive cecum and large, transverse, and small intestines should be examined.	—
21. Remove spleen and examine external surface. Make several parallel cuts to examine parenchyma. Collect histopathologic sample.	—
22. Remove the liver.	—
23. Inspect capsule and external surface of liver.	—
24. Cow: Incise gallbladder. Examine mucosal surface. Collect histopathologic or culture samples.	24a. Horses do not have a gallbladder.
25. Make several parallel incisions in liver. Examine parenchyma for abnormalities.	25a. Collect histopathologic or culture samples.
26. Examine portal and hepatic veins for thrombosis.	—
27. Remove reproductive and urinary systems.	—
28. Examine external surface of kidneys. Make longitudinal incision to inspect renal pelvis.	28a. Collect culture sample. Histopathologic sample should include cortex and medullary areas.
29. Make several parallel cuts to examine parenchyma. Incise along ureters opening to bladder. Incise bladder and examine mucosal surface.	29a. Obtain culture and histopathologic samples.
30. Remove adrenal glands. Examine external surface and parenchyma.	30a. Adrenal glands are found on anterior poles of kidneys.
	30b. Collect histopathologic samples.
31. Examine external surface of testicles or uterus and ovaries. Make several longitudinal incisions to examine parenchyma of testicles and ovaries. Incise uterus and examine mucosal surface.	31a. Collect culture and histopathologic samples.

TECHNICAL ACTION	RATIONALE/AMPLIFICATION
32. Incise along medial aspect of mandibles to free the tongue.	32a. This is much more difficult in the horse. Separation of the mandibular symphysis will facilitate removal.
33. Pull tongue downward toward trachea. Cut as needed to free tongue and trachea.	—
34. Pull trachea caudally to begin removing trachea, lungs, and heart *in situ*.	34a. The aorta, vena cava, and esophagus will require transection.
	34b. This allows the veterinarian to remove the entire pluck consisting of the trachea, thyroid, parathyroid, heart, lungs, thymus, tongue, and esophagus.
35. Locate and examine thymus.	35a. The thymus normally atrophies with age.
	35b. Collect histopathologic sample.
36. Locate thyroid glands on anterior aspect of trachea. Note size.	36a. Collect histopathologic sample.
37. Incise along lumen of esophagus. Examine mucosal surface.	37a. Collect histopathologic sample.
38. Examine external surface of heart, noting condition of pericardium and presence of normal fat in the cardiac groove.	38a. It is difficult to remove the equine heart with intact pericardium.
39. Incise heart to examine valves.	39a. Aortic, pulmonary, right and left atrial and ventricular chambers, and wall size.
40. Examine greater vessels for evidence of stenosis or aneurysm. Examine atrial and ventricular septa for defects.	—
41. Incise myocardium several times, using parallel cuts, to examine condition of myocardium.	41a. Collect culture and histopathologic samples.
42. Examine external surface of lungs.	—
43. Incise trachea to level of mainstem bronchi.	—
44. Palpate lung tissue consistency. Incise lungs, making several parallel incisions.	44a. Do not handle lung tissue intended for histopathologic study, because it will create excessive amounts of artifact.
	44b. Collect culture and histopathologic samples.

TECHNICAL ACTION	RATIONALE/AMPLIFICATION
45. Examine teeth and hard palate.	—
46. Collect cerebral spinal fluid from atlantooccipital joint using an 18g × 3-inch needle and 12-cc syringe.	**46a.** Flexing the head toward the chest will open this joint.
47. Remove the head at this time if needed. Extend head and transect through neck musculature and atlantooccipital joint.	—
48. To remove the brain, remove skin and musculature from dorsal aspect of cranium. Cut through cranium using saw or axe.	**48a.** Fig. 5–23.
49. Using screwdriver or similar object, remove skull cap using a prying motion. Cut through dura mater. Transect tentorium cerebelli.	**49a.** Transecting the tentorium cerebelli allows the cerebellum and cerebral hemispheres to be removed together.
50. Obtain culture sample if needed.	**50a.** The optic chiasm provides a good sampling site.
51. Cut cranial nerves and remove the intact brain.	**51a.** The pituitary gland typically remains in the cranium and must be removed using forceps.
	51b. Submit sections for histopathologic analysis.
	51c. Alternatively, the entire brain can be placed in 3 L of formalin.

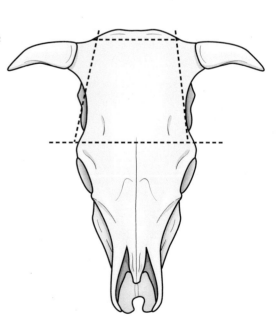

FIG. 5-23 *Dashed lines* depict the cuts to be made through a cow's skull to gain access to the brain.

Special Notes

- Necropsies performed more than 48 hours after death are not diagnostic.
- Samples submitted for culture or histology must not be frozen.
- Samples submitted for histopathologic analysis should be placed in 10% formalin. Tissue thickness should not exceed ¾ inch.
- All samples should be submitted in leak-proof containers.
- Use a ruler to measure and record size of all gross lesions.
- Sharpen instruments as needed during the necropsy.
- Animals undergoing necropsy for insurance purposes should be photographed before the procedure. Confirm any brands or tattoos in ears or lips. Contact the insurance company for details before doing the necropsy.

REVIEW QUESTIONS

1. State appropriate venipuncture and arterial puncture sites of all large animal species.
2. Describe differences in technique used for sample collection of routine venous blood samples and those used for **coagulation studies**.
3. State possible complications of fecal collection.
4. State the most common reason for performing **abdominocentesis** in the horse.
5. Describe the two techniques used to perform abdominocentesis. Which technique is most likely to cause a laceration of an abdominal structure?
6. Name the tests typically performed on urine collected via the bovine digital method.
7. State the most common test performed on a **rumenocentesis** sample.
8. Describe the techniques used to assess placement of a nasogastric tube.
9. State reasons for avoiding heavy sedation during a transtracheal wash.
10. Identify the location for performing a bovine liver biopsy.
11. Compare and contrast dermal aspiration and biopsy.
12. Explain reasoning for pinching skin and plucking hair during a routine skin scrape.
13. State two culture techniques that are performed routinely on the skin.
14. State the most common reason for performing a CMT.
15. Identify the recumbency positions used during horse and cow necropsy, and explain why these positions are used.

BIBLIOGRAPHY

Ackerman, L. (1989). *Practical equine dermatology* (2nd ed.). Goleta, CA: American Veterinary Publications.

Auer, J. (1992). *Equine surgery*. Philadelphia: W. B. Saunders.

Crow, S., & Walshaw, S. (1997). *Manual of clinical procedures in the dog cat and rabbit* (2nd ed.). Philadelphia: Lippincott, Williams & Wilkins.

Faerber, C. (1999). *Dairy production medicine and management.* Animal Health Publications.

Fowler, M. (1998). *Medicine and surgery of South American camelids* (2nd ed.). Ames: Iowa State University Press.

Herman, H., Mitchell, J., & Doak, G. (1994). *The artificial insemination and embryo transfer of dairy and beef cattle* (8th ed.). Danville, Illinois: Interstate Publishers.

Koterba, A., Drummond, W., & Kosch, P. (1990). *Equine clinical neonatology.* Philadelphia: Lea & Febiger.

McCurrin, D. (1998). *Clinical textbook for veterinary technicians* (4th ed.). Philadelphia: W. B. Saunders.

Morrow, D. (1986). *Current therapy in theriogenology 2.* Philadelphia: W. B. Saunders.

Nordlund, G. (1994). Rumenocentesis: A technique for the diagnosis of subacute rumen acidosis in dairy herds. *Bov Pract, 28,* 109–112.

Oakley, G., Jones, D., Harrison, J. A., Wade G. E., et al. (1980). A new method of obtaining arterial blood samples from cattle. *Vet Rec* Vol 106, pg. 460.

Oehme, F. (1974). *Textbook of large animal surgery* (2nd ed.). Baltimore: Williams & Wilkins.

Pearce, S. G., Firth, E. C., Grace, N. D., & Fennessy, P. F. (1997). Liver biopsy techniques used for adult horses and neonatal foals to assess copper status. *Aust Vet J, 75,* 194–198.

Pratt, P. (1998). *Principles and practice of veterinary technology* (4th ed.). St. Louis: C. V. Mosby.

Smith, B. (1990). *Large animal internal medicine.* St. Louis: C. V. Mosby.

CHAPTER 6

Clinical Procedures

KEY TERMS

balling gun
beef quality assurance
catheter
intramuscular
intravenous
intubation
nasolacrimal duct
ocular
parenteral
sheath
stylet
subcutaneous
syringe gun

OBJECTIVES

- Describe equipment and techniques used in oral, **parenteral**, and **ocular** drug administration.
- Identify purpose and methodology for routine large animal clinical procedures.
- Compare and contrast clinical techniques used in the various large animal species.

A fellow went into a restaurant and asked, "What's the special of the day?"

The waiter replied, "Beef tongue."

The fellow said, "Ugh! I don't want anything coming out of a cow's mouth…Fry me up a couple of eggs."

ADMINISTERING ORAL MEDICATIONS

Substances orally administered to large animals are numerous and can include antibiotics, antiinflammatory agents, deworming medications, probiotic or other nutritional supplements. Most animals are amenable to having substances placed in the oral cavity; however, the degree of restraint needed will vary with each animal species.

ADMINISTERING PASTES

Purpose

- Administer oral medication or deworming agents

Complications

- Spitting out paste
- Oral trauma
- Personnel injury
- Buccal ulceration
- Aspiration pneumonia

Equipment

- Dosing syringe containing medication

Procedure for Administering Oral Pastes to the Horse

TECHNICAL ACTION	RATIONALE/AMPLIFICATION
1. Halter the horse.	1a. A stock can be used if desired but usually is not needed.
2. Ensure that oral cavity is clear of grain and hay.	2a. If feed is present in the mouth, the horse will easily spit out the medication.
3. Stand lateral to horse and place one hand over bridge of nose.	3a. Most horses will lift the head when the syringe is introduced into the oral cavity.
4. Insert syringe at commissure of lips through interdental space.	4a. Never use a balling gun with the horse.
	4b. Fig. 6–1.
5. Administer paste on the tongue. Lift head.	5a. Do not place paste in cheek pouch. This can result in ulceration from prolonged medication contact time.
	5b. Lifting the head discourages the horse from spitting out the medication.

FIG. 6-1 Administering oral medication to a horse.

Procedure for Administering Oral Pastes to the Goat

TECHNICAL ACTION	RATIONALE/AMPLIFICATION
1. Secure goat by straddling neck or holding collar and pushing against wall.	—
2. Ensure that oral cavity is empty. Grasp mandible and elevate head slightly above horizontal.	2a. Elevating the head helps retain liquid medication in the mouth.
3. Insert syringe tip into commissure of lips until goat opens mouth. Once mouth is open, direct syringe to caudal aspect of oral cavity and administer medication.	3a. Do not place medication in cheek pouch, because this can cause buccal ulceration. 3b. If goat begins coughing or showing signs of distress, lower the head.

Procedure for Administering Oral Pastes to the Llama

TECHNICAL ACTION	RATIONALE/AMPLIFICATION
1. Halter llama. Place in stock if available.	—
2. Ensure that mouth is clear of feed.	—
3. Snub head as close as possible to post.	3a. Llamas that are handled frequently probably will not require snubbing.
4. Place syringe in lip commissure and advance through interdental space.	—
5. Gently advance syringe to caudal aspect of mouth and administer medication. Lift head.	5a. Do not place paste in cheek pouch. This can result in ulceration from prolonged medication contact.
	5b. Lifting the head discourages spitting out the medication.

BALLING GUNS

Purpose

- Administer oral medication or magnet

Complications

- Laryngeal trauma
- Aspiration

Equipment

- Balling gun
- Calf-size gun for goats

Procedure for Using a Balling Gun

TECHNICAL ACTION	RATIONALE/AMPLIFICATION
1. Place cow in chute. Straddle neck of goat.	1a. Calves are also straddled over the neck.
2. Allow cow or goat to swallow any feed currently in the mouth.	2a. Having a clear oral cavity expedites the procedure.

TECHNICAL ACTION	RATIONALE/AMPLIFICATION
3. Place fingers in commissure of lips and insert fingers or hand into interdental space to open mouth.	3a. Animals may resist by tossing head. Take care to prevent your hands from being hit against the chute.
4. Insert balling gun and direct caudally in mouth. As the base of the tongue is reached, the tip may need to be directed slightly dorsally to pass over the tongue base.	4a. Cow and goats frequently will attempt to chew on or push out gun. 4b. The gun does not need to be lubricated. 4c. Fig. 6–2.
5. Depress plunger to administer bolus.	5a. Use plastic balling guns when administering magnets. If a metal gun must be used, place hay around the magnet to prevent adherence to the gun.

FIG. 6–2 Administering oral medication to a ruminant using a balling gun.

PARENTERAL ADMINISTRATION OF DRUGS

Parenteral drug, implant, or vaccine administration is a cornerstone of veterinary practice. The importance of developing good injection techniques cannot be overemphasized. In addition to minimizing the animal's immediate pain or distress, appropriate injection techniques decrease future economic loss from tissue damage. The technician's role in adherence to quality assurance protocols is pivotal.

Before administering a substance, confirm that you have the correct patient, dosage, route, drug or biologic agent, time, and frequency of administration. Adherence to this routine will ensure a safe, effective therapy.

COMMON NEEDLE SIZES

Needle size selection varies with the species, age, location of injection, and substance injected. Table 6–1 lists appropriate needle sizes for each species.

TABLE 6–1 COMMON NEEDLE SIZES

SPECIES	NEEDLE GAUGE AND LENGTH	LOCATION	COMMENTS
Horse			
Adult	18–20g × 1.5 inch	IM, IV	Viscous fluids require a larger gauge needle. Vaccines can be administered using a 1-inch needle. 20g × 1.5-inch is the most common size for IM injection smaller than 10 cc.
Foal or pony	20g × 1 inch	IM, IV	—
Cow			
Adult	16g × 1 inch	IM, SQ, IV	16g × 1 inch: Vaccination of adult cows.
	16–18g × 0.5 inch		16–18g × 1.5-inch needle for mature cattle; Gauge depends on fluid volume and viscosity. Use 1-inch needle for all SQ injections. Use 1.5-inch needle for IV injections.
Calf	18g × 1 inch	IM, SQ, IV	18g × 1: Vaccination of calves.
	20g × 0.5–1 inch		20 g × 1: Routine calf injections.
Pig			
Piglets	18–20g × 0.5 inch	IM, SQ	—
	25–25g × 1 inch	IV	
Finishers	16g × 1 inch	IM, SQ	—
	20g × 1 inch	IV	
Adult	14–16g × 1–1.5 inches	IM, SQ	Needle length depends on back fat depth and method of restraint.
	18–20g × 1 inch	IV	Gauge depends on fluid viscosity and volume of injection.
Goat			
Adult	20g × 1 inch	IM, SQ, IV	—
Kid	22g × 1 inch	IM, SQ, IV	—
Llama			
Cria	22–20g × 1 inch	IM, SQ, IV	—
Adult	18–20g × 1 inch	IM, SQ	Gauge depends on fluid viscosity.
	18–20g × 1.5 inches	IV	

IM, intramuscular; *IV*, intravenous; *SQ*, subcutaneous.

ADMINISTERING INTRAMUSCULAR INJECTIONS

Purpose

- Administer biologic agents or other drugs into the muscle

Complications

- Myositis
- Iatrogenic infection
- Inadvertent injection into bloodstream
- Trauma to personnel

Equipment

- Isopropyl alcohol 70%
- Needle
- Syringe
- Cotton

Procedure for an Intramuscular Injection in the Horse

TECHNICAL ACTION	RATIONALE/AMPLIFICATION
1. Halter horse and place in stock.	1a. Stocks help to minimize motion and therefore discomfort during the injection.
	1b. Never administer any injection without haltering the horse, even if confined to a stall.
2. Locate muscle for injection.	2a. Fig. 6–3 shows injection locations.
	2b. Table 6–1 lists a selection of appropriate needle sizes.
	2c. Table 6–2 lists appropriate sites for intramuscular injection
3. Wipe injection site using alcohol.	3a. Alcohol removes surface debris and minimizes contamination.
4. Remove needle from syringe. Grasp hub of needle tightly. Hold needle perpendicular to horse and insert rapidly into muscle. Ensure that entire length of needle is inserted.	4a. Occasionally technicians will hold the needle and tap the muscle firmly 3–4 times with the back of the hand (same hand holds and taps). On the last tap, the hand is turned over and the needle inserted. This technique typically is used by inexperienced technicians

TECHNICAL ACTION	RATIONALE/AMPLIFICATION
	having difficulty placing the needle swiftly into the muscle. This technique will alert the horse to the pending injection, thereby causing muscle contraction and increased pain upon injection.
	4b. Hubbing the needle, or inserting the entire length of the needle into the horse, ensures that the needle will not advance after aspiration and inadvertently enter a vessel.
	4c. Many individuals will use the free hand to apply a hand twitch on the neck while placing the needle. This distraction technique works well.
5. Attach syringe to needle.	**5a.** Horses frequently move after a needle is inserted. To avoid bent needles and significant muscle trauma, syringes are attached after needle placement.
	5b. Fig. 6–4.

FIG. 6–3 Intramuscular injection sites in the horse.

TECHNICAL ACTION	RATIONALE/AMPLIFICATION
6. Aspirate to ensure that a blood vessel has not been entered.	6a. Aspiration means to pull back on the plunger of the syringe while watching the hub of the needle for the appearance of blood. 6b. Never inject into the muscle without aspirating. Many biologic agents can cause death if injected into the bloodstream. 6c. If blood is noted upon aspiration simply remove the needle, obtain a new needle, and inject 1–2 inches from first site.
7. Inject at a moderate rate of speed.	7a. The maximum volume per injection site is 15 cc. Neonates should receive a maximum of 5–10 cc per site.

TABLE 6-2 INTRAMUSCULAR INJECTION SITES FOR THE HORSE

MUSCLE	COMMENTS
Neck	Fig. 6–3 shows sites used for intramuscular injection. Site is used for smaller volumes or occasional large-volume injection. Site is safer for the technician as the horse is unlikely to kick. External boundaries of this site include nuchal ligament, cranial border of scapula, and jugular groove. Always palpate to ensure that injection is not placed directly over palpable wing of cervical vertebra.
Triceps and pectorals	Site is used primarily for vaccinations. These muscles compose the chest and area just cranial to the girth. Site is safer for the technician as the horse is unlikely to kick. The larger muscle can accommodate larger volumes or repeated injections.
Semimembranosus or semitendinosus	This is the most commonly used site in foals. Technicians should exercise caution to avoid being kicked by standing on the side opposite to which they are injecting. Never inject into the ligament area between the two muscle bellies. These muscles are located on the caudal rump area.
Gluteus	This large muscle can accommodate large volumes or repeated injections. Drainage is difficult to establish should an abscess develop. These muscles comprise the rump of the horse.

TECHNICAL ACTION	RATIONALE/AMPLIFICATION
8. Remove needle and massage injection site.	8a. Massaging helps alleviate injection site pain.
	8b. Occasionally bleeding from the injection site is noted upon removal of the needle. This is most likely caused by disruption of vessels in the subcutaneous tissue. It does not mean that the injection entered a vessel. Wipe blood off using cotton or gauze.

FIG. 6-4 Administration of an intramuscular injection to a horse.

Procedure for Intramuscular Injection in the Cow

TECHNICAL ACTION	RATIONALE/AMPLIFICATION
1. Place cow in chute or stock.	1a. Stock helps to minimize motion and therefore discomfort during the injection.
2. Locate cervical muscles for injection. The cervical vertebrae run roughly between the base of the ear and point of the shoulder. Keep injections above this line, approximately 4 inches below the crest of the neck and cranial to the scapula.	2a. Neck muscles should be used if at all possible. Alternative sites for injection include gluteals, semimembranosus, and semitendinosus muscles.
	2b. **Beef quality assurance** issues dictate that injections be administered in the cervical area whenever possible. Even small

TECHNICAL ACTION	RATIONALE/AMPLIFICATION
	injection volumes can adversely affect meat quality years after injection. Use of gluteals or other muscles should be limited to neonates, multiple injection sites, or very large injection volumes. See Beef Quality Assurance section in text for recommendations.
	2c. Intramuscular injections administered in front of the head catch are most likely administered too far cranially in the neck.
	2d. Table 6–1 gives a selection of appropriate needle sizes.
3. Wipe injection site using alcohol.	**3a.** Alcohol removes surface debris and minimizes contamination; however, this action is performed only on hospitalized animals. Cattle that are being processed are not wiped with alcohol.
4. Hold syringe and keep needle perpendicular to neck. Rapidly insert needle to the hub.	**4a.** Hubbing the needle, or inserting the entire length of the needle into the cow, ensures that the needle will not advance after aspiration and inadvertently enter a vessel.
	4b. Cow will move rapidly forward or backward in the chute upon penetration of needle.
	4c. If large volumes are being administered, many individuals will place the needle and then attach the syringe. This is done when one or two cows are being treated. This technique would never be used when processing a large number of animals.
	4d. Fig. 6–5.
5. Aspirate and inject.	**5a.** Aspiration means to pull back on the plunger of the syringe while watching the hub of the needle for the appearance of blood.
	5b. Unfortunately, many intramuscular injections in cattle are performed without aspirating.

TECHNICAL ACTION	RATIONALE/AMPLIFICATION
	This situation typically arises when processing very large numbers or using an automatic **syringe gun**.
	5c. Automatic syringe guns increase efficiency, because the technician does not need to draw up vaccine after injecting each cow. Each loaded gun can hold 10–25 doses. Aspiration cannot be done with a gun like it can with a syringe.
	5d. The maximum volume per injection site is 10 cc. Neonates should receive a maximum of 5 cc per site.
6. If a second injection must be administered, place the injection 3–4 inches from the first site.	**6a.** Placing two vaccines within 2–3 inches will increase the likelihood of abscess formation.
7. Massage injection site if feasible.	**7a.** Massaging helps alleviate injection site pain. Massage is not feasible if processing large numbers or handling wild cattle.

FIG. 6-5 Injection sites used in the cow. (**A**) Preferred site for all subcutaneous and intramuscular injections. (**B**) Intramuscular injection sites (use is not recommended under BQA guidelines).

Beef Quality Assurance Recommendations

Given that most cattle (beef and dairy) are eventually consumed, the following recommendations have been assembled to minimize muscle damage and increase carcass quality.

- Use muscles on the neck cranial to the shoulder for **intramuscular** injections if at all possible.
- Administer all **subcutaneous** injections cranial to the shoulder.
- Use clean, sharp needles, and change needles every 8–10 injections when processing large numbers of cattle.
- Use 16–18g × 1-inch needles for vaccinations.
- Administer a maximum of 10 cc per injection site.
- Properly restrain animals during injection.
- Ensure that injection sites are at least 4 inches apart.
- Keep vaccination bottles (or any drug) clean, and use a new sterile needle when drawing product from the bottle.
- If possible, administer injections subcutaneously instead of intramuscularly.

Care of Automatic Syringe Guns

TECHNICAL ACTION	RATIONALE/AMPLIFICATION
1. Discard any unused biologic agent or medicine appropriately.	1a. Consult Material Safety Data Sheets (MSDS) for appropriate disposal methods.
2. Cap, remove, and discard needle.	2a. Needles should be placed in an appropriate sharps container.
	2b. Removing needles from automatic guns can prove difficult. Using pliers or another gripping device is helpful. Recap needles before removal.
3. Disassemble gun and wash in a mild soap detergent.	3a. Use care not to lose the rubber washers.
	3b. Fig. 6–6.
4. Thoroughly rinse with warm water.	4a. Even traces of soap residue can inactivate modified live vaccines.
5. Inspect all parts for damage, paying close attention to barrel and washers.	5a. Cracking of barrels and rubber washers is a very common problem.

TECHNICAL ACTION	RATIONALE/AMPLIFICATION
6. Reassemble gun.	6a. Do not overtighten barrel or tension of plunger.
	6b. Washers frequently are forgotten. The gun will not function without the washer.
	6c. If needed, a light coating of petroleum jelly (Vaseline) can be applied to the rubber plunger before assembly.

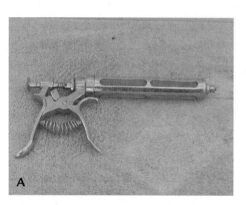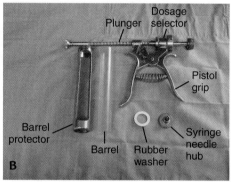

FIG. 6–6 (A) An automatic pistol-grip syringe gun. (B) Disassembled syringe gun. Note the barrel protector, the barrel, plunger, pistol grip, syringe needle hub, dosage selector, and the rubber washer.

Procedure for Intramuscular Injection in the Goat

TECHNICAL ACTION	RATIONALE/AMPLIFICATION
1. Grasp goat by collar or place in stanchion.	1a. Encouraging the goat to stand against a wall will facilitate the procedure if the goat is being held by the collar.
2. Localize muscle for injection.	2a. Intramuscular injections in the goat are performed most frequently in the semimembranosus and semitendinosus muscles. Triceps and neck muscles can be used, but their use is limited to smaller volumes. Lumbar muscle injection should be avoided.
	2b. Never inject into the ligament area between muscle bellies.
	2c. Table 6–1 gives a selection of appropriate needle sizes.

CHAPTER 6 Clinical Procedures **207**

TECHNICAL ACTION	RATIONALE/AMPLIFICATION
3. Wipe injection site using alcohol.	3a. Alcohol removes surface debris and minimizes contamination.
4. Hold syringe and keep needle perpendicular to muscle. Rapidly insert needle to the hub.	4a. Hubbing the needle, or inserting the entire length of the needle into the goat, ensures that the needle will not advance after aspiration.
	4b. Goats typically will vocalize at this point.
	4c. Fig. 6–7.
5. Aspirate to ensure that a blood vessel has not been entered.	5a. Aspiration means to pull back on the plunger of the syringe while watching the hub of the needle for the appearance of blood.
	5b. Never inject into the muscle without aspirating. Many biologic agents can cause death if injected into the bloodstream.
	5c. If blood is noted upon aspiration simply remove the needle, obtain a new needle, and inject 1–2 inches from first site.
6. Inject at a moderate rate of speed.	6a. Do not inject more than 5 cc per injection site.
7. Remove needle and massage area.	7a. Massage alleviates pain and improves drug absorption.

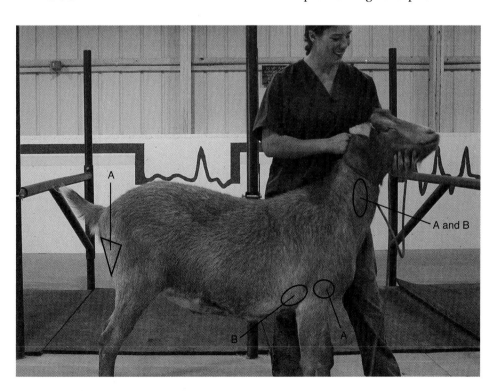

FIG. 6-7 Injection sites used in the goat. Intramuscular (**A**) and subcutaneous (**B**).

Procedure for Intramuscular Injection in the Pig

TECHNICAL ACTION	RATIONALE/AMPLIFICATION
1. Restrain pig using snare, crate, or chute.	1a. Grower and finisher hogs can be restrained by holding the hind limbs in the air.
	1b. Neonates are lifted off the ground.
	1c. Pigs will vocalize loudly during this procedure. Technicians and veterinarians are advised to use ear protection devices.
2. Identify injection site just caudal and ventral to ear on lateral neck area. Ensure that injection site is clean and dry.	2a. Fig. 6–8.
	2b. Meat quality assurance issues are extremely important in pigs. Do not use other muscles for injection.
	2c. Avoid injecting into fat, because drug absorption will be delayed dramatically.
	2d. Table 6-1 gives a selection of appropriate needle sizes.
3. Grasp skin and pull cranially. Keep needle perpendicular to skin and insert rapidly into muscle.	3a. The needle and syringe are kept attached.
	3b. Pulling the skin cranially helps to seal the site and prevents leakage from the injection site.
4. Aspirate and inject.	4a. Aspiration means to pull back on the plunger of the syringe while watching the hub of the needle for the appearance of blood.
	4b. If blood is noted upon aspiration simply remove the needle, obtain a new needle, and inject 1 inch from first site.
	4c. Inject a maximum volume of 5–10 cc per site in adults. Inject a maximum volume of 1–2 cc per site in piglets.
	4d. Excessive volume injected into one site will cause pressure necrosis and delayed absorption.

CHAPTER 6 Clinical Procedures **209**

TECHNICAL ACTION	RATIONALE/AMPLIFICATION
5. Withdraw needle and massage.	5a. If using an automatic syringe gun, change needles after every 10 pigs. Immediately change needles that become bent, burred, or contaminated.

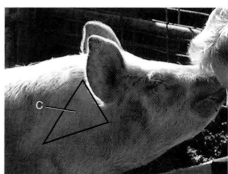

FIG. 6–8 Subcutaneous and intramuscular injection sites used in the pig. (**A**) Preferred subcutaneous site. (**B**) Alternative subcutaneous site. (**C**) Intramuscular site.

Procedure for Intramuscular Injection in the Llama

TECHNICAL ACTION	RATIONALE/AMPLIFICATION
1. Halter and secure llama.	1a. Minimizing motion during injections decreases trauma and pain of injection.
2. Localize and clean injection site using alcohol wipe. The semimembranosus and semitendinosus muscles are used most commonly.	2a. Avoid using neck muscles. 2b. Triceps muscles are also commonly used. 2c. Fig. 6–9.

TECHNICAL ACTION	RATIONALE/AMPLIFICATION
3. Keep needle attached to syringe. Hold them perpendicular to skin and insert smoothly and rapidly into muscle.	3a. Table 6–1 gives a selection of appropriate needle sizes.
4. Aspirate and inject at a moderate rate of speed.	4a. Aspiration means to pull back on the plunger of the syringe while watching the hub of the needle for the appearance of blood.
	4b. Never inject into the muscle without aspirating. Many biologic agents can cause death if injected into the bloodstream.
	4c. If blood is noted upon aspiration simply remove the needle, obtain a new needle, and inject 1 inch from first site.
5. Remove needle and massage.	5a. Massage alleviates pain and improves drug absorption.

FIG. 6–9   Injection sites used in the llama. Intramuscular (**A**) and subcutaneous (**B**).

ADMINISTERING INTRAVENOUS INJECTIONS

Purpose

- Rapid absorption of medication

Complications

- Inadvertent arterial puncture
- Extravascular injection
- Hematoma
- Adverse reaction to medication

Equipment

- Isopropyl alcohol 70%
- Cotton or gauze
- Medication or syringe drawn into appropriate size syringe and needle (Table 6–1)

Procedure for Intravenous Injection in the Horse

TECHNICAL ACTION	RATIONALE/AMPLIFICATION
1. Halter horse and place in stock. Alternatively an assistant standing on the same side as the collector can restrain the horse.	1a. Never stand directly in front of horse or under neck.
2. Identify jugular furrow.	—
3. Wet furrow using alcohol-soaked cotton swab, then remove needle form syringe.	3a. The needle can remain attached to the syringe if administering small volumes (less than 5 cc).
	3b. Table 6–1 gives a selection of needle sizes.
4. Occlude jugular vein.	—
5. Hold needle bevel side up and place in distended jugular vein.	5a. If bright red streaming or pulsating blood is noted, an inadvertent arterial puncture of the carotid artery probably has occurred. Remove needle immediately and apply compression for 5 minutes.
	5b. The smallest diameter needle that can be used safely in the equine is 20 gauge. Smaller diameter needles will not allow the technician to detect accidental arterial puncture.

TECHNICAL ACTION	RATIONALE/AMPLIFICATION
6. Attach syringe and aspirate to confirm intravenous placement.	**6a.** Blood will flow into the syringe confirming intravenous placement.
	6b. Fig. 6–10.
7. Inject at a moderate rate of speed.	—
8. Remove needle and apply compression for 1–2 minutes.	**8a.** Pressure prevents hematoma formation.

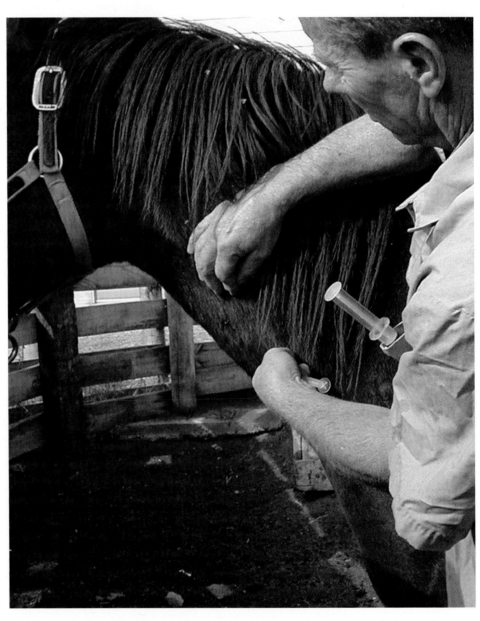

FIG. 6–10 Intravenous (jugular vein) injection in the horse. Needle in vein, verifying flash of venous blood in hub of needle.

Special Notes

Needle may be placed in a downward direction, toward the heart. This technique will lessen the chance of inadvertent carotid puncture. Alternatively, needle may be placed in an upward direction, away from the heart. This technique minimizes the amount of air entering the vein.

Procedure for Intravenous Injection in the Cow

TECHNICAL ACTION	RATIONALE/AMPLIFICATION
1. Place cow in chute or stock.	—
2. Secure head to side of stock using halter.	2a. Head can be secured to either side.
3. Identify jugular furrow.	—
4. Wet furrow using alcohol-soaked cotton swab and occlude vein.	4a. Alcohol assists in raising vein.
5. Grasp needle by hub with thumb and forefinger. Thrust needle through skin at a 45–90 degree angle to the vein.	5a. Fig. 6–11. 5b. Table 6–1 gives a selection of needle sizes.
6. Confirm vein penetration by witnessing blood come from hub of needle.	6a. Cows will have a stream of blood from the hub, horses will have a drip. 6b. If needle has only penetrated skin and not entered the vein, reidentify the vein location and thrust needle into it without removing from skin. 6c. If the needle has been placed too deep and penetrated through the vein, pull back slowly.
7. Thread the needle by laying it parallel to the jugular and advancing it into the vein. Thread needle to its hub.	—
8. Attach syringe and administer medication at a moderate rate.	—
9. Remove needle and apply pressure for 20 seconds.	9a. Pressure prevents hematoma formation.

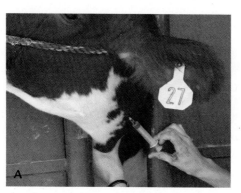

FIG. 6-11 (A) Intravenous (jugular vein) injection in the cow. (B) Administration of an intravenous (jugular vein) infusion.

Procedure for Intravenous Injection in the Goat

TECHNICAL ACTION	RATIONALE/AMPLIFICATION
1. Place goat in stanchion; halter and secure head to the side. Alternatively an assistant can back goat into a corner and then straddle neck.	1a. If the goat is too large to straddle, the assist can back animal into corner or against fence.
2. Grasp mandible and elevate head while turning head slightly to the side.	2a. Fig. 6–12.
	2b. Table 6–1 gives a selection of needle sizes.
3. Identify jugular furrow and wipe with alcohol.	3a. Alcohol will remove dirt and assist with visualization, especially in goats with long or thick hair.
4. Occlude vein at the level of the lower third of the neck or thoracic inlet.	4a. Occlude and releasing vein several times can help with visualization.
5. Attach needle to syringe. Insert needle bevel up at a 30-degree angle to the skin, parallel to the jugular vein.	—
6. Aspirate blood into syringe to confirm placement.	—
7. Inject at a moderate rate of speed.	—
8. Remove needle and apply pressure for 20 seconds.	8a. Pressure prevents hematoma formation.

CHAPTER 6 Clinical Procedures

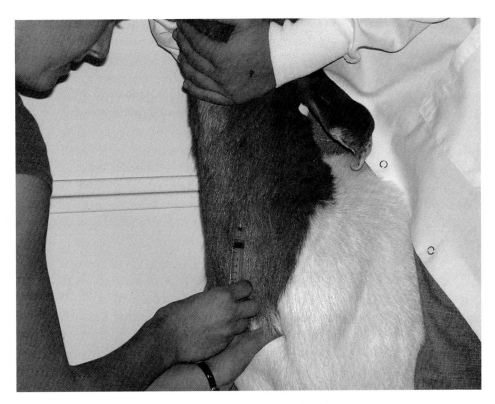

FIG. 6-12 Intravenous (jugular vein) injection in the goat.

Procedure for Intravenous Injection in the Pig

TECHNICAL ACTION	RATIONALE/AMPLIFICATION
1. Secure pig using snare, chute, or assistant holding against body.	1a. Restraint technique will depend on pig size.
2. Marginal ear veins typically are visible without occluding. Three veins are present on the ear, the lateral or central being the easiest to visualize.	2a. Fig. 6-13. 2b. Pigs are unable to sweat, so the ears are an important component of temperature regulation. Increasing the ambient temperature will result in vasodilation and assist with vein visualization.
3. Assistant should compress the base of the ear to occlude vein. A pumping motion can be used to help raise the vein.	3a. Alternatively the collector can use the free hand to occlude the vein.
4. Attach needle to syringe. Insert needle along vessel at a very shallow angle of penetration.	4a. Table 6-1 gives a selection of appropriate needle sizes. 4b. Needles must be attached to the syringe before insertion, because ear veins are very fragile.

TECHNICAL ACTION	RATIONALE/AMPLIFICATION
5. Aspirate to confirm vascular placement. Ask assistant to release occluding pressure. Inject at a moderate rate of speed.	—
6. Remove needle and apply compression for 1 minute.	6a. Compression reduces incidence of hematomas.

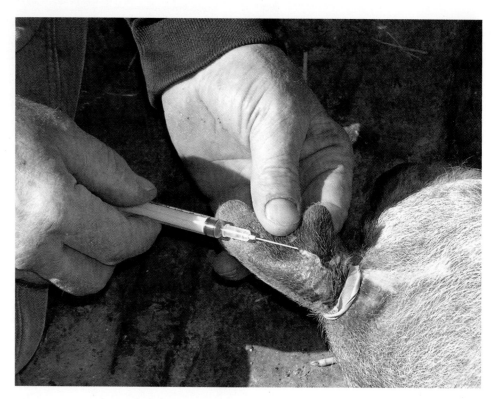

FIG. 6-13 Intravenous (ear vein) blood collection in the pig. If injecting, do not use an elastic band to occlude the vein.

Procedure for Intravenous Injection in the Llama

TECHNICAL ACTION	RATIONALE/AMPLIFICATION
1. Place llama in stock or other restraint device to prevent forward motion.	—
2. Elevate head.	—
3. In lower third of neck, palpate for enlarged transverse processes of the sixth cervical vertebra.	3a. Fig. 6-14.
	3b. The jugular vein lies just medial to this process.
	3b. The pulsating carotid artery can be palpated medial to the transverse process.

TECHNICAL ACTION	RATIONALE/AMPLIFICATION
4. Occlude vein. Confirm jugular vein penetration point by occluding and releasing vein several times.	4a. Watch for jugular vein distention between the fifth and sixth cervical vertebrae. 4b. Do not clip animal unless permission is obtained from owner. Regrowth of fleece can take up to 18 months. If permission is obtained, clipping can assist with visualization.
5. Attach needle to syringe and insert slightly medial to the process, aiming toward the center of the neck.	5a. Attaching the syringe before insertion aids in collection. 5b. Table 6-1 covers needle size selection.
6. Aspirate to confirm vascular placement. Inject at a moderate rate.	—
7. Remove needle and apply compression for 1 minute.	7a. Compression reduces incidence of hematomas.

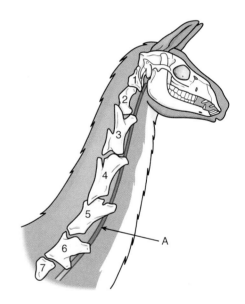

FIG. 6-14 Location for intravenous (**A**, jugular vein) injection in the llama.

ADMINISTERING SUBCUTANEOUS INJECTIONS

Purpose

- Administer medication, vaccinations, or fluids into subcutaneous tissue

Complications

- Inadvertent vascular penetration
- Iatrogenic abscess

Equipment

- Isopropyl alcohol 70%
- Cotton
- Needle (size per Table 6–1)
- Syringe

Procedure for Subcutaneous Injection in the Cow

TECHNICAL ACTION	RATIONALE/AMPLIFICATION
1. Secure cow in stock or chute.	1a. Adequately securing animal minimizes motion, thereby decreasing the amount of tissue damage caused by injection.
2. Two-handed technique. Grasp skin 4 inches cranial to the shoulder. Lift skin to form a skin tent.	2a. A line from the ear base to the point of the shoulder effectively outlines the cervical vertebrae. Injections should be made dorsal to this line and 4 inches below the crest of the neck.
	2b. See beef quality assurance recommendations.
	2c. Fig. 6–5.
3. Hold needle at a 20-degree angle to skin and insert at base of tent.	3a. Lifting the needle tip slightly to raise skin from underlying tissue will help confirm that needle is not in muscle.
	3b. Use care not to stab or inject the hand that is holding the skin.
	3c. Do not use a two-handed technique when administering live zoonotic agents such as brucellosis.
	3d. See Table 6–1 for selection of needle size.
4. Technique for the one-handed injection is essentially the same except that the skin is not tented using the second hand.	4a. The one-handed technique should be used when the safety of the person injecting would be compromised by the two-handed technique.
	4b. The one-hand technique is much quicker and safer for the operator.
5. Aspirate and inject.	5a. Aspiration involves pulling on the plunger while watching the needle hub for evidence of blood.
	5b. Aspiration often is not performed if many cattle are being processed. Aspiration is impossible with syringe guns.

Procedure for Subcutaneous Injection in the Goat and Llama

TECHNICAL ACTION	RATIONALE/AMPLIFICATION
1. Halter llama; collar goat.	1a. A stock (llama) or a stanchion (goat) can also be used.
2. Identify injection site just cranial to shoulder.	2a. Alternative injection sites include caudal to the elbow (llama) and lateral chest area 3 inches caudal to the shoulder (goat).
	2b. These areas are characterized by a loose skin attachment.
3. Ensure that injection site is clear of dirt or debris.	3a. Individual animals often are swabbed with alcohol, but this is not performed when processing a herd.
	3b. Figs. 6–7, 6–9.
4. Grasp and lift skin with free hand.	4a. This forms a skin tent.
5. Holding needle at a 20-degree angle to the skin, insert at base of tent.	5a. Lifting the needle tip slightly to raise skin from underlying tissue will help confirm that needle is not in muscle.
	5b. Use care not to stab or inject the hand that is holding the skin.
	5c. Llamas frequently will kick out and goats will vocalize at this point.
6. Aspirate and inject.	6a. Aspiration involves pulling on the plunger while watching the needle hub for evidence of blood.
7. Massage area of injection.	7a. Massage decreases pain and distributes drug.

Procedure for Subcutaneous Injection in the Pig

TECHNICAL ACTION	RATIONALE/AMPLIFICATION
1. Restraint technique and location of injection are dependent upon pig size.	—
2. Piglet injections: Hold piglet up by rear legs to expose the injection site on the inside of flank along the abdominal wall.	2a. Fig. 6–8 gives location of injection.
	2b. The skin fold caudal and medial to the elbow can also be used.
	2c. This method of injection can be used until the pig is too large to secure by holding the hind limbs. Most grower and finisher size animals can be restrained in this manner.

TECHNICAL ACTION	RATIONALE/AMPLIFICATION
3. Larger pig injections: Use hog snare or chute, which allows access to area of loose skin caudal to the ear.	—
4. Ensure that injection site is clean and dry.	4a. Contamination of the injection site will increase the frequency of abscess formation.
5. Grasp skin and pull dorsally. Insert needle at a shallow angle (approximately 10 degrees) to the skin.	5a. Inserting at a shallow angle prevents entry into the muscle.
6. Aspirate and inject at a moderate rate of speed.	6a. Aspiration involves pulling on the plunger while watching the needle hub for evidence of blood. 6b. A maximum volume of 10 cc should be injected per site in adult pigs.
7. Remove needle and massage.	7a. Massage helps to disperse medication and alleviate pain.

GROWTH HORMONE IMPLANTS

Purpose

- Feed utilized more efficiently by cattle

Complications

- Abscess formation
- Expelled implants
- Implant becomes embedded in cartilage
- Implants crushed or broken

Equipment

- Knife
- Stiff brush
- Chlorhexidine solution
- Bucket
- Implant gun with needles
- Implants

Procedure for Placing Growth Hormone Implants

TECHNICAL ACTION	RATIONALE/AMPLIFICATION
1. Place cow in chute.	—
2. If the ear is clean and dry, go directly to step 4. If needed, scrape manure or debris off back side of ear using knife.	2a. The scrape, brush, and disinfect method was initiated to minimize infections associated with contaminated implants.
3. Brush ear using chlorhexidine solution.	3a. A two-sided brush with one brass and one nylon bristle side is used most commonly.
	3b. The brush should be placed back in the chlorhexidine between animals.
4. Insert needle subcutaneously on back side in the middle third of the ear. Insert the full length of the needle. Squeeze trigger and withdraw needle slowly as trigger is being depressed.	4a. Correct insertion will result in a smooth row of implanted pellets.
	4b. The only approved location for implant administration is the back side of the ear. This location minimizes human exposure to the implants because ears are discarded at the time of slaughter.
	4c. Rotating the needle slightly and using a light touch will help to avoid placing implants in ear cartilage.
	4d. Use only sharp needles. Dull needles make implanting much more difficult and increase the infection rate.
5. Inspect and palpate implant site. Correctly placed implants are not bunched and are slightly movable within the subcutaneous tissue.	—

INTUBATION

Gastric **intubation** serves many purposes. A longtime mainstay of colic therapy and required for successful relief of bloat, gastric intubation remains one of the most valuable large animal medical procedures. Although the method of accessing the stomach may vary by species (oral versus nasal), the basic techniques of intubation remain constant. Technicians should invest adequate time developing this skill. It is a procedure meriting the time and effort that is required for mastery.

NASOGASTRIC INTUBATION

Purpose

- Administer fluids, medication, or nutrients, or relieve gastric pressure

Complications

- Epistaxis
- Aspiration
- Esophageal trauma

Equipment

- Nasogastric tube
- Bucket
- Stomach pump
- Lubrication gel or oil
- Twitch

Procedure for Nasogastric Intubation of the Horse

TECHNICAL ACTION	RATIONALE/AMPLIFICATION
1. Halter and place horse in stock.	1a. Many horses are easier to intubate when a twitch is applied. Some horses may require tranquilization.
2. Lightly coat the tube end with lubricant or oil.	2a. Lubrication facilitates tube passage.
	2b. Keep a towel handy, because the technician's hand can become slippery during the procedure.
	2c. The length of the tube often is draped over the shoulder, with the free end placed in the technician's mouth during the intubation.
3. Stand lateral to horse. Insert tube into ventral medial aspect of nostril. Use forefinger or thumb to ensure that tube is directing ventrally.	3a. Placing one hand over the bridge of the nose will help to stabilize the head. Be certain not to occlude airway when this is done.
	3b. Directing the tube in a ventromedial direction minimizes trauma to nasal turbinates and therefore decreases the incidence of epistaxis.

TECHNICAL ACTION	RATIONALE/AMPLIFICATION
	3c. If epistaxis should occur during intubation, wait 5–10 minutes and then proceed using the other nostril. This is a minor complication and does not warrant discontinuing the procedure.
	3d. Horses typically will resist (head toss) until the first 6 inches of tube is passed. After this point, most animals tolerate the procedure very well.
4. Slowly advance tube until reaching the level of nasopharynx. Continue to apply soft pressure and allow horse to swallow the tube.	**4a.** Swallowing can be facilitated by flexing the neck. Other horses can be stimulated to swallow by slowly rotating the tube.
	4b. Fig. 6–15.
5. Confirm that tube has entered the esophagus by detection of negative pressure or palpation of the tube in the esophagus.	**5a.** To assess for negative pressure, suck on the tube end.
	5b. Palpate the left side of the neck dorsal to jugular groove. Palpation is assisted by moving the tube. It is easiest to keep fingertips in a fixed position on the

FIG. 6–15 Nasogastric intubation of the horse. Note use of thumb to direct the tube ventrally.

TECHNICAL ACTION	RATIONALE/AMPLIFICATION
	cranial third of neck and then palpate for passage of tube under fingertips.
	5c. It is often easier to detect negative pressure than to palpate the tube.
6. Blow into the tube and slowly advance it into the stomach.	6a. Blowing into the tube dilates the esophagus, facilitating atraumatic tube passage.
7. Check for gastric reflux by attaching the stomach pump and priming the tube with warm water. Lower the tube end and allow fluid to drain out.	7a. Priming means to fill the tube with water. This will help to create a siphon.
	7b. Horses are unable to vomit and are therefore susceptible to gastric rupture. Before administering fluids, the technician should be certain that gastric pressure is normal and that excessive amounts of reflux (fluid) is not present.
	7c. Pulling the tube out rapidly 3–7 inches once the tube is primed can facilitate the siphon effect.
	7d. Record volume of reflux removed in patient chart. Confirm with veterinarian if fluids are to be administered when reflux is present.
8. Attach stomach pump and administer fluids or medication.	8a. Administer oral fluids at a maximum rate of 6 liters every 2 hours.
9. To remove, blow tube clear of all fluids. Place thumb over tube end and kink tube. Withdraw tube in one fluid motion.	9a. Clearing tube and kinking it helps to prevent aspiration during tube removal.
	9b. In cases of potential gastric rupture, nasogastric tubes can be secured temporarily to the halter using tape.

OROGASTRIC INTUBATION

Purpose

- Administer fluids, medication, or nutrients, or relieve gastric pressure

Complications

- Aspiration
- Esophageal trauma
- Regurgitation (llama)

Equipment

- Tube (size per Step 3 of Orogastric Intubation)
- Frick speculum or mouth gage (cow)
- Oral speculum (llama, goat)
- Bucket
- Stomach pump

Procedure for Orogastric Intubation of Cows, Llamas, and Goats

TECHNICAL ACTION	RATIONALE/AMPLIFICATION
1. Place cow or llama in stock. Secure goat by collar.	1a. Small llamas can be placed in sternal recumbency with the collector on his knees straddling the llama.
	1b. Goats usually are straddled.
2. Place Frick speculum in oral cavity of cow. Place oral speculum in llama and goat.	2a. Permanently attaching a nose tong to the end of a Frick speculum allows it to be self-retaining. This permits the use of both hands by the operator (Fig. 6–16).
	2b. An 8-inch segment of rubber garden hose works very well for a llama speculum. Alternatively, polyvinyl chloride (PVC) pipe wrapped in adhesive tape also does nicely.

FIG. 6–16 (A) Placing a Frick speculum for orogastric intubation of a cow. (B) Confirming the presence of negative pressure. (C) Kinking the tube off for removal.

TECHNICAL ACTION	RATIONALE/AMPLIFICATION
	2c. A speculum is required, because the sharp molars of both species will lacerate tubes.
	2d. An old roll of tape or a piece of polyvinyl chloride (PVC) pipe wrapped in tape makes an excellent goat speculum (Fig. 6–17).
3. Pass tube through speculum into esophagus. Gentle pressure is required to pass through the oral pharynx area. Rotating tube in circular motion can facilitate swallowing.	**3a.** Tube sizes: Cattle: 1-inch internal diameter or larger Goats: Small foal tube

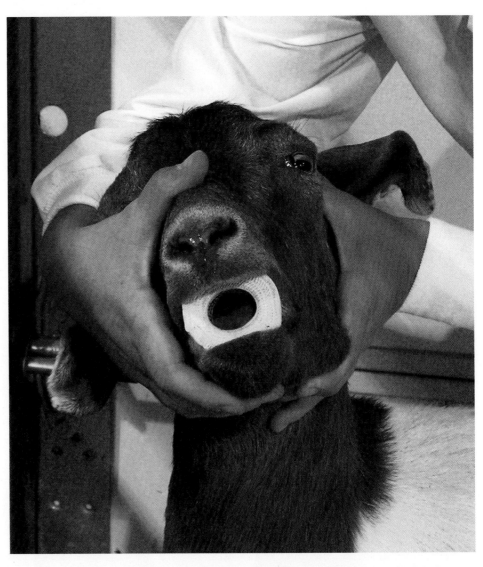

FIG. 6-17 Placement of a speculum for orogastric intubation of a goat.

TECHNICAL ACTION	RATIONALE/AMPLIFICATION
	Llamas:
	10–20 lb: 18–22 French, ¼-inch outside diameter
	65–200 lb: 30–40 French, ½-inch outside diameter (foal or small equine tube)
	200 or more lb: 40–45 French, ½ to ⅝-inch outside diameter (small equine size)
	3b. Passage of tube in llamas and goats is facilitated by flexing head slightly.
	3c. Never force tube. If resistance is met, reposition speculum or withdraw tube slightly and begin again.
4. Suck on the end of the tube as it is passed into the esophagus. Negative pressure indicates that the tube is placed correctly.	**4a.** Sucking on the end of the tube allows the operator to assess if the tube is correctly placed in the esophagus or incorrectly placed in the trachea (Special Notes).
	4b. Llamas may regurgitate at this time. Although it is not very dangerous, it may require that the procedure be started over.
5. The operator can also palpate for the end of the tube along the left ventral side of the neck as it passes through the esophagus.	**5a.** This is an excellent means of confirming placement.
6. Once tube is confirmed in esophagus, blow into tube to dilate esophagus and facilitate atraumatic passage.	**6a.** A strong grass or grain odor is typically noted once the tube has entered the rumen.
7. Attach stomach pump and administer fluids or medication.	**7a.** Fluids should be warmed to body temperature.
8. Before removing tube, blow forcefully into the tube to clear all contents within the tube. Place thumb over tube end or kink the tube.	**8a.** These techniques minimize the chance of accidental aspiration during removal.
9. Remove tube using a steady downward, continuous, pulling motion.	—
10. Remove speculum.	—

Special Notes

Negative pressure means that the operator will not be able to pull air in through the tube. The operator's cheeks will cave inward when sucking. To simulate this feeling, place a straw in your mouth. Cover the end with your index finger and try to suck air into your mouth. This feeling indicates negative pressure. A tube that has been positioned incorrectly in the trachea will demonstrate positive pressure when sucked on. To simulate this feeling, place a straw in your mouth and suck in air. This is called positive pressure.

In addition to assessing negative pressure, tube placement is confirmed by smelling for gastric content and listening for motility sounds through the tube.

CATHETERS

Intravenous catheterization ensures rapid administration of drugs or fluids into the venous vascular space. It minimizes the potential for inadvertent extravascular injection and serves to minimize vascular trauma when repeated intravenous injections are anticipated. The decision to place an intravenous catheter typically reflects the overall fluid volume the animal will receive and the duration of treatment. Catheter type often depends up the medical team's experience, personal preference, and the anticipated duration of use (Tables 6–3 and 6–4). In contrast to intravenous catheters, which are used for a myriad of reasons, intraarterial catheters are used primarily in the anesthetized patient.

CATHETER TYPES

Fig. 6–18 shows two types of catheters.

TABLE 6–3 CATHETER TYPES

CATHETER TYPE	USES AND ADVANTAGES	DISADVANTAGES
Butterfly needle	Rapid placement	Short-term limited use
		Lacerated vessels
Over-the-needle catheter	Low cost	Less flexible
	Easy placement	Higher incidence of thrombosis, phlebitis
Through-the-needle catheter	Very flexible	Expensive
	Decrease incidence of phlebitis	Increased risk of lacerating catheter
	Long-term use	More difficult to place

CATHETER SIZES FOR VARIOUS ANIMAL SPECIES

TABLE 6-4 CATHETER SIZES FOR VARIOUS ANIMAL SPECIES

EQUIPMENT NAME	IV OR IA	VESSEL NAME	SIZE
Horse			
Adult	IV	Jugular	16–18g × 6 inches
	IA	Facial Greater metatarsal	18–20g × 1.5–2 inches
Foal	IV	Jugular	18–20g × 2 inches
	IA	Facial Greater metatarsal	20–22g × 1 inch
Cow			
Adult	IV	Jugular	14–16g × 6 inches
Calf	IV	Jugular	18–20g × 2 inches
Goat			
Adult	IV	Jugular	18–20g × 2 inches
Kid	IV	Jugular	20–22g × 1–2 inches
Llama			
Adult	IV	Jugular	16–18g × 6 inches
Cria	IV	Jugular	18–20g × 2 inches
Pig			
Adult	IV	Auricular	18g × 2 inches
Piglet	IV	Auricular	20–22g × 1–2 inches

IA, intraarterial; *IV,* intravenous.

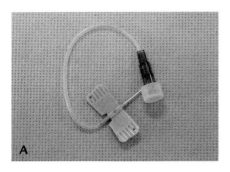 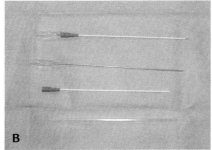 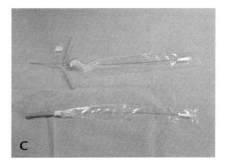

FIG. 6-18 (A) Butterfly catheter. (B) Over-the-needle catheter assembled and disassembled showing the catheter and stylet. (C) A through-the-needle catheter assembled and opened, showing the catheter passing through the needle and the needle guard.

Special Note

Intraarterial catheters can be placed in all animal species. However, placement in species other than equine is rather limited. Intraarterial catheter size in other species would be comparable to that used in horses.

GENERAL INTRAVENOUS CATHETER PLACEMENT AND CARE RECOMMENDATIONS

Certain principles apply to all catheters. The following recommendations should be adhered to regardless of the species, vessel selected, or type of catheter placed.

- To prepare a catheter for placement, visually inspect the stylet and catheter tip for burrs or other manufacturing defects. Flush using heparinized saline.
- When inserting a catheter, hold both the stylet and catheter hub firmly together. This prevents the catheter from sliding forward over the stylet.
- Handle only the hub of the catheter. To prevent contamination, never touch the shaft of the catheter.
- Never advance a catheter over a stylet into a vessel and then attempt to pull the catheter back onto the stylet. This action can result in severing of the catheter, thereby causing a foreign body embolus (death).
- When injecting through a catheter cap, use a needle that is a smaller gauge than the catheter itself. Always wipe the catheter cap with alcohol before inserting a needle.
- The catheter system and method of securing chosen for each patient depends on the type and volume of fluids given, duration of catheterization, and potential for thrombosis.
- All catheters should be removed immediately if any swelling, pain at catheter site, thrombosis, or fever of unknown origin is noted.
- Catheters should be flushed with heparinized saline every 8–12 hours.
- Catheters should be inspected for patency or problems 2 to 3 times per day.
- Catheters should be changed every 72 hours unless otherwise indicated.
- Application of topical lidocaine after the three-step surgical preparation will minimize catheter placement discomfort in all species.
- Heparinized saline flush solution often is made at the hospital by adding heparin to 1 liter of sterile saline. The final solution should contain 2 units of heparin per ml of saline.
- Fig. 6–19 shows catheter parts.

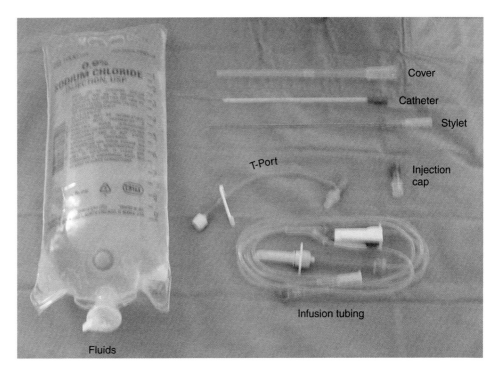

FIG. 6-19 Set-up for intravenous fluid administration.

INTRAVENOUS CATHETERS

Purpose

- Provide a safe, effective means of administering fluids, drugs, or anesthetic agents

Complications

- Thrombophlebitis
- Infection
- Hematoma
- Subcutaneous hemorrhage
- Inadvertent arterial puncture
- Nonpatent catheter

Equipment

- Sterile gloves
- Clippers with #40 blade
- Povidone-iodine scrub and solution

- Isopropyl alcohol 70%
- Lidocaine (topical and injectable)
- Number 15 blade or 14g needle
- Catheter (Table 6–4)
- One-inch tape
- Superglue or tissue adhesive (methoxypropyl cyanoacrylate)
- Suture (nonabsorbable size 0 suture on a cutting needle)
- Scissors
- Catheter cap
- Extension set
- Fluids
- Syringes and needles
- Protective wrap materials

Procedure for Placing an Over-the-needle Intravenous Catheter in the Horse

TECHNICAL ACTION	RATIONALE/AMPLIFICATION
1. Halter horse and place in stock.	1a. Fig. 6–20.
2. Wash hands and prepare catheter.	2a. See instructions listed under general catheter care recommendations.
	2b. If the catheter will be sutured in, apply butterfly tape at this point (Step 12 in this procedure).
3. Clip a 4 × 4-inch area over the jugular groove in the middle to proximal third of neck.	3a. The left or right side of the neck can be used. In emergencies the lateral thoracic or cephalic veins are used.
4. Perform a standard three-step preparation.	4a. Directions for the three-step preparation can be found in Chapter 8. Sterile placement will help minimize iatrogenic infections.
5. Apply sterile gloves. Apply topical lidocaine to clipped area.	5a. Sterile gloves help to minimize iatrogenic infection.
	5b. Topical lidocaine minimizes placement discomfort.
6. Occlude and distend jugular vein at level of thoracic inlet or proximal portion of neck using the hand that will not be inserting the catheter.	6a. This vessel is easy to discern. However inserting the catheter in some ponies, mules, or thick-necked horses will prove more difficult.

TECHNICAL ACTION	RATIONALE/AMPLIFICATION
7. Grasp catheter and hub firmly together. Hold catheter at a 60-degree angle to the skin and advance rapidly through skin and subcutaneous tissue into jugular vein.	7a. Catheters are inserted toward the heart, not directed toward the head. 7b. The length of the catheter may cause some awkwardness for novice technicians. This will resolve with experience.
8. Confirm catheter placement by visualizing blood in hub.	8a. With correct placement, blood will drip from catheter or be visualized in hub. This occurs once 1–1.5 inches of the catheter is in the horse. Do not advance additional length of the catheter into the horse searching for the vessel.
9. Once blood is visualized, advance stylet and catheter $\frac{1}{4}$ inch further into vessel to seat the catheter.	9a. Seating the catheter ensures that both the catheter and stylet are within the vessel.

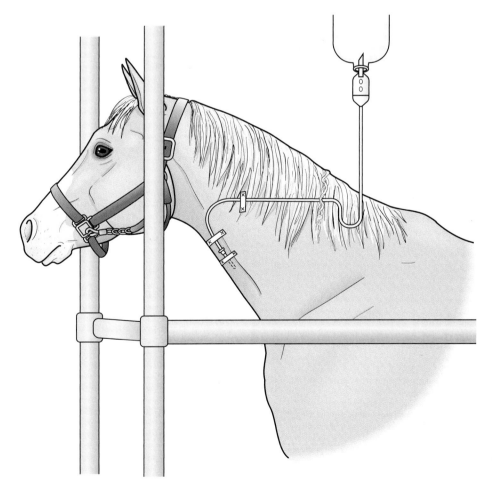

FIG. 6-20 Administration of continuous intravenous fluids to a horse in a stock.

TECHNICAL ACTION	RATIONALE/AMPLIFICATION
10. Hold the stylet very still and slowly advance the catheter into the vessel.	10a. Catheters should insert smoothly with minimal resistance. If resistance is encountered, the catheter is most likely in the extravascular space.
11. Hold catheter hub and remove stylet. Attach catheter cap or extension set.	11a. Ensure that cap is secured tightly. 11b. Extension sets provide a flexible length of tubing that facilitates access for injections. If an extension set is used, place the cap at the end of the extension set.
12. Secure catheter using one of the following methods.	12a. Figs. 6–21d and e.
13. Adhesive method: Place a dab of adhesive glue on the medial aspect of the catheter hub. Press firmly against neck for 1 minute.	13a. This method is much quicker and often is better tolerated by the animal than suturing. 13b. Extension sets can not be used with this method. All medications or fluids will have to be administered through the cap. 13c. It is much easier for an animal to remove a catheter secured by this method. Thus, this is typically reserved for routine, rapid procedures, not with lengthy surgeries or debilitated animals.
14. Suture method: a. Use adhesive tape to form a butterfly over catheter hub. b. Inject 1 cc of lidocaine into subcutaneous tissue ventral to butterfly. c. Suture both wings of butterfly to skin using a simple interrupted pattern.	14a. The tape butterfly provides a suturing surface. 14b. Lidocaine decreases discomfort during suture placement. Wait 3 minutes after injection before proceeding. 14c. Draping the extension set over the neck during the securing process will help maintain catheter placement. Do not allow the set to hang, or it will pull out the catheter. 14d. See Chapter 8 for placing simple interrupted sutures.

TECHNICAL ACTION	RATIONALE/AMPLIFICATION
15. If an extension set is used, apply a second butterfly 3 inches from the catheter. Secure to neck using suture method. Coil the remaining tube, secure with a third butterfly, and suture to lateral neck area.	**15a.** This second butterfly should be placed in the jugular groove area to maintain alignment and prevent kinking of the catheter.
16. Place a protective wrap using gauze and bandage.	**16a.** Catheters secured using the adhesive method may remain unwrapped. **16b.** All catheters with extension sets must be wrapped.

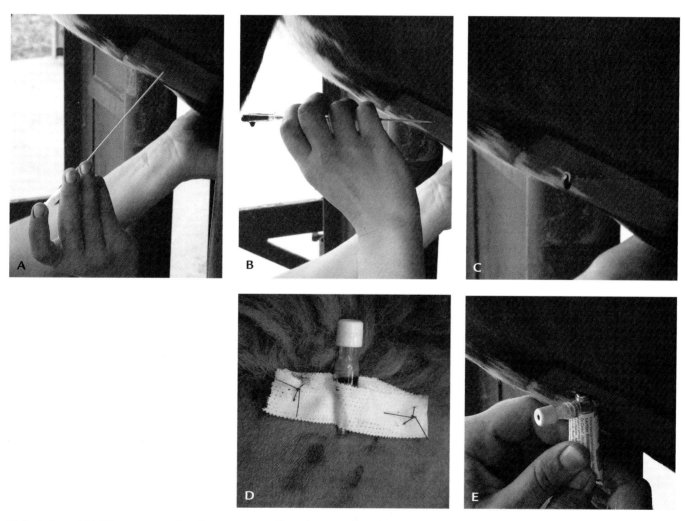

FIG. 6–21 (A) Placement of an intravenous (jugular vein) catheter in the horse. Area of clip and preparation. Insertion of stylet and catheter through the skin. (B) Advancing the catheter into the vein. (C) Showing the flash of blood in the hub of the catheter. (D) Catheter secured using the butterfly and suture technique. (E) Securing the catheter using tissue adhesive.

Special Note

Catheters are inserted commonly in foals. Use an 18-gauge or smaller catheter. Length varies from 2 to 5¼ inches depending on foal size and catheter needs.

Procedure for Placing a Through-the-needle Intravenous Catheter in the Horse

TECHNICAL ACTION	RATIONALE/AMPLIFICATION
1. Complete steps 1–5 of equine over-the-needle catheter placement procedure.	1a. The preparation procedure is identical for both types of catheters.
2. Expose the catheter needle by sliding plastic ring back onto hub of needle guard.	2a. Do not discard the plastic ring or remove the protective clear plastic catheter cover at this point.
3. Occlude and distend jugular vein at level of thoracic inlet or proximal portion of neck using the hand that will not be inserting the catheter.	3a. This vessel is easy to discern. However, the vessel in some ponies, mules, or thick-necked horses will prove more difficult.
4. Grasp needle and hold it at a 40–60 degree angle to the skin. Advance through the skin and subcutaneous tissue into the jugular vein.	4a. Catheters are placed toward the heart, not directed toward the head.
	4b. The bevel of the needle should be up.
	4c. You will not see blood to confirm placement of catheter, because the guide wire (stylet) obstructs the lumen of the catheter.
5. Once vessel has been entered, feed the catheter fully into vein by thumbing through plastic sleeve.	5a. The catheter should feed easily. There should be no resistance. If resistance is encountered, remove both the needle and catheter simultaneously.
	5b. Never pull the catheter back through the needle or it can sever the catheter and create an embolus (leading to death).
6. Connect the catheter hub to the hub of the needle guard. Remove plastic sleeve.	6a. The plastic sleeve ensures the sterility of the catheter as it is being advanced through the needle.
7. Withdraw needle and catheter together to expose 1.5 inches of the catheter. Close the needle guard.	7a. Needle guards prevent inadvertent severing of the catheter.

TECHNICAL ACTION	RATIONALE/AMPLIFICATION
8. Remove catheter guide wire (stylet) and apply catheter cap.	8a. The guide wire (stylet) gives the catheter rigidity, thus facilitating passage through the needle.
9. Secure guard to the skin using a simple interrupted suture pattern.	9a. A butterfly of adhesive tape may be needed to provide a suturing surface for the guard. 9b. Lidocaine can be injected to decrease discomfort during suture placement. Wait 3 minutes after injection before proceeding.
10. Flush catheter using heparinized saline. Place protective wrap or attach extension set.	10a. If an extension set is used, remove the catheter cap and attach extension set directly to catheter. Place the cap on the end of the extension set. 10b. Complete Step 15 in equine over-the-needle intravenous catheter placement if using an extension set.

Procedure for Placing an Intravenous Catheter in Cows and Llamas

TECHNICAL ACTION	RATIONALE/AMPLIFICATION
1. Restrain animal appropriately.	1a. Place cow in chute. Halter and secure head to side. 1b. Halter and place llama in stock. Secure head. 1c. If a stock is not available, stand llama such that one side is against a wall.
2. Wash hands and prepare catheter.	2a. Refer to General Intravenous Catheter Placement and Care Recommendations.
3. Prepare catheter site. Cow: Clip a 4 × 4-inch square in the midcervical area, centering on the jugular vein. Llama: Clip a 4 × 4-inch area on the lower third of neck, centering on the transverse process of the sixth cervical vertebra.	3a. Always inform owners that the animal will be clipped. Llama wool is very slow growing, and the catheter site may be visible for up to 1 year. 3b. There is no discernible jugular groove in the llama. Occasionally the vessel can be detected with percussion. This involves briskly stroking the vessel with one hand while feeling for fluid movement with the occluding hand.

TECHNICAL ACTION	RATIONALE/AMPLIFICATION
	3c. Although the right jugular vein is easier to access in llamas, the left one can be catheterized as well.
	3d. In emergency situations the cephalic or antecubital veins of llamas can be used.
	3e. Tape can be used to secure wool out of clipped area.
	3f. Refer to intravenous blood draw diagram for depiction of llama jugular vein, which lies just medial to the vertebral transverse process.
	3g. Fig. 6–22.
4. Perform a three-step surgical preparation over catheter site.	4a. Directions for the three-step surgical preparation can be found in Chapter 8. Sterile placement will help minimize iatrogenic infections.
5. Apply sterile gloves.	5a. Sterile gloves help to minimize iatrogenic infection.
6. If the operator elects to perform a stab incision, inject 1–2 cc lidocaine subcutaneously over the jugular vein. Puncture skin over vessel using a #15 blade or a 14g needle.	6a. Stab incisions facilitate passage of the stylet through thick-skinned animals. It is important to avoid puncturing the underlying vessel.
	6b. Watch for jugular vein distention between the fifth and sixth cervical vertebrae in the llama.
7. Occlude jugular vein. Grasp catheter and stylet at the hub. Holding catheter perpendicular to skin, thrust rapidly through the subcutaneous tissue into the vessel. If a stab incision was made, a minimal amount of pressure is required.	7a. Stab incisions are not required. If the operator chooses to forgo the step, stronger pressure will be needed to advance stylet through intact skin.
	7b. Many individuals will find it awkward to use a 6-inch catheter for the first time. This will resolve with practice.
	7c. Wait 3 minutes after injecting lidocaine. Lidocaine minimizes discomfort of stab incision.
8. Once blood is visible in hub, align the catheter parallel to the jugular vein. Hold the stylet still and slowly advance the catheter into vessel.	8a. Catheters should enter smoothly, with minimal resistance. If resistance is encountered, the catheter is most likely in the extravascular space.

TECHNICAL ACTION	RATIONALE/AMPLIFICATION
9. Hold catheter hub and remove stylet. Apply catheter cap or extension set.	9a. Ensure that cap is secured tightly.
	9b. Extension sets provide a flexible length of tubing that facilitates access for injections. If an extension set is used, place the cap at the end of the extension set.
10. Secure catheter using one of the following methods.	—
11. Adhesive method:	11a. Place a dab of adhesive glue on the medial aspect of the catheter hub. Press firmly against neck for 1 minute.
	11b. This method is much quicker and is often better tolerated by the animal.
	11c. Extension sets can not be used with this method. All medications or fluids will need to be administered through the cap.

FIG. 6-22 Preparing for placement of an intravenous (jugular vein) catheter in the cow. Note the clipped area.

TECHNICAL ACTION	RATIONALE/AMPLIFICATION
	11d. It is much easier for an animal to remove a catheter placed with this method.
12. Suture method:	
a. Use adhesive tape to form a butterfly over catheter hub.	**12a.** The tape butterfly provides a suturing surface.
b. Inject 1 cc lidocaine into subcutaneous tissue ventral to butterfly.	**12b.** Lidocaine decreases discomfort during suture placement. Wait 3 minutes after injection before proceeding.
c. Suture both wings of butterfly to skin using a simple interrupted pattern.	**12c.** Draping the extension set over the neck during the securing process will help maintain catheter placement. Do not allow the set to hang, or it will pull out the catheter.
13. If an extension set is used, place a second butterfly 3 inches from the catheter. Secure to neck using suture method. Coil the remaining tube, secure with a small piece of tape, and affix to wool or skin.	—
14. Place a protective wrap using gauze and veterinary wrap.	**14a.** Catheters secured using the adhesive method may remain unwrapped.
	14b. All catheters with extension sets must be wrapped.

Special Notes

Intravenous catheters can be inserted readily in calves and cria. Do not perform a stab incision, because neonate skin is very thin and delicate. The adhesive method of securing catheter is recommended.

Procedure for Placing an Intravenous Catheter in the Goat

TECHNICAL ACTION	RATIONALE/AMPLIFICATION
1. Ask assistant to secure goat by straddling neck and lifting head.	**1a.** Securing both the collar and the mandible with the same hand will remove the collar from the area of interest.

TECHNICAL ACTION	RATIONALE/AMPLIFICATION
	1b. Goats can also be restrained using a head catch and halter.
2. Wash hands and prepare catheter.	**2a.** Refer to General Intravenous Catheter Placement and Care Recommendations.
3. Clip a 2 × 2-inch square over jugular vein in middle of neck. Perform a standard three-step surgical preparation.	**3a.** The goat jugular vein usually is easy to visualize.
	3b. Directions for the three-step preparation can be found in Chapter 8. Sterile placement will help minimize iatrogenic infections.
	3c. Fig. 6–23.

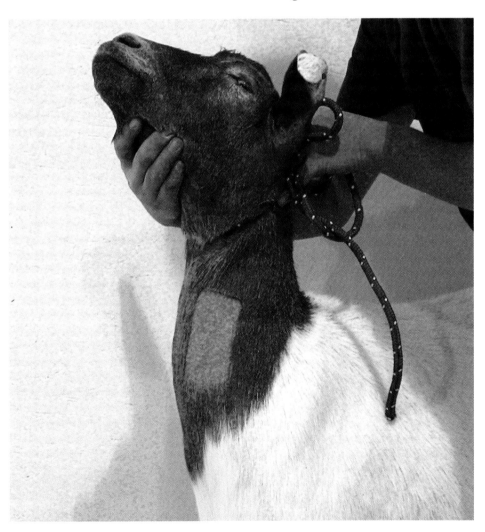

FIG. 6–23 Preparing for placement of an intravenous (jugular vein) catheter in the goat. Note the clipped area.

TECHNICAL ACTION	RATIONALE/AMPLIFICATION
4. Apply sterile gloves.	4a. Sterile gloves help to minimize iatrogenic infection.
5. Kneeling in front of goat, occlude jugular vein with the hand that is not holding the catheter.	5a. Goats are easier to catheterize when the technician is kneeling in front of the animal. This is not advised with other species due to the risk of injury.
6. Grasp stylet and catheter at hub. Holding catheter at a 45-degree angle to skin. Advance quickly through skin and subcutaneous tissue into vein.	6a. Stab incisions are not performed in goats.
7. Once blood is visible, align catheter parallel to jugular vein. Hold stylet still and advance catheter.	7a. Catheters should insert smoothly with minimal resistance. If resistance is encountered, the catheter is most likely in the extravascular space.
8. Remove stylet and place catheter cap.	8a. Retain index finger over catheter hub as stylet is removed and cap is attached. Goats are extremely sensitive to air emboli.
9. Secure catheter by placing a dab of tissue adhesive or superglue gel on the medial aspect of the catheter hub. Press firmly against neck for 1 minute.	9a. The suture method of securing catheters is not recommended in goats. 9b. Protective wraps are generally not needed, but this will depend on the temperament of the individual goat.

Procedure for Placing an Intravenous Catheter in the Pig

TECHNICAL ACTION	RATIONALE/AMPLIFICATION
1. Secure pig using hog snare.	1a. Small pigs can be restrained manually. This is much preferable to using a snare, because the pig will vocalize continually when the snare is used.

TECHNICAL ACTION	RATIONALE/AMPLIFICATION
2. Wash hands and prepare catheter.	2a. Refer to General Intravenous Catheter Placement and Care Recommendations.
3. Identify catheter site along dorsal aspect of pinna and perform a three-step surgical preparation.	3a. Directions for the three-step preparation can be found in Chapter 8. Sterile placement will help minimize iatrogenic infections. 3b. Clipping is not usually performed. During the winter months, excessively hairy pigs may warrant a quick clip.
4. Apply sterile gloves.	4a. Sterile gloves help to minimize iatrogenic infection.
5. Ask assistant to occlude vein at the base of the ear.	5a. If an assistant can not manually occlude the auricular vein, a large rubber band can be used. Secure the rubber band with hemostats.
6. Grasp catheter and stylet at the hub. Hold the catheter at a 30-degree angle to the skin and insert into vessel.	6a. The bevel of the stylet should always face upward when entering a vessel. 6b. Compared with those of other large animal species, pig vessels are very delicate.
7. Advance the stylet ¼ inch into the vessel. Hold the stylet in position and slowly advance the catheter into the vessel.	7a. Advancing the stylet ¼ inch is known as seating the catheter. 7b. Catheters should enter smoothly, with minimal resistance. If resistance is encountered, the catheter is most likely in the extravascular space.
8. Hold catheter hub and remove stylet. Apply catheter cap.	8a. Ensure that cap is tightly secured. 8b. Extension sets are not used. Pigs do not tolerate them. 8c. Remove tourniquet if used.
9. Place a dab of adhesive glue on the medial aspect of the catheter hub. Press firmly against ear for 1 minute.	9a. Catheters are not secured using a suture method.
10. Place a roll of 2-inch tape (or cylinder of similar size) on ventral aspect of pinna. Place a small square of Telfa or gauze directly over the catheter. Then secure cylinder to ear using adhesive tape.	10a. Use of wrap is optional and often depends on temperament of pig.

Removal of Intravenous Catheters

TECHNICAL ACTION	RATIONALE/AMPLIFICATION
1. Secure animal appropriately.	1a. Method of securing is determined by animal species.
2. Use suture scissors to cut securing sutures.	2a. If the adhesive method was used, proceed to Step 3.
3. Grasp catheter hub and remove catheter in one fluid, swift motion.	3a. Catheters secured using suture will slide out easily.
	3b. If the catheter was secured with adhesive, a significant amount of force will be needed to pull the hub free from the skin. Once the attachment is broken, the catheter will slide out easily.
4. Apply direct digital pressure to catheter site for 2–3 minutes.	4a. Digital pressure helps to minimize hematoma formation.
5. Inspect catheter site for signs of swelling, infection, or phlebitis.	—
6. Apply a small dab of antibiotic ointment to catheter site.	6a. Wraps are not usually applied but may be warranted if abnormalities are noted.
	6b. Dimethyl sulfoxide (DMSO) often is applied topically to vessels exhibiting abnormalities.

INTRAARTERIAL CATHETERS

Purpose

- Provide a mechanism to obtain blood for serial blood gas analysis or monitor arterial blood pressure

Complications

- Hematoma
- Thrombosis
- Catheter occlusion
- Iatrogenic infection

Equipment

- Catheter 18–20g $\times$ 1.5–2 inches
- One-inch tape

- Sterile heparinized saline
- 12-cc syringe
- Extension set
- 3-way stopcock
- Monitoring equipment with connecting tubing
- 18g needle
- Clippers
- Povidone-iodine scrub and solution
- Isopropyl alcohol 70%
- Gauze or cotton

Procedure for Placing an Intraarterial Catheter

TECHNICAL ACTION	RATIONALE/AMPLIFICATION
1. Access appropriate limb or head. Animals requiring arterial catheters are typically under anesthesia.	1a. Horses are most commonly catheterized during anesthesia.
2. Clip a 2 × 2-inch square over the facial or greater metatarsal artery.	2a. The facial artery crosses the ventral border of the mandible and runs along the cranial border of the masseter (cheek) muscle.
	2b. The greater metatarsal artery is accessible distal to the hock on the lateral aspect of limb.
	2c. In addition to the arteries noted above, the transverse facial, caudal auricular, and lateral and medial palmar digital arteries can be used.
	2d. Fig. 6–24.
3. Perform a standard three-step surgical preparation.	3a. Directions for the three-step preparation can be found in Chapter 8.
4. Pull skin to the side of the artery and pierce skin using a 16–18g needle.	4a. The skin is pulled to the side to avoid accidental puncture of the artery.
	4b. A stab incision facilitates passage of the catheter.
	4c. Once the skin is released, the hole should lie directly over the artery.

TECHNICAL ACTION	RATIONALE/AMPLIFICATION
5. Occlude artery using thumb of hand that will not be holding the catheter. In addition to occluding the artery, apply slight traction to the skin to help immobilize the vessel.	—
6. Holding catheter and stylet together at a 20-degree angle to the skin, advance through the stab incision and insert into artery.	6a. Given the elastic nature of the arterial wall, small controlled thrusts are required to penetrate the vessel. 6b. Blood within the hub verifies that the stylet is within the artery. Blood will stream out rapidly.
7. Seat the catheter by advancing stylet $\frac{1}{16}$ of an inch.	7a. The facial and greater metatarsal arteries have very small diameters. Piercing the far wall and thus exiting the artery is not uncommon. Use caution to ensure this does not occur when seating the catheter.
8. Hold stylet still and advance catheter into vessel.	8a. There should be no resistance.

FIG. 6-24 Location of an intraarterial (lateral facial artery) catheter in the horse.

TECHNICAL ACTION	RATIONALE/AMPLIFICATION
9. Remove stylet and attach extension set that has been primed with sterile heparinized saline and capped with a 3-way stopcock.	9a. Priming helps to minimize vessel occlusion. 9b. A 3-way stopcock facilitates easy flushing and attachment of monitoring equipment.
10. Tape catheter and extension set in place or use a small amount of tissue adhesive to secure.	10a. Catheters are maintained only during the anesthetic period and therefore do not warrant suturing in.
11. Flush catheter every 3 minutes using 5 cc of heparinized saline to ensure patency. Attach a 12-cc syringe to the stopcock for flushing.	11a. Refer to General Intravenous Catheter Placement and Care Recommendations for preparing heparinized flush.
12. Attach monitoring equipment tubing to stopcock.	12a. Equipment must be at the level of the heart to maintain accuracy of measurement.
13. Collect blood sample or record pressures as needed.	—
14. Upon removal, maintain digital pressure over catheter site for a minimum of 5 minutes.	14a. Pressure minimizes hematoma formation. 14b. A temporary pressure bandage can be applied to the site if needed.

BANDAGING

Bandaging is a fundamental medical procedure integral to large animal practice. The equine patient is the primary beneficiary of this procedure. The number of techniques is enormous and reflects the many needs and functions of the bandage. Remembering that bandages are classified as having primary (closest to wound), secondary (absorption and padding), and tertiary (outer protection) layers will help the technician decide how best to adapt a bandage to the circumstances.

BASIC LOWER LIMB WRAPS

Purpose

- Provide support, promote wound healing, protect wound from desiccation or contamination, absorb exudates, or minimize tissue swelling

Complications

- Wound maceration
- Pressure necrosis

Equipment

- Primary layer (nonadherent dressing pad)
- Roll gauze
- Secondary layer (roll cotton or cotton sheet or quilted pad)
- 2-inch tape
- Tertiary layer (veterinary wrap, track wrap, or polo wrap)

Procedure for Applying a Lower Limb Wrap

TECHNICAL ACTION	RATIONALE/AMPLIFICATION
1. Halter horse. Tie or have assistant hold animal.	1a. Do not place in a stock. Stocks interfere with bandaging.
2. Brush limb thoroughly.	2a. If a wound is present, take care to keep brush away from wound.
	2b. Wounds should be assessed at every bandage change and appropriate treatment provided before reapplication.
3. Place primary layer if wound is present. If no wound is present, proceed to step 4.	3a. Primary layers can be absorbent, nonabsorbent, wet, dry, adherent, and nonadherent. A very common primary layer is the nonadherent Telfa.
	3b. Sweat wraps are used to increase tissue temperature and circulation. Sweat typically is generated using a poultice followed by a plastic wrap material. The plastic material would be considered the primary layer.
	3c. Equine leg wraps are applied frequently to support limbs that do not have open wounds.
4. Apply roll or quilted cotton snugly around the limb. Be certain to avoid wrinkling or unevenness. The cotton layer should extend over the hoof to the ground.	4a. Large animal bandages should always extend to the ground surface, or excessive bandage slipping and distal edema can occur.
	4b. It is often difficult to conform this layer to the leg, especially over the fetlock. Practice will improve technique.
	4c. Wrinkles will cause uneven pressure and tissue trauma.
5. Beginning midcannon, apply roll gauze. Wrap snugly and evenly	5a. The amount of tension applied depends on the amount of

CHAPTER 6 Clinical Procedures **249**

TECHNICAL ACTION	RATIONALE/AMPLIFICATION
toward the ground, then proceed toward the knee, then toward ground again. Wrap should finish midcannon.	cotton or padding used. The thicker the padding, the greater the tension that can be applied.
	5b. Gauze will help conform the cotton to the limb. Do not twist or wrinkle gauze during application. The gauze should overlap the previous layer by 50% as it is applied.
	5c. Snugly encircle the bulb of the heel or the wrap will flip upward.
	5d. Ideally the wrap should finish midcannon.
6. Apply the tertiary layer in the same manner as the roll gauze.	**6a.** Common equine tertiary layer materials include veterinary wrap, polo wraps, or track wraps. Bovine wraps typically are made of Elastikon. All these materials have a stretch factor. It is extremely important to have adequate padding when using these materials.
	6b. Many equine support wraps use only a quilted cotton pad and track wrap. These materials can be washed and reused indefinitely.
7. Apply tape at bandage end to ensure security of wrap end.	**7a.** Tape or Elastikon can be applied at the proximal end of the bandage to prevent slippage or debris from entering wrap. Be certain that no tension is applied around the limb with these materials.
	7b. Fig. 6–25.

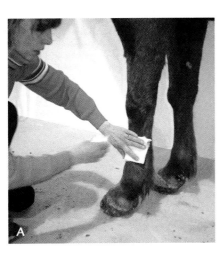 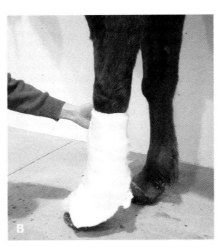 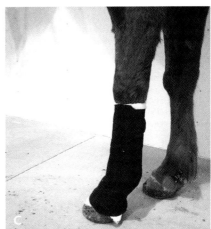

FIG. 6–25 (A) Application of the primary layer of a horse lower limb bandage. (B) Application of the secondary (padding) layer. (C) Application of the tertiary layer.

TAIL WRAPS

Purpose

- Protect tail or prevent tail hairs from entering rectum or vagina during clinical procedures

Equipment

- Brown gauze (temporary wrap)
- One-inch tape
- Veterinary wrap or bandage

Complications

- Injury to personnel
- Pressure necrosis

Procedure for Placing a Tail Wrap

TECHNICAL ACTION	RATIONALE/AMPLIFICATION
1. Place horse in a stock or have assistant hold.	1a. Stand to the side of the rump to avoid being kicked. 1b. Fig. 6–26.
2. Gently lift tail. Begin wrap 3–4 inches from base of tail. Wrap snugly toward tail base, then continue distally toward end of tail.	2a. A variety of wrap material can be used depending on length of time wrap is required. Brown gauze is used for rectals or other quick obstetrical work. Track wraps or veterinary wrap can be used if the tail will be wrapped for a significant period.
3. Wrap must cover 18 inches of tail and can continue to further if desired.	3a. Wraps should not extend distal to coccygeal vertebrae.
4. Secure wrap end using tape.	4a. If using brown gauze, the end of the gauze is torn and tied in a bow instead of securing with tape.

FIG. 6-26 Application of a tail wrap.

ABDOMINAL WRAPS

Purpose

- Apply pressure after abdominal surgery to prevent swelling or edema of suture line

Complications

- Suture line irritation
- Bandage slippage

Equipment

- Ace bandages or Elastikon (many rolls)
- Absorbent pad
- 2-inch tape

Procedure for Applying an Abdominal Wrap

TECHNICAL ACTION	RATIONALE/AMPLIFICATION
1. Halter and hold horse.	1a. The horse can be tied or cross-tied, but do not place in a stock.
	1b. A stock will interfere with this procedure.
2. Brush horse. Do not brush directly over suture line. Ensure that suture line is clean.	—
3. One person should stand on each side of horse.	3a. Bandage materials will be passed over and under the horse.
4. Place absorbent pad on suture line.	4a. Roll cotton is not suitable, because it sticks to the sutures.
5. Wrap snugly by passing wrap around abdomen. Each person should ensure that no wrinkles or twists occur.	5a. At the end of each bandage roll, the next roll must overlap to prevent unraveling of the previous wrap.
6. The wrap should extend 4–6 inches cranial and caudal to the suture line.	—
7. If ace wraps are used, secure both the front and hind margins of the wrap with tape.	7a. Taping the wrap to the hair helps to minimize slippage.

REPRODUCTIVE TREATMENTS

Sheath cleaning is a component of all good equine husbandry protocols. It is a straightforward, easy-to-master procedure that improves the genital health of both geldings and stallions. Owners should be encouraged to perform this procedure either on their own or with the help of a veterinary paraprofessional.

SHEATH CLEANING

Purpose

- Clean penis and sheath for breeding purposes or routine husbandry

Complications

- Trauma to technician
- Residual soap irritation

Equipment

- Mild soap (Ivory or commercial sheath cleaner)
- Warm water

- Obstetrical sleeve or glove
- Gauze squares
- Bucket
- Tranquilizer if needed

Procedure for Cleaning the Sheath

TECHNICAL ACTION	RATIONALE/AMPLIFICATION
1. Halter horse and place in stock.	1a. A stock used for this procedure should have easy side access.
	1b. Routine husbandry sheath cleaning should be performed every 6–12 months. Owners should be instructed on how to complete this procedure.
2. Sedate horses using acepromazine or xylazine if necessary.	2a. Most horses will learn to accept sheath cleaning, so tranquilization becomes unnecessary. Horses not accustomed to the procedure may kick.
	2b. Acepromazine often facilitates dropping down of the penis which makes cleaning easier; however, this tranquilizer has been associated with paraphimosis.
3. Apply an obstetrical sleeve or gloves to prevent smegma from adhering to hands. Thoroughly lubricate hands using 2–3 tablespoons of soap.	3a. Smegma is a thick, black, waxy material containing sebaceous gland secretions and cells. It has a foul odor.
	3b. Do not use iodine scrubs or chlorhexidine for sheath cleaning. These agents are more irritating to the sheath.
4. Insert hand into the sheath and gently begin to wash walls of sheath and penis.	4a. The sheath is also called the prepuce.
	4b. It is much easier to wash the penis when it is distended, but cleaning can be accomplished when the penis is still within the sheath.
	4c. The cleaner's arm will need to be inserted to the level of the forearm if the penis is not dropped.

TECHNICAL ACTION	RATIONALE/AMPLIFICATION
5. Once the area is thoroughly soaped, wet gauze can be used to help remove the smegma.	5a. Use additional soap, water, and gauze as needed.
6. Check the urethral fossa on the head of the penis for a hard ball of smegma (bean).	6a. The fossa is located adjacent to the urethral orifice and the smegma must be scooped out using the index finger (Fig. 6–27).
	6b. The fossa is a few centimeters deep.
	6c. Smegma balls are a common cause of sheath swelling or urine flow obstruction.
7. Continue cleaning until all smegma is removed. Thoroughly rinse soap using warm water.	7a. Soap residue will cause irritation to the penis and sheath.
	7b. Horses accustomed to sheath cleaning will often tolerate a low-pressure, warm water flush using a hose. This makes rinsing much faster.

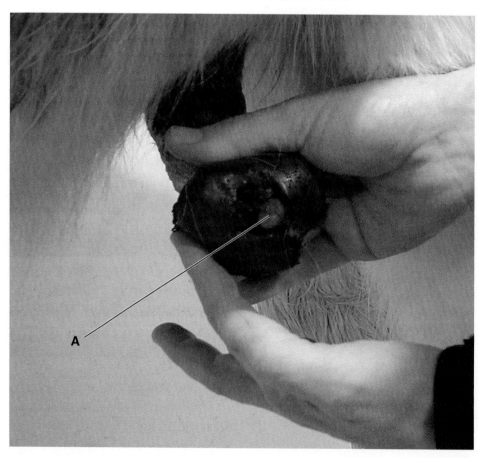

FIG. 6-27   Distal end of the horse penis showing the urethral orifice (**A**) and fossa, which contains a smegma bean.

INTRAMAMMARY INFUSION

Purpose

- To prevent (dry-off treatment) or treat mastitis

Complications

- Iatrogenic infection

Equipment

- Antiseptic swab or Povidone-iodine scrub
- Isopropyl alcohol 70%
- Paper towels
- Medication with cannula

Procedure for Performing an Intramammary Infusion

TECHNICAL ACTION	RATIONALE/AMPLIFICATION
1. Restrain animal appropriately.	1a. Method of restraint is determined by species. Cattle are placed in stocks or lock-up. Goats are placed on stanchions or held by the collar.
	1b. Infusion of the cow is most common. Sheep and goats are treated occasionally. Horses and llamas are not usually given infusions.
2. Strip udder.	2a. See Chapter 5 "Procedure for obtaining a milk culture" for milking procedure.
	2b. Removal of milk permits better entry of medication.
3. Wash and dry teat using paper towel and antiseptic.	3a. It is important to dry the teat thoroughly.
4. Swab teat sphincter with alcohol and allow to air dry.	—
5. Grasp teat at base and insert cannula into teat orifice. Administer medication.	—
6. Occlude teat orifice and gently massage teat (upward) and quarter to half of udder to distribute medication.	—
7. Appropriately mark the animal to ensure that milk is not consumed.	7a. It is extremely important to follow milk withdrawal times. Milk must be discarded until appropriate withdrawal time is completed.

OCULAR PROCEDURES

Most of the ocular procedures described are performed on the equine patient. Although cattle experience many ocular problems, treatment that requires attention twice or 4 times a day frequently is limited to the more expensive breeding animals.

ADMINISTERING OCULAR MEDICATION

Purpose

- Administer medication to eye

Complications

- Ocular trauma

Equipment

- Eye drops or ointment

Procedure for Administering Ocular Medication

TECHNICAL ACTION	RATIONALE/AMPLIFICATION
1. Secure animal as needed.	1a. Place cattle in chute and secure head to side using halter.
	1b. Halter horse or llama and place in a stock.
	1c. Secure goat by collar.
2. Open lids of eye using thumb and forefinger.	2a. Always administer ocular solutions before ointments. Ointments will prevent corneal penetration of the solution.
	2b. If multiple solutions are administered, wait 3–5 minutes after each medication.
3. Allow the hand that is holding the medication to rest against the animal's head.	3a. Resting the hand against the head will help to minimize ocular trauma should the animal move rapidly.
4. Squeeze a ribbon of ointment along upper sclera or lower palpebral border. Do not contact any portion of the eye with tip of medication dispenser.	4a. Drops are placed in the same location.
	4b. Contacting the tip of the dispenser can cause severe ocular trauma.
	4c. Medication is dispersed across the eye as the animal blinks.

FLUORESCENCE STAINING

Purpose

- Detect corneal ulceration and confirm nasolacrimal duct patency

Complications

- Iatrogenic infection
- Ocular trauma

Equipment

- Sterile fluorescein stain strips
- Sterile saline
- Gauze

Procedure for Performing a Fluorescein Stain

TECHNICAL ACTION	RATIONALE/AMPLIFICATION
1. Secure species as needed.	1a. Place cattle in chute and secure head to side using halter.
	1b. Halter horse or llama and place in a stock.
	1c. Secure goat by collar.
2. Wet the tip of sterile dye strip with saline. Place a few drops of the dye into the eye.	2a. Artificial tear solutions can be used in place of saline.
	2b. Alternatively, a dye strip can be placed in a syringe filled with some saline. The resultant dye is then dropped into eye.
	2c. Rose bengal stains can be used to detect devitalized surface cells and will detect much more subtle lesions.
	2d. Staining should occur after Schirmer tear tests and cultures have been performed.
3. Allow animal to blink several times, and then flush thoroughly with saline.	3a. Horses experiencing painful eyes may object strongly to manipulation of the palpebrae. General tranquilization or auriculopalpebral or frontal nerve blocks may be necessary.

TECHNICAL ACTION	RATIONALE/AMPLIFICATION
4. Examine cornea for signs of ulceration. Ulcers will appear as green dots.	4a. Examination can be facilitated by using a dimly lit room and penlight.
5. Examine nasal puncta for presence of dye.	5a. The punctum is located on the medial, ventral aspect of the nostril.
	5b. Dye should appear within 5 minutes to confirm patency of the nasolacrimal duct.

NASOLACRIMAL FLUSHING

Purpose

- Clear blockage of the nasolacrimal duct

Complications

- Inability to open duct
- Iatrogenic infection

Equipment

- Polyethylene tubing size 5 French
- Proparacaine 0.5%
- Sterile lubrication
- 60-cc syringe
- Sterile saline

Procedure for Flushing the Nasolacrimal Duct

TECHNICAL ACTION	RATIONALE/AMPLIFICATION
1. Halter horse and place in a stock.	—
2. Apply 0.5% proparacaine to eye and nasolacrimal opening in nostril.	—
3. Inspect catheter tip for any irregularities, and flush with saline. Apply light coat of sterile lubricant.	3a. Catheter should be placed on sterile surface.
4. Advance catheter 5–6 inches into nasolacrimal opening.	4a. Horses typically will toss their heads during this portion. Tranquilize if necessary.
	4b. The nasolacrimal opening is located on the medial aspect of the nostril.
5. Attach syringe and flush with sterile saline.	—

NASOLACRIMAL CANNULATION OR INDWELLING INFUSION TUBE

Purpose

This procedure provides a means to administer ocular medication without contacting the eye. This usually is placed when medication is required at least 3 times per day or the horse is extremely temperamental.

Complications

- Premature cannula removal
- Occluded cannula
- Iatrogenic infection

Equipment

- 3-0 nonabsorbable suture
- Number 15 scalpel blade
- Polyethylene tubing size 5 French
- Topical anesthetic (0.5% proparacaine)
- 2-inch tape
- 3-cc syringe with 20g × 1-inch needle
- 12-cc syringe
- Lidocaine (injectable)
- Needle drivers
- Scissors
- Extension set
- Sterile saline
- Sterile gauze
- Sterile lubrication

Procedure for Cannulating the Nasolacrimal Duct

TECHNICAL ACTION	RATIONALE/AMPLIFICATION
1. Halter horse and place in stock.	1a. Sedation may be necessary.
2. Inspect catheter tip for any irregularities, flush with saline, and set aside.	2a. Catheter should be placed on sterile surface.
3. Apply 0.5% proparacaine to eye and nasolacrimal opening in nostril.	3a. Fig. 6–28.

TECHNICAL ACTION	RATIONALE/AMPLIFICATION
4. Inject 1–2 cc lidocaine subcutaneously in external dorsal lateral aspect of nostril. Wait 3–5 minutes.	—
5. Use a #15 blade to make a stab incision through external dorsal lateral aspect of nostril.	—
6. Pass catheter through stab incision. Wipe catheter end using sterile gauze. Apply a very light coat of lubricant.	6a. Canine male urinary catheters of appropriate size are used most commonly.
7. Insert catheter into nasolacrimal opening in nostril and advance catheter 5–6 inches.	7a. The nasolacrimal opening is located on the medial aspect of the nostril.
8. Apply tape to catheter in butterfly fashion 2–3 inches from stab incision dorsal to the nostril. Suture both wings of butterfly to skin.	8a. Horses tolerate this better if a local lidocaine block is used.
9. A second butterfly is attached inside nostril on ventral floor near nasolacrimal opening.	—
10. Finish securing catheter by placing a third butterfly over forehead.	10a. It is important that the catheter remain close to the forehead. This will help to prevent premature removal by the horse.

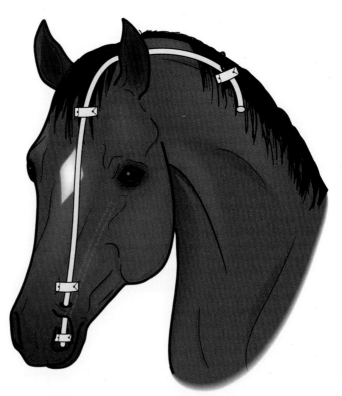

FIG. 6–28 Nasolacrimal cannulation of the horse.

TECHNICAL ACTION	RATIONALE/AMPLIFICATION
11. Attach primed extension set to catheter. Bring extension set between ears and secure to dorsal neck 4–6 inches caudal to ear.	**11a.** The extension set frequently is taped or braided in the forelock and mane. **11b.** A primed extension set contains saline. **11c.** It is important to know the exact volume needed to prime the extension set. After administering eye drops, the extension set is flushed with just enough saline to deliver the medication to the eye.
12. Administer medication as directed.	**12a.** Ointments can not be delivered using this method. **12b.** A catheter should remain in place a maximum of 2 weeks. **12c.** Inspect catheters daily and remove at first sign of redness, abnormal discharge, swelling, or pain.

IDENTIFICATION TECHNIQUES

Appropriate identification minimizes animal theft and assists with routine husbandry issues such as calving dates, milk production, medication administration, and tracking. The decision to apply a permanent or removable means of identification is dependent on the animal species, primary function of the identification, and owner preference.

HOT BRANDING

Purpose

- Provide permanent individual animal identification

Complications

- Hide damage
- Excessive thermal damage caused by hide moisture

Equipment

- Heat source
- Extension cord and outlet if using electric brands
- Branding iron

Procedure for Hot Branding

TECHNICAL ACTION	RATIONALE/AMPLIFICATION
1. Place cow in chute or horse in stock.	1a. Always confirm with owner the exact location of brand. Common sites include shoulder, hip, or rib.
	1b. It is recommended that horses be tranquilized for this procedure.
2. Place iron in heat source.	2a. Propane is used most commonly.
	2b. Irons that are ready to apply are ash gray. Red irons are too hot and black irons are too cold.
	2c. Application of overheated irons will result in hair fire, excessive hide damage, and animal distress.
	2d. Irons are composed of iron or steel. Less intricate, simple patterns make for better brands.
3. Ensure that coat is not wet or covered with excessive debris.	3a. Never brand wet animals. Wet hair will transfer heat over a large area, causing excessive hide damage and blotched brands.
4. Apply brand for 3–5 seconds, using a slight rocking motion to ensure uniformity of application.	4a. Brand area should appear leather brown in color.
	4b. Do not allow branding iron to slide or move from site during application.
	4c. Hot branding destroys hair follicles, thus resulting in a permanent bald scar.
	4d. The length of time irons are applied will vary dramatically with the animal age and species. Calves, goats, and horses require much less time than adult cattle.
5. Place iron back in heat source.	—

FREEZE BRANDING

Purpose

- Provide permanent individual animal identification that is easy to visualize

Complications

- Inadequate brands that are difficult to read
- Blotched brands

Equipment

- Freeze brand iron
- Coolant (liquid nitrogen or dry ice and 70% isopropyl alcohol mix)
- Cooler
- Clippers with #40 blade
- Stiff brush
- Squeeze bottle filled with 70% isopropyl alcohol
- Heavy gloves and safety goggles

Procedure for Freeze Branding

TECHNICAL ACTION	RATIONALE/AMPLIFICATION
1. Secure cow in chute or horse in stock.	1a. Compared with hot branding, freeze branding causes less damage to the hide and is less painful.
	1b. Always confirm with owner the exact location of brand. Common sites are shoulder, hip, or rib.
2. Cool branding iron for 20 minutes before use on first animal by placing the iron in freezing mixture.	2a. Alcohol and dry ice coolant mixture for branding 20–30 head requires 2–5 gallons alcohol and 1 lb dry ice per animal.
	2b. Coolant should cover iron head by at least 1 inch. Bubbling of the alcohol should stop before application of the brand.
	2c. Frost lines on the handle of the iron indicates that it is ready.
	2d. Old 20-quart plastic coolers work very well.
	2e. Liquid nitrogen requires less contact time than the alcohol mix (30 seconds versus 1 minute). Use extreme care when working with liquid nitrogen.
	2f. Freeze-branding irons are composed of copper or bronze. Steel irons used with hot branding do not cool sufficiently.

TECHNICAL ACTION	RATIONALE/AMPLIFICATION
3. Clip animal over branding area.	3a. Freeze branding kills the melanocyte pigment–producing cells in the hair follicle. Thus, freeze brands appear white. If the iron is applied too long, the hair follicle is destroyed and the brand will appear similar to a hot brand.
4. Use stiff brush to remove any debris.	—
5. Saturate area with alcohol using a squeeze bottle.	5a. Use sufficient alcohol. Area must be thoroughly soaked.
6. Apply brand. Hold iron firmly against skin and apply a gentle rocking motion top to bottom and left to right.	6a. Use gloves when holding branding irons. 6b. The rocking motion ensures equal pressure for all iron contact points. Do not permit iron to move from brand spot while rocking.
7. Length of application depends on skin thickness, animal age, coolant, and type of metal used.	7a. Typical application times for cattle: Alcohol and dry ice: 40–60 seconds; liquid nitrogen: 20–30 seconds. 7b. Horses have shorter application times. Liquid nitrogen application time is typically 15 seconds. 7c. The brand area initially will appear indented. Subsequent swelling will cause the brand to rise within a few minutes. The area will scab and peel 3–4 weeks after branding. Future hair growth will be white.
8. Place iron back in coolant immediately. Irons must be cooled a minimum of 2 minutes between animals.	—

EAR TATTOOING

Purpose

- Provide permanent identification for breed registration or verification of brucellosis vaccination status

Complications

- Illegible tattoo
- Transmission of papilloma virus

Equipment

- Gauze or cotton
- Isopropyl alcohol 70%
- Tattoo pliers
- Numbers or letters for pliers
- Tattoo ink or paste
- Toothbrush
- Paper and pen

Procedure for Tattooing the Ear

TECHNICAL ACTION	RATIONALE/AMPLIFICATION
1. Place cow in chute.	**1a.** Goats and pigs may also be tattooed in this manner. Secure as needed.
	1b. Some horses may have a lip tattoo, but this is performed only by breed registry personnel.
2. Clean ear of wax using alcohol.	**2a.** The right ear is reserved for brucellosis tattooing, which may be performed only by licensed veterinarians. Use the left ear for breed or owner tattoos.
3. Test accuracy of tattoo by clamping a sheet of paper.	**3a.** Inexperienced operators may reverse numbers or place upside-down in the pliers.
	3b. The identification number and letter sequence will be provided by owner if tattooing for breed registry purposes. Be certain to record any other identification numbers used on the cow at the time of tattoo application. For example, ear tag numbers probably will not correspond with tattoo numbers.
	3c. Brucellosis tattoos include the letter R, state shield, and the last digit of the year.
4. Apply ink or paste to ear. Tattoos are placed between second and third rib of the ear.	**4a.** For pigs the thinner part of the lower ear (inside or outside) works well.
5. Apply and firmly clamp pliers.	**5a.** Excessive bleeding at the tattoo site causes the ink to wash out, resulting in a poor quality tattoo.

TECHNICAL ACTION	RATIONALE/AMPLIFICATION
6. Apply ink or paste a second time, using toothbrush to ensure adequate penetration of ink.	6a. Tattoo legibility is extremely important, because it will directly affect animal marketability. Illegible tattoos can result in significant monetary losses.
7. Place tattoo pliers in chlorhexidine solution between animals.	7a. Pliers are a common vector for transmitting the papilloma (wart) virus.

EAR TAGS

Purpose

- Provide an affordable, easy-to-read, removable identification system

Complications

- Premature tag removal
- Inadvertent spread of papilloma virus

Equipment

- Ear tags
- Tag gun
- Knife
- Paper and pen for recording identification numbers

Procedure for Attaching Ear Tags

TECHNICAL ACTION	RATIONALE/AMPLIFICATION
1. Place cow in chute.	1a. Secure goats, sheep, and pigs as needed.
2. Place ear tag in gun.	2a. Ensure that tag is placed in gun such that number will be facing forward upon application (Fig. 6–29).
	2b. Radio frequency identification devices (RFIDs) are gaining popularity. These devices can store information about the individual animal and are placed in the same location as the older style tags. An RFID dramatically increases the efficiency or ability to track animal movements (sales).

TECHNICAL ACTION	RATIONALE/AMPLIFICATION
3. Use knife to remove old tag if necessary.	3a. Tag cutters are made specifically for this purpose.
4. Identify center between second and third rib of left ear.	4a. If a previous tag was removed the old hole or piercing site can be used, provided it is not significantly stretched out.
	4b. The right ear should be reserved for brucellosis vaccination tattoo.
5. Place tag and record number or other animal identification information as requested by owner.	—

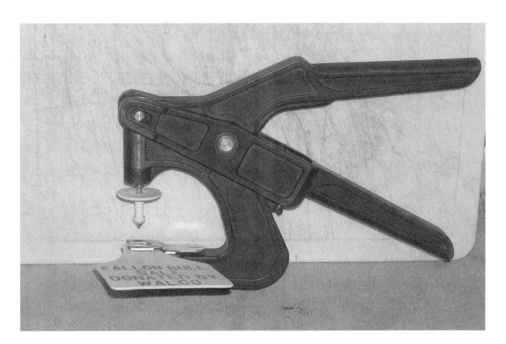

FIG. 6-29 Ear tag gun loaded with button and tag.

EAR NOTCHING

Purpose

- Provide permanent, inexpensive, individual pig identification

Complications

- Notches applied in incorrect area

Equipment

- Notching pliers
- Chlorhexidine solution

Procedure for Notching a Pig Ear

TECHNICAL ACTION	RATIONALE/AMPLIFICATION
1. Secure piglet.	1a. It is often easiest to have one person hold and a second person notch. However, very experienced individuals hold and notch simultaneously.
	1b. Notching is performed in the farrow house along with teeth trimming, tail docking, and iron administration. Ideally this occurs when the animal is less than 3 days old. Perform notching after all other processing has been completed, because piglets find this procedure least desirable.
2. Apply notching pliers to appropriate quadrants of ear to identify pig.	2a. Many different marking patterns can be used. Each notch represents a number. Typically the right ear signifies the sow number and the left ear is the piglet's number within the litter (Fig. 6–30).
	2b. Notches placed at the ear base must be deep enough to ensure that they do not diminish with time. Pigs under 25 lb should have a notch that is $3/16$–$1/4$ inch deep.
	2c. Notches at ear tip should be slightly shallower to prevent drooping of the ear.
	2d. Leave at least $1/4$ inch between notches. Avoid making notches too close to the head.
	2e. Bleeding is minimal and will stop spontaneously.
3. Clean notching pliers before use on the next pig.	3a. Place in chlorhexidine.

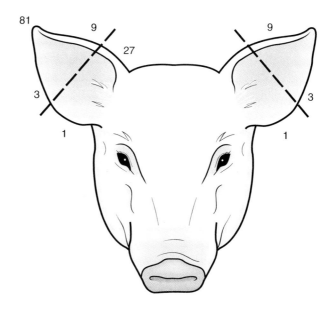

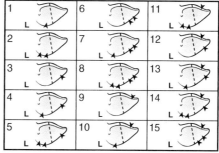

FIG. 6-30 Ear notching identification system used in pigs.

REVIEW QUESTIONS

1. State the complications associated with administering oral pastes.
2. Identify equine muscles suitable for intramuscular injection.
3. List specific actions and recommendations technicians can follow during bovine injections to minimize tissue trauma.
4. Explain why automatic syringe guns must be rinsed thoroughly using water.
5. Define hubbing a needle and explain the importance of this action.
6. Identify appropriate locations for intramuscular, subcutaneous, and intravenous injections in the pig.
7. List the complications associated with growth hormone implants in cattle.
8. Discuss the function and importance of assessing gastric reflux in a colicky horse.

9. State three indications for removing intravenous catheters.
10. Draw and label the following intravenous catheter parts: stylet, catheter, hub, bevel, guide wire, and needle guard.
11. Explain why stab incisions are used frequently during placement of a bovine intravenous catheter.
12. Define smegma and bean as they relate to sheath cleaning.
13. List the equipment required to insert an indwelling nasolacrimal infusion tube.
14. Compare and contrast hot and freeze branding.
15. State the rationale for using the left ear when placing bovine identification tags.

BIBLIOGRAPHY

Fowler, M. (1998). *Medicine and surgery of South American camelids* (2nd ed.). Ames: Iowa State University Press.

Hall, B., Greiner, S. P., & Gregg, C. (2004). *Cattle identification: Freeze branding.* Virginia Cooperative Ext. Blacksburg, Publication number 400–301. Retrieved January 12, 2006, from http://www.ext.vt.edu/. Click on Educational Programs and Resources; Livestock, Poultry & Dairy; Beef.

McCurnin, D. (1998). *Clinical textbook for veterinary technicians* (4th ed.). Philadelphia: W. B. Saunders.

Smith, B. (1990). *Large animal internal medicine.* St. Louis: C. V. Mosby.

(2000). *Cattle and horse branding.* Saskatchewan Agriculture, Food and Rural Revitalization. Retrieved January 23, 2006, from www.agr.gov.sk.ca

(2002). Sheffield, England: 5M Enterprises. Retrieved January 23, 2006, from www.thepigsite.com

CHAPTER

Neonatal Clinical Procedures

If people were superior to animals, they'd take better care of the world.
— WINNIE THE POOH

KEY TERMS

colostrum
cria
enema
kid
meconium

nasogastric intubation
neonate
orogastric intubation
parenteral nutrition
placenta

OBJECTIVES

- Discuss techniques used to provide neonatal supplemental nutrition.
- Identify clinical procedures and processing techniques commonly performed on the neonatal patient.
- Compare and contrast neonatal care provided to each species immediately after birth.

BOTTLE FEEDING

Bottle feeding provides the newborn with the nutritional support that is fundamental to life. In addition to nutritional support, it provides a positive interaction between the **neonate** and medical care team. It should be an unhurried time of comfort for the baby, not an episode of intense stress and anguish. Remember, when dealing with neonates, patience and persistence are essential.

PURPOSE

- Provide nutrition to orphaned neonates or those unable to nurse the dam

COMPLICATIONS

- Aspiration
- Refusal of neonate to suck

EQUIPMENT

- Human baby bottle with nipple (foal, llama, pig)
- Calf bottle with sheep or goat nipple (goat, foal)
- Calf bottle with calf nipple (calf)
- Towel

Procedure for Bottle Feeding a Foal

TECHNICAL ACTION	RATIONALE/AMPLIFICATION
1. Mix milk replacer according to instructions.	1a. Always use warm water for mixing.
	1b. If feeding a foal for first time, widen the nipple hole so that it drips at 2 drops per second when inverted.
	1c. Newborn foals should be fed every 30–40 minutes. (Normal foals will nurse the dam 7 times per hour.) Foals 1 month and older should be fed a minimum of 4 times a day.
	1d. Foal should consume 1 ounce of milk for every 10 lb body weight per hour.
	1e. If no replacer is immediately available, the following mixture can be used on a very short-term basis. Mix together 1 can evaporated milk, 2 cans water, and 50 cc of 50% glucose.

TECHNICAL ACTION	RATIONALE/AMPLIFICATION
2. Cradle head and neck in left arm, with foal and handler facing the same direction. Insert nipple in mouth and compress bottle to place small (5 cc) amount of milk in mouth.	2a. These instructions assume that foal is nursing for first time or having difficulty nursing. Foals accustomed to bottle feeding are much less labor intensive. Just hold the bottle at wither level for them. 2b. Covering (draping) the foal's eyes with a towel will often facilitate sucking. 2c. Do not hold bottle above the level of the foal's withers. This will increase the risk of aspiration.
3. Foal should begin sucking at this point. Continue to support head loosely while foal sucks.	3a. Many foals will resist initially. Be patient and repeat the process slowly. 3b. Foals may nurse standing or in sternal recumbency. Never feed in lateral recumbency. 3c. Poor suckle reflexes are associated with increased risk of aspiration pneumonia.
4. Wipe chin and nose to remove spilled milk. Record amount consumed and thoroughly wash all equipment.	4a. Foals that are bottle fed are highly susceptible to enteric infections. Cleanliness of feeding equipment is essential.

Procedure for Bottle Feeding a Calf

TECHNICAL ACTION	RATIONALE/AMPLIFICATION
1. Mix milk replacer according to instructions.	1a. Milk should be warm. 1b. Newborn calves should be fed 3 times per day. Healthy older calves usually are fed 2 times per day. 1c. Calves consume about 2 quarts per feeding. Sick animals should receive smaller feedings more frequently.
2. Hold bottle at height that allows calf to maintain head in natural position. This is usually just below the level of the calf's shoulder.	2a. Fig. 7–1.

TECHNICAL ACTION	RATIONALE/AMPLIFICATION
3. Hold nipple in front of calf. Most calves will latch on readily.	—
4. If calf does not latch on, straddle neck of calf and place nipple in mouth. Hold mouth closed over nipple. Move nipple back and forth in mouth.	4a. Most nipples have a very small aperture. This hole can be widened so that a steady drip is noted when the bottle is inverted.
5. Wipe chin and nose to remove spilled milk. Record amount consumed and thoroughly wash all equipment.	5a. Calves that are taught to feed from buckets often experience fewer enteric problems. This is a sanitation issue reflecting poor cleaning of the bottle and nipple by the caretaker. Unless the calf has been drinking from the bucket for a significant amount of time, it should be allowed to nurse from a bottle. Nursing from the bottle will facilitate closure of the esophageal groove.

FIG. 7-1 Bottle feeding a calf.

Procedure for Bottle Feeding a Kid

TECHNICAL ACTION	RATIONALE/AMPLIFICATION
1. Mix milk replacer according to instructions.	1a. Kids are the easiest of all species to bottle feed. Actively nursing kids will wag their tails.
	1b. Offer a maximum of 1 pint per feeding. Milk should be warm.
2. Offer bottle.	2a. Kids that have nursed several times on the doe will often refuse a bottle initially.
3. If kid does not latch on immediately, hold it under left arm with kid facing forward. Cup mandible in left hand. Place nipple in mouth and gently hold mouth closed around nipple for 1 minute.	3a. Encourage kid to nurse only if a suckle reflex is present.
	3b. The nipple can be rocked gently back and forth in the mouth to encourage sucking.
	3c. The nipple hole can be enlarged if needed. It should not stream when inverted but can drip at a rate of 1–2 drops per second.
4. Wipe chin and nose to remove spilled milk. Record amount consumed and thoroughly wash all equipment.	4a. Fig. 7–2.

FIG. 7–2 Bottle feeding a kid.

Procedure for Bottle Feeding a Pig

TECHNICAL ACTION	RATIONALE/AMPLIFICATION
1. Mix milk replacer according to instructions.	1a. If pig milk replacer is unavailable, goat milk replacer can be used temporarily.
	1b. Ensure that milk is warm.
	1c. If another sow is available, piglets should be fostered.
2. Allow piglet to remain standing. Cup one hand over dorsum of piglet and offer bottle.	2a. Piglets typically nurse readily from the bottle.
	2b. At 1 day of age, piglets consume 2–3 tablespoons each feeding. This increases with age.
	2c. In addition to milk replacer supplement, offer solid feed at 10 days of age.
3. If piglet does not grasp nipple, place it in mouth and hold mouth closed for a minute.	3a. The nipple can be rocked gently back and forth in the mouth to encourage sucking.
	3b. The nipple hole can be enlarged if needed. It should not stream when inverted but can drip at a rate of 1–2 drops per second.
4. Wipe chin and nose to remove spilled milk. Record amount consumed and thoroughly wash all equipment.	4a. Newborn piglets have no ability to maintain body temperature, so an external heat source must be supplied for several weeks
	4b. Pigs typically are weaned at 28 days of age (12–14 lb).

Procedure for Bottle Feeding a Cria

TECHNICAL ACTION	RATIONALE/AMPLIFICATION
1. Mix replacer according to package directions.	1a. Llamas are similar to foals in that it is fairly difficult to establish nursing from a bottle. Crias rarely learn to drink milk from a bucket.
	1b. Crias should be offered 120 cc (or as much as they will consume) every 2–3 hours.
	1c. Llamas frequently are fed goat milk replacers. Two to three ounces of live-culture yogurt can be added.

CHAPTER 7 Neonatal Clinical Procedures

TECHNICAL ACTION	RATIONALE/AMPLIFICATION
	1d. Milk replacers containing antibiotics will cause diarrhea in cria.
2. Hold bottle level with head. Place left arm over neck and cup mandible in left hand. Snug head slightly against your leg or abdomen if needed.	**2a.** Fig. 7–3.
3. Place nipple in mouth and move back and forth (½ inch) to encourage sucking. Continue to support head while sucking.	—
4. Wipe chin and nose to remove spilled milk. Record amount consumed and thoroughly wash all equipment.	—

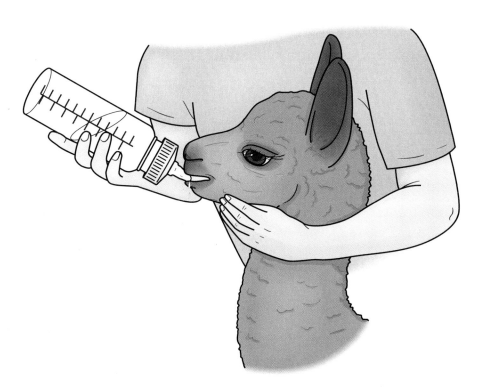

FIG. 7–3 Bottle feeding a cria.

MILK REPLACERS AND COLOSTRUM

Production of commercial milk replacers has made it much easier to meet the nutritional needs of the neonate (Table 7–1). If at all possible, use a species-appropriate milk replacer that does not contain antibiotics, and be meticulous about cleaning the bottles and nipples after use.

COMMON MILK REPLACER COMPONENTS

TABLE 7-1 COMMON MILK REPLACER COMPONENTS

Calf	20–22% Crude protein (all of which is milk derived)
	20% Crude fat
	Feed 3 times a day
Foal	18–22% Crude protein
	16% Crude fat
	Feed 16 times per day on day 1, and 8 feedings per day by day 8
Kid	21–24% Crude protein
	24–30% Crude fat
	Feed 4 times a day for 1 week, then twice a day
Pig	25% Crude protein
	10% Fat
	Feed every 2 hours for first week, then feed 4 times a day

Special Note

Milk replacers containing antibiotics should be avoided.

COLOSTRUM RECOMMENDATIONS

Many large animal species are born immunologically naive; they obtain antibodies after birth, not during gestation. **Colostrum** contains antibodies that are essential to fight off infection. Also called the first milk, colostrum is truly one of life's golden bullets.

- Foals should consume 1.5 liters of colostrum during the first 24 hours. Administer 250 cc every hour for the first 6–8 hours. If horse colostrum is not available, use cow colostrum.

- Calves should receive 10–15% of body weight during the first feeding immediately after birth. A second feeding of colostrum should occur 8–10 hours later. Depending on size of calf, this is typically 2–4 liters.

- A **kid** should receive 500 cc to 1 liter within the first 24 hours. Kids typically will consume 200 cc with each feeding. A minimum of four feedings should occur during the first 24 hours.

- Pig colostrum is difficult to obtain. Substitute cow colostrum and feed piglets for the first 3–4 days of life.

- A **cria** should consume 10% of body weight, given over an 18-hour period. If llama colostrum is not available, use cow or goat colostrum. Do not use sheep colostrum.

INTUBATION

Intubation involves inserting a tube into an organ or body cavity. Intubation of the neonate is most commonly performed to place nutritional supplements directly into the stomach. Foals are intubated through the nose. This is termed **nasogastric intubation**. Claves, cria and kids are intubated through the mouth. This is termed **orogastric intubation**. Although it may seem rather difficult at first, intubation is easily mastered with practice.

NASOGASTRIC INTUBATION OF FOALS

Purpose

- Administer fluids
- Evaluate for reflux or alleviate gas distention

Equipment

- Foal nasogastric tube
- 1-inch tape
- Bucket
- Pump
- Warm water

Complications

- Aspiration
- Secondary aspiration pneumonia
- Pharyngeal irritation
- Esophagitis
- Epistaxis
- Secondary guttural pouch infection
- Gastrointestinal intolerance (colic, bloat, diarrhea)

Procedure for Nasogastric Intubation of Foals

TECHNICAL ACTION	RATIONALE/AMPLIFICATION
1. Measure tube by holding against foal from mouth to 12th rib. Mark tube using tape.	1a. Remember to allow for the curvature of the neck. 1b. Marking the tube provides an easy method of determining when to stop advancing it.
2. Warm tube in water.	—
3. Lubricate tube end with KY jelly.	3a. Lubrication minimizes esophageal trauma.

TECHNICAL ACTION	RATIONALE/AMPLIFICATION
4. Standing cranial to foal, insert tube into ventral medial aspect of nostril. Place index finger dorsal to tube to assist with ventral medial placement.	4a. Tube must enter ventral nasal meatus. 4b. Foals typically will snort and resist as tube passes into nostril. This will stop once tube is inserted 4–5 inches inside the nostril. 4c. Epistaxis can occur at this point and is associated most commonly with irritation or trauma to the meatus epithelium. Bleeding should stop within 10–15 minutes. Once bleeding has ceased, begin again using the other nostril. This is not typically a serious complication of **nasogastric intubation**. 4d. Do not attempt to stop bleeding by applying pressure to nostrils.
5. Advance tube into pharyngeal area; encourage swallowing by rotating the tube slightly. Continue advancing tube into esophagus.	5a. Foals should swallow naturally to prevent tube from entering trachea.

6. Verify tube placement via:
  a. Palpating cranial third of neck on the left side. Operator should be able to feel the tube pass under the fingers.
  b. Check for negative pressure.
	6a. It is much easier to palpate the tube when it is in motion. Sometimes moving the tube back and forth will assist with palpation.
	6b. See Chapter 6 for discussion on assessing negative pressure.

7. Advance tube to level of tape.	—
8. Attach pump and administer fluids.	8a. Tube may be retained temporarily using tape to secure the tube to a halter. Foals should be under continual observation while tube is secured in this manner. 8b. If tube is to remain in place, the end should be plugged or capped to prevent excess air from entering foal's stomach.
9. To remove tube, clear it by blowing fluid through, kink tube, and remove it in one smooth, fluid motion.	9a. Clearing and kinking the tube prevents inadvertent aspiration during removal.

OROGASTRIC INTUBATION

Purpose

- Administer medication or nutrients

Complications

- Aspiration
- Esophageal trauma

Equipment

- Magrath fluid feeder (calf)
- Tube with 10–13 mm outside diameter, 24 French (cria)
- Red rubber feeding tube (kid)
- 60-cc catheter-tip syringe
- Towel

Procedure for Orogastric Intubation of Calves

TECHNICAL ACTION	RATIONALE/AMPLIFICATION
1. Straddle calf.	1a. Straddle neck area, enabling you to pin head between thighs if needed.
	1b. Large fractious calves can be intubated in chute using adult cow techniques.
2. Kink feeder tube at base of tube.	2a. Kinking prevents fluid from entering the tube prematurely, causing aspiration.
	2b. Fig. 7–4.
3. Grasp calf's head with left hand and gently insert tube into mouth. Advance slowly into pharynx.	3a. Tube may encounter slight resistance as it goes over the tongue.
4. Continue advancing tube gently, allowing calf to swallow.	4a. The tube almost always will pass into the esophagus, even if the calf is bawling.
5. Palpate left cranial third of neck approximately 4 inches from the mandible and confirm tube placement. Insert length of orogastric tube.	5a. It is easiest to palpate bulbous tube end when in motion. Placing palpating fingers on neck while gently moving tube up and down will assist in detection.
	5b. Never administer fluids without confirming tube placement.

TECHNICAL ACTION	RATIONALE/AMPLIFICATION
6. Invert bottle and allow fluid to enter by gravity flow.	**6a.** If the fluid does not run readily, loosen cap slightly (1–2 twists). Do not remove tube when you do this. Withdrawing tube 1–2 inches can also facilitate flow.
7. Once fluids have been administered, kink tube end again and withdraw in one fluid motion.	**7a.** Kinking helps prevent aspiration of fluids during tube removal.

FIG. 7-4 Orogastric intubation of a calf.

Procedure for Orogastric Intubation of Kids

TECHNICAL ACTION	RATIONALE/AMPLIFICATION
1. Place kid between legs or under arm.	**1a.** Position is determined by size and excitability of kid.
2. Approximate the length of tube that must be passed by holding it against the kid. Measure from mouth to 12th rib. Tape can be applied if a visual aid is needed.	**2a.** Some individuals prefer a feeding probe over a tube. A probe is basically a rigid stainless steel tube with a smooth ball on the end.
3. Lightly lubricate the tube using KY jelly or water.	**3a.** Lubrication minimizes esophageal trauma.
4. Place left arm over neck or thorax of kid and grasp ventral mandible.	**4a.** Fig. 7–5.
5. Insert tube into mouth and slowly advance until desired length is reached.	**5a.** Speculums are not used.

TECHNICAL ACTION	RATIONALE/AMPLIFICATION
	5b. Checking for negative pressure is not done.
	5c. Entering the trachea is extremely uncommon because of the kid's pharyngeal anatomy.
6. Attach syringe and administer fluids.	—
7. To remove tube, kink the end, and remove in one fluid motion.	**7a.** Kinking the tube prevents aspiration during removal.

FIG. 7-5 Orogastric intubation of a kid.

Procedure for Orogastric Intubation of Cria

TECHNICAL ACTION	RATIONALE/AMPLIFICATION
1. Measure tube to level of thoracic inlet. Straddle cria, and if recumbent straddle in kneeling position.	**1a.** Do not use tube with diameter larger than 13 mm, because it can interfere with cardiac function.
2. Lubricate end of tube with water.	—
3. Flex head 40 degrees and insert tube into mouth.	**3a.** It is important to flex the head to facilitate passage into esophagus.
	3b. Do not use a speculum.

TECHNICAL ACTION	RATIONALE/AMPLIFICATION
4. Advance gently and allow cria to swallow, then palpate left cervical area for correct placement.	4a. The tube can be palpated next to the trachea. It is easiest to palpate the tube when it is moving; thus it is helpful to move tube slightly up and down to recognize passage under fingers.
5. Advance tube to level of thoracic inlet.	5a. Do not place tube directly in the stomach. The llama stomach is composed of 3 compartments. The third compartment is termed the true stomach. Ideally the fluid passing through the distal esophagus will induce closure of the esophageal groove, causing milk to flow past the first and second compartments directly into the third.
6. Attach catheter-tip syringe and administer fluids.	6a. Administer a maximum of 250 cc.

EARLY NEONATAL CARE CHECKLIST

Delivering babies is one of the most rewarding aspects of veterinary medicine. Watching a foal attempt to stand or seeing a calf take its first breath somehow just never gets old. It is during these first few minutes to hours, however, that simple actions taken by the medical team can make the difference between life and death.

Purpose

- Assist the neonate during the immediate postpartum period

Complications

- Neonatal death
- Asphyxia
- Hypothermia
- Failure of passive antibody transfer
- Umbilical hemorrhage
- Dermatitis surrounding umbilicus

Equipment

- Towels
- Stethoscope

- Thermometer
- Umbilical tape or suture
- Povidone-iodine tincture, solution, or chlorhexidine solution (species dependent)
- Enema materials

Procedure for Immediate Postpartum Care of All Species

TECHNICAL ACTION	RATIONALE/AMPLIFICATION
1. Clear airway. Remove fluid from nose and mouth.	1a. Newborns with excessive fluid accumulation should be lifted by their hind limbs to facilitate fluid removal.
	1b. Sticking a finger up the nostril will often stimulate snorting, which helps clear fluid from the nostrils.
2. Determine pulse and respiratory rate.	2a. Doxapram (Dopram) can be administered to facilitate ventilation.
	2b. If warranted, the temperature can also be taken at this time.
3. Confirm that there is no hemorrhage from umbilicus.	3a. If bleeding is noted, clamp or tie using umbilical tape.
4. Examine mucous membranes and determine capillary refill time.	4a. Membrane color and capillary refill time improve rapidly within the first few minutes.
5. Assist mother with drying neonate, if warranted.	5a. Drying neonate with towels will help maintain temperature. This is critical during winter birthing.
	5b. If everything appears normal, the best option is to leave the mother and offspring alone. Unnecessary interventions typically cause more problems than they prevent.
6. Ensure that colostrum is consumed.	6a. Colostrum is the first milk consumed by a neonate. Rich in antibodies, this milk is very thick and yellow.
	6b. Colostrometers provide an accurate means of assessing immunoglobulin (antibody) content.

Procedure for Postpartum Care of the Foal

TECHNICAL ACTION	RATIONALE/AMPLIFICATION
1. Perform steps outlined in Procedure for Immediate Postpartum Care of All Species.	1a. Equine labor and delivery is very rapid compared with that of other domestic species. Horses usually deliver during early morning hours. Waxing of the teats, vulvar laxity, and relaxation of the sacroiliac ligaments are all signals that the pregnancy is about to end.
	1b. Average parturition times follow. Stage one: 1–4 hours, stage two: 10–20 minutes, stage three: 30 minutes to 2 hours.
	1c. Fig. 7–6.
2. The umbilical stump should be pinched off or clamped 3–4 inches from the abdominal wall if bleeding. Apply iodine tincture or solution to navel. Navel should be iodized twice a day for 48 hours.	2a. The umbilical cord usually will tear on its own. Do not tear unless there is a significant risk of contamination or tearing close to body wall. If foal is attempting to walk and the **placenta** is still attached, the assistant should intervene.

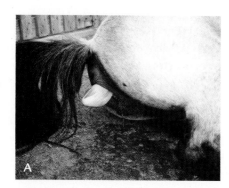  

FIG. 7-6 (A) Stage 2 labor. Appearance of the foal's feet. (B) End of stage 2 labor. Foal on ground still attached to dam by umbilical cord. (C) Stage 3 labor. Dam passing the placenta.

TECHNICAL ACTION	RATIONALE/AMPLIFICATION
	2b. Never cut the umbilical cord using scissors. Tie using umbilical tape 3 inches from abdominal wall, and tear cord distal to tape.
	2c. Use a minimum of 2% iodine. If using iodine tincture, do not allow it to contact skin or a severe dermatitis will result.
3. Administer enema. Ensure passage of meconium.	**3a.** Some clinics routinely administer enemas to all foals. Other facilities administer enemas only if difficulty passing meconium is noted.
	3b. Meconium is the first stool passed by the foal. This fecal material is very sticky, which often makes it difficult to pass. Meconium typically is passed within the first 6–10 hours.
	3c. Examine the stall for evidence of meconium. It is often the size of an average dog bowel movement and very sticky. Consistency is similar to that of toothpaste.
4. Examine placenta.	**4a.** Placenta should be examined to ensure that it was passed in entirety. The mare will become extremely ill if any portion of the placenta is retained.
	4b. Do not allow mare to consume placenta. This often results in colic.
	4c. If the mare has not passed the placenta within 1 hour of birth, place an obstetrical sleeve over exposed hanging placenta (using tape to secure top). Placenta can be tied in a knot if hanging very low. Placing several knots in the placenta that is hanging out of the vulva will reduce the chance that the mare will step on it. Do not attempt to pull the placenta out.
5. Administer 1,500 units of tetanus antitoxin to the foal, if the mare has not received a toxoid booster during last 2 months of gestation.	—

Procedure for Postpartum Care of the Calf

TECHNICAL ACTION	RATIONALE/AMPLIFICATION
1. Perform steps outlined in Procedure for Immediate Postpartum Care of all Species.	1a. Average parturition times follow. Stage one: 30 minutes to 10 hours, stage two: 30 minutes to 6 hours, stage three: 30 minutes to 8 hours. Heifers experience longer labors than cows.
	1b. Normally delivery will occur 1 (cow) to 3 (heifer) hours after the appearance of the water sac (amnion).
	1c. Dripping milk and mucoid vaginal discharge often signal that labor may begin within 12 hours.
2. The umbilical cord usually will break a few inches from the body wall. If this does not occur, tear umbilicus 3–4 inches from body wall.	2a. Fig. 7–7.
3. Apply iodine tincture to umbilical stump.	3a. Minimize tincture contact with skin. Tincture can be irritating and can cause dermatitis.
	3b. Ideally iodine is applied once a day for 3 days.
4. Calf should nurse within 1 hour of birth.	—

FIG. 7–7 (A) Stage 2 labor. Appearance of amniotic sac. (B) Stage 3 labor. Cow passing the placenta. (C) Appearance of a normal umbilicus on a newborn calf.

Procedure for Postpartum Care of the Kid

TECHNICAL ACTION	RATIONALE/AMPLIFICATION
1. Perform steps outlined in Procedure for Immediate Postpartum Care of all Species.	1a. Kids are very lively and typically attempt to stand within 15–30 minutes of birth. 1b. Kids should attempt to nurse within 1 hour of birth. 1c. Average parturition times follow. Stage one: 30 minutes to 6 hours, stage two: 30 minutes to 2 hours, stage three: 30 minutes to 8 hours.
2. If the umbilical cord does not tear naturally, apply pressure for 3–5 minutes and then tear cord 2 inches from body wall. Apply tincture of iodine.	2a. Use care when applying iodine, and minimize contact with skin. A small squirt bottle works nicely.
3. Do not administer enema.	—

Procedure for Postpartum Care of the Piglet

TECHNICAL ACTION	RATIONALE/AMPLIFICATION
1. Perform steps outlined in Procedure for Immediate Postpartum Care of all Species.	1a. Sows typically show vulvar enlargement and a clear vaginal discharge 24–48 hours before farrowing. 1b. Average parturition times follow. Stage one: 2–12 hours, stage two: 3 hours. The time from onset of active labor to delivery of the first piglet is 30 minutes to 1 hour. The average time between piglets is 15 minutes. Stage three: irregular times. 1c. Sows usually farrow in lateral recumbency. Piglets can be born head or tail first.
2. If any bleeding is noted from umbilical cord, tie using suture. Iodine is not usually applied to piglet umbilical cords.	2a. Umbilical cord bleeding is rare. 2b. Do not permit sow to consume placenta.

TECHNICAL ACTION	RATIONALE/AMPLIFICATION
3. Observe to ensure that piglets nurse within 1 hour of birth.	3a. Piglets generally feed every hour and consume 30–50 cc at each feeding.
4. Processing of piglets occurs at 24 hours of age. This includes clipping needle teeth, weighing, tail docking, ear notching, and iron administration.	4a. Technical actions for these processing procedures are provided in this chapter and in Chapter 6.

Procedure for Postpartum Care of the Cria

TECHNICAL ACTION	RATIONALE/AMPLIFICATION
1. Perform steps outlined in Procedure for Immediate Postpartum Care of all Species.	1a. Llamas do not usually nudge and lick offspring. Females will make humming sound near cria. Many llamas deliver standing up. Early morning is the most common time for delivery.
	1b. Cria should stand within 1 hour and nurse 2–3 times per hour. Nursing sessions last only 1–2 minutes. If the cria has not nursed within 5 hours, initiate supportive nutrition.
2. If umbilical cord does not tear naturally, apply pressure for 3–5 minutes and then tear cord 6 inches from body wall. Do not ligate cord. Spray cord with chlorhexidine solution.	2a. Chlorhexidine is much preferred over iodine tincture. Cria will experience severe dermatitis if skin is exposed to iodine.
	2b. Chlorhexidine solution: Mix 5 cc chlorhexidine with 15 cc water. Ideally the solution is applied 2 times per day for 2–3 days.
	2c. The placenta should pass within 45 minutes to 2 hours. Do not permit llama to consume the placenta.
3. Do not administer an enema.	3a. Meconium passes within first 18–24 hours.

PARENTERAL NUTRITION OF FOALS

Parenteral nutrition often is reserved for expensive neonates. Although the supplies and equipment are nominal, neonates receiving this therapy are very debilitated and require extensive medical intervention.

Purpose

- Provide nutritional support to debilitated foals

Complications

- Sepsis
- Phlebitis
- Electrolyte imbalances

Equipment

- Catheter supplies (Chapter 6)
- Intravenous fluid
- Isopropyl alcohol 70%
- Heparinized saline

Procedure for Administering Parenteral Nutrition to the Foal

TECHNICAL ACTION	RATIONALE/AMPLIFICATION
1. Mix intravenous solution. Ensure solution is body temperature.	1a. Solution mixture: 1000 cc of 50% glucose, 1000 cc of 8.5% amino acids, and 500 cc of 10% lipid emulsion.
	1b. Mix all solutions under strict sterile conditions. Use of a laminar flow hood is recommended.
2. Insert intravenous catheter.	2a. See Chapter 6 for directions.
	2b. A 3.5-inch catheter can be used if it will be left in for less than 3 days. Long-term therapy requires using a central venous catheter.
	2c. Monitor several times per day for evidence of phlebitis.

TECHNICAL ACTION	RATIONALE/AMPLIFICATION
	2d. Intravenous catheters placed for total parenteral nutrition (TPN) should be used only for that purpose.
3. Administer solution at a rate of 2 ml/kg/hr. Flush catheter with saline after each treatment. Wipe catheter cap with alcohol.	3a. New intravenous lines and extension sets should be hung every 8–12 hours.
	3b. Strict adherence to aseptic technique during delivery of TPN is essential.
4. Check blood glucose level every 6 hours.	4a. If blood glucose value exceeds 250 mg/dl, decrease glucose administration immediately.
	4b. Monitoring for electrolyte imbalances is also recommended. Watch potassium closely. If needed, supplement at 20–40 mEq/L of intravenous fluid.
	4c. Other electrolytes frequently monitored include calcium, phosphorus and magnesium.
	4d. TPN should be discontinued over a 12–24 hour period. Rapid removal is inadvisable.

NASAL OXYGEN

Critically debilitated neonates are often hypoxic. Administration of oxygen serves to raise blood oxygen content, thereby facilitating tissue oxygenation. The benefits of improved oxygenation are enormous and warrant the technician's time that is required to administer this therapy.

Purpose

- Alleviate hypoxia

Complications

- Epistaxis
- Nasal irritation
- Inadvertent removal of tube

Equipment

- Oxygen catheter or nasopharyngeal catheter (14 French, 40 cm)
- Oxygen tank
- Humidifier

Procedure for Administering Nasal Oxygen to the Foal

TECHNICAL ACTION	RATIONALE/AMPLIFICATION
1. Using soft flexible catheter, advance tube into nasal cavity to the level of the medial canthus of the eye.	1a. Nasal oxygen supplementation often is initiated if foal's respiratory rate is less than 30 or more than 80 breaths per minute. Other indications include labored breathing, cyanotic mucous membranes, and abnormal chest radiograph results or abnormal blood gas values. 1b. Increased abdominal contraction is often noted with laborious respiration. 1c. Do not advance catheter too far back into the pharynx. 1d. Catheter tip should be fenestrated. 1e. Face masks can be used if oxygen administration is very short term.
2. Administer oxygen at a rate of 5–7 liters per minute.	—
3. Humidify oxygen if therapy lasts longer than 30 minutes.	3a. Dry oxygen will damage delicate mucous membranes, causing significant discomfort. 3b. Sterile in-line humidification chambers are commonly used.

ENEMAS

Meconium impaction is the bane of the newborn foal. This first fecal material is very sticky, making passage exceptionally difficult. Because of the high frequency of meconium impactions, many facilities administer enemas routinely at birth. The decision to administer enemas as a routine practice or upon visualization of clinical signs should be determined by the veterinarian and farm managers.

Purpose

- Remove first fecal material (meconium) from rectum

Complications

- Rectal tears
- Failure to pass meconium
- Apparent abdominal discomfort

Equipment

- Enema solution (warmed) and delivery device
- Lubrication gel

Procedure for Administering an Enema to a Foal

TECHNICAL ACTION	RATIONALE/AMPLIFICATION
1. Ask assistant to restrain foal by placing arms around cranial aspect of shoulder and neck area.	1a. See Chapter 2 for foal restraint and cradling techniques.
2. Stand lateral to hindquarters and lift tail.	2a. Do not stand directly behind foal. This can result in injury.
3. Lubricate and insert enema tip into rectum 2–3 inches.	3a. Foals often will move forward or kick during insertion.
	3b. Enema equipment commonly includes a 60-cc syringe with a 20 French catheter attached, bulb syringes, or prepackaged enema solution.
4. Administer enema, allowing fluid to fill rectum slowly. Remove catheter.	4a. Most enemas are 120–180 cc in volume. If the anus bulges and fluid no longer flows by force of gravity alone, then sufficient volume probably has been administered.
	4b. Enema solution should be warmed to body temperature.
	4c. Common enema solutions include warm water, docusate sodium (DSS) and water mixture, or a mild soap and water mixture.

TECHNICAL ACTION	RATIONALE/AMPLIFICATION
5. Meconium usually is passed within 15–20 minutes.	5a. Meconium is often dark and dry or sticky. As foal nurses, the fecal material will become yellow and pasty.
6. Monitor foal for signs of abdominal discomfort.	6a. Abdominal discomfort in the foal is manifested as tail wagging, kicking at abdomen, looking at flanks, or repeatedly getting up and lying down.

Special Note

More severe cases of meconium impaction have been resolved using an enema solution containing acetylcysteine (Mucomyst). This mixture is composed of 1.5 tablespoons of baking soda, 200 cc of water, and 8 grams of acetylcysteine. Use a 30 French Foley catheter with balloon end. Insert catheter 1–2 inches, inflate balloon, and administer fluid. Keep catheter in position for 15 minutes, and then deflate the balloon and remove the catheter.

ROUTINE NEONATAL PROCESSING PROCEDURES

Routine processing procedures are designed to improve animal gains, marketability, and quality of life. These procedures address husbandry issues and reflect industry needs for efficiency. The timing of many procedures is age and species dependent. As a general rule, processing animals at an earlier age minimizes the procedural stress and incidence of complications.

CLIPPING NEEDLE TEETH

Purpose

- Remove piglet needle teeth to prevent injury to litter mates and to the sow's teats

Complications

- Tongue laceration
- Stomatitis

Equipment

- Tooth clippers or cutting pliers

Procedure for Clipping Needle Teeth

TECHNICAL ACTION	RATIONALE/AMPLIFICATION
1. Contain sow in another pen.	1a. Sows are very dangerous when they are concerned for the welfare of their piglets.
	1b. See Chapter 6 for ear notching procedure.
2. Hold piglet's head and press the corner of mouth to open. Tilt head so that clipped teeth will fall out of the mouth.	2a. Clipping needle teeth can be done as soon as 15 minutes after birth.
	2b. This procedure usually is performed between 1 and 7 days of age.
	2c. See Chapter 6 for ear notching procedure.
3. Clip tooth as close as possible to the gum line, being careful not to injure the tongue.	3a. Clip upper and lower teeth.
	3b. Fig. 7–8.
4. Return piglet to sow or continue other processing measures.	—
5. Clean equipment before processing the next piglet.	5a. Pliers can be placed in a chlorhexidine solution.

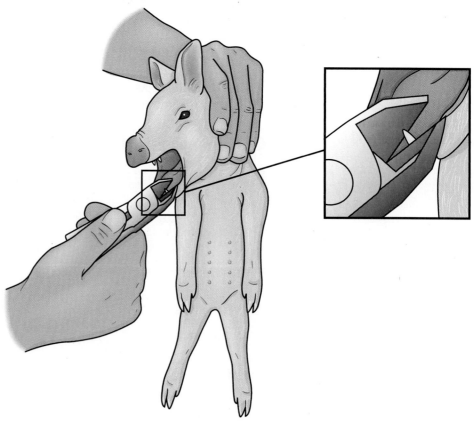

FIG. 7–8 Diagram of clipping needle teeth.

DISBUDDING KIDS

Purpose

- Remove goat horns during neonatal period

Complications

- Cauterization of inappropriate areas
- Spur horn growth

Equipment

- Cautery iron
- Chair

Procedure for Disbudding Kids

TECHNICAL ACTION	RATIONALE/AMPLIFICATION
1. Sit on chair or bucket and place kid between legs. The kid and the technician should be facing the same direction. The kid's head should be secured between the technician's thighs.	1a. Kids should be disbudded between 3 and 10 days of age using the dehorning hot iron. Older animals require a Barnes dehorner or knife to remove the horn tissue, in addition to application of the hot iron. 1b. Older animals require different restraint techniques. 1c. Fig. 7–9.
2. Palpate and localize horn bud.	2a. Horn buds are easily palpable on the top of the head.

FIG. 7–9 (A) Location of horn buds on a young kid. (B) Positioning of kid for disbudding.

TECHNICAL ACTION	RATIONALE/AMPLIFICATION
3. Place iron directly over horn. Maintain firm pressure and apply a slight rocking motion for 15–30 seconds.	3a. The rocking motion ensures that all horn base areas are contacted. 3b. The kid will vocalize loudly during this procedure. Do not attempt to hold mouth closed. 3c. Holding the ear opposite to the side you are dehorning can facilitate restraint of the head.
4. Ensure that entire area is leather brown in appearance.	—

LAMB TAIL DOCKING

Purpose

- Remove tail to facilitate sanitation and improve marketability of lambs

Complications

- Infection
- Fly blow
- Tetanus
- Rectal prolapse

Equipment

- Equipment will vary with technique selected
- Elastrator with bands
- Hot docking iron
- Emasculator
- Fly repellent

Procedure for Docking Lambs' Tails

TECHNICAL ACTION	RATIONALE/AMPLIFICATION
1. Assistant should restrain lamb with hind end facing technician.	1a. An experienced technician can restrain and dock the lamb simultaneously. This is performed while seated, with lamb grasped between the technician's legs.

TECHNICAL ACTION	RATIONALE/AMPLIFICATION
2. Several techniques for docking are used. All techniques involve removing the tail 1.5 inches from the tail base. Select one of the following methods.	2a. Tail docking should occur before the kid reaches 2 weeks of age (2–3 days of age is preferable).
	2b. The caudal tail folds on the ventral aspect of the tail can serve as a landmark. The tail is removed where the tail folds end. This corresponds to 1.5 inches from the base or the third joint.
	2c. Purebred show or 4-H sheep tails frequently are docked shorter, at the level of the first joint.
	2d. Lambs with excessively short tail docks experience increased incidences of rectal prolapse as adults.
3. Elastrator method: Use elastrator gun and place band.	3a. Elastrator bands eliminate the blood supply distal to the band. The tail will shrivel and drop off within 10–14 days.
	3b. Advantages of this method include no blood loss, minimal complications, rapid to perform, and ease of band application.
	3c. Disadvantages include increased incidence of fly blow and tetanus.
	3d. All lambs docked in this manner should receive tetanus vaccination.
4. Hot docking iron method: Apply hot iron to tail until the tail separates from the lamb.	4a. The lamb's tail can be inserted through a hole drilled in a wooden board. This prevents accidental burning of inappropriate areas and facilitates uniform docking length.
	4b. Heat iron to a cherry red color. Irons can be electrically (best) or propane heated. The iron will cut and cauterize simultaneously.
	4c. Post-docking infection and bleeding rates are higher using this method. The wound will also take longer to heal.

TECHNICAL ACTION	RATIONALE/AMPLIFICATION
5. Emasculator method: Ensure that crushing edge (not the cutting edge) of the emasculator is nearest to the tail base. Use emasculator to crush, then cut, tail.	5a. The emasculator simultaneously crushes the vascular supply and cuts off the distal tail. If the emasculator is applied backward, the tail will be cut off without crushing the blood supply. 5b. This method is the least desirable and is associated with the most complications.
6. Apply fly repellent after procedure for all methods.	6a. Repellent decreases the likelihood of fly strike.

REVIEW QUESTIONS

1. State the preferred bottle feeding frequency for a newborn foal.
2. Identify the function of colostrum.
3. List the solutions that can be used for foal enemas.
4. Describe the complications associated with nasal oxygen administration.
5. List complications associated with nasogastric intubation of foals.
6. Describe the techniques used to verify esophageal placement of an orogastric tube.
7. List parameters and identify procedures that should be performed on every neonate immediately after birth.
8. Name the solution that should be used to care for the cria's umbilical stump.
9. Compare and contrast the three tail docking techniques used on lambs.
10. Identify the purposes for clipping piglet needle teeth.

BIBLIOGRAPHY

Fowler, M. (1998). *Medicine and surgery of South American camelids* (2nd ed.). Ames: Iowa State University Press.

Haskell, S. *Neonatology of camelids and small ruminants.* College of Veterinary Medicine, University of Minnesota.

Jones, T. (2002). *Complete foaling manual.* Stallside Books.

Madigan, J. (1991). *Manual of equine neonatal medicine* (2nd ed.). Woodland, CA: Live Oak Publishing.

McCurnin, D. (1998). *Clinical textbook for veterinary technicians* (4th ed.). Philadelphia: W. B. Saunders.

Smith, B. (1990). *Large animal internal medicine.* St. Louis: C. V. Mosby.

SECTION FOUR

Surgical, Radiographic, and Anesthetic Preparation

Chapter 8
Surgical Preparation

Chapter 9
Selected Lower Limb Radiograph Procedures

Chapter 10
Anesthesia

CHAPTER 8

Surgical Preparation

Surgeons must be very careful

When they take the knife

Underneath their fine incisions

Stirs the Culprit—Life!

EMILY DICKINSON 1859
NO. 108

KEY TERMS

antiseptic
aseptic technique
autoclave
castration
enucleation
laparotomy
obstetrical
ovariectomy
parturition
prolapse
sepsis
sterile

OBJECTIVES

- Identify the roles a veterinary technician may have in performing routine surgical procedures on the horse, cow, llama, goat, and pig.
- Describe general standards for application and maintenance of **aseptic technique** in surgery.
- Identify suturing materials and patterns commonly employed in large animal surgery.
- Identify routine surgical procedures for the horse, cow, llama, goat, and pig; how to prepare for them and common complications that may occur.

SURGERY TEAM

The surgery team consists of the surgeon, the anesthetist, one or more sterile assistants (scrub nurses), and one or more non-sterile assistants (circulators). Because the surgeon must be a veterinarian, that role will not be discussed in this book. The duties of the anesthetist are covered in Chapter 10. All persons within the surgery suite should wear caps, masks, and scrub clothing at a minimum.

SCRUB NURSE

Duties

- Assist the surgeon in performing surgery in a sterile manner
- Drape the patient using aseptic technique
- Pass instruments to the surgeon
- Maintain the sterile field
- Retract or hold organs for the surgeon
- Assist in suturing or ligating

Complications

- Iatrogenic infection from breach of aseptic technique
- Injury to patient as a result of failure to follow surgeon's instructions

Equipment

- Supplies to scrub and gown up

CIRCULATOR

Duties

- Prepare the patient and surgery site
- Set out and open sterile instrument and suture packs
- Gather and provide sterile supplies required by the surgeon
- Help to maintain surgical asepsis

Complications

- Contamination of sterile field or items
- Injury to patient because of failure to follow surgeon's orders

Equipment

- Scrub clothes
- Surgical cap and mask

PERSONAL PREPARATION FOR SURGERY

Although many large animal surgical procedures take place outside the surgery suite, aseptic personal preparation is still appropriate for a number of surgeries performed in the field or those done in a chute or stock. Use of cap, mask, sterile gown, and sterile gloves protects both the patient and the operator from contamination or sepsis.

Purpose

- Maintain asepsis within the surgical suite
- Remove contaminants from the surgeon or surgical assistant's hands and forearms
- Cover the face and nose with a filter to decrease airborne contaminants emanating from the surgical personnel
- Cover the head of surgical personnel to protect the patient from hair and debris shed by the surgical personnel
- Cover the surgeon's and surgical assistant's bodies in sterile gowns to protect the patient from contaminants on the clothing
- Cover the surgeon's and surgical assistant's hands in sterile gloves to protect the patient from contaminants and to protect the operators from exposure to pathogens carried by the patient

Equipment

- Mask (cloth or disposable)
- Surgery cap (cloth or disposable)
- Shoe covers
- Sterile gown pack
- Sterile surgical gloves
- Surgical scrub soap (povidone-iodine or chlorhexidine)

Complications

- Allergic reaction to latex-containing gloves
- Allergic reaction to surgical scrub soap
- Inadequate scrubbing of hands and arms
- Improper technique in gowning or gloving resulting in contamination of sterile field

Procedure for Personal Preparation for Surgery

TECHNICAL ACTION	RATIONALE/AMPLIFICATION
1. Remove jewelry from hands and neck.	1a. They may harbor bacteria and other contaminants.
2. Apply cap and mask, making all adjustments to mask before scrubbing hands, because you cannot touch mask afterward.	2a. Put cap on first, and then place mask over nose and mouth with the wire on the bridge of the nose.
	2b. Squeeze wire to fit snugly over bridge of nose.
	2c. Tie top ties in a bow behind head.
	2d. Pull lower portion of mask down to chin and tie lower ties behind neck.
3. Open sterile gown and glove packs on a clean, dry surface near the wash area.	3a. Place them in an area where they won't be touched accidentally once they are open.
	3b. Be certain not to touch anything within the packs until you have performed the surgical scrub.
4. Bring surgical scrub sponge to sink and turn on water.	4a. Surgical scrub is a soap that usually is made from povidone-iodine or chlorhexidine. Disposable scrub brush-sponges are available already impregnated with the soap. Reusable brushes may be autoclaved.
5. Thoroughly wash hands starting from fingertips, going to wrists.	—
6. Perform surgical scrub (5–7 minutes) using the single brush technique. Opinions vary on how long a surgical scrub should take; the important thing is that all surfaces of your fingers, hands, wrists, and forearms be scrubbed thoroughly.	6a. Clean under fingernails with pick.
	6b. Keep hands elevated above elbows at all times to prevent contaminated water from running back down to your clean hands.
	6c. Scrub each finger, one at a time, from tip to palm, all four sides at least 10 times (Fig. 8–1).
	6d. Scrub back of hand at least 10 times.
	6e. Scrub palm at least 10 times.
	6f. Scrub all four sides of forearm to elbow at least 10 times.
	6g. Rinse from fingertips to elbow, drain to elbow.
	6h. Repeat procedure on other hand.

TECHNICAL ACTION	RATIONALE/AMPLIFICATION
	6i. Keeping hands clasped together in front to your chest, walk over to gown and gloves.
7. Dry hands thoroughly with the sterile towel that is wrapped in the package with the gown, starting at the fingertips and working toward the elbow.	**7a.** Never go back to hands once the towel has touched the arm.
	7b. Let the towel drop to the floor or counter so that you do not risk touching anything else.
8. Put on the gown using aseptic technique.	**8a.** Grasp gown with neck facing your neck, inside surface toward your chest.
	8b. Without touching the outside surfaces of the gown, place hands in the arm holes of the gown and pull it on.
	8c. Have an assistant tie the neck and inner gown ties without touching the outside surface of the gown.
	8d. Tie the outer tie after gloving up.
9. Put on sterile gloves using aseptic technique. What follows is the open gloving technique.	**9a.** Pick up one glove by the turned-down cuff and pull it onto the hand with the cuff left turned down.
	9b. Right-handed people usually begin with the right glove.

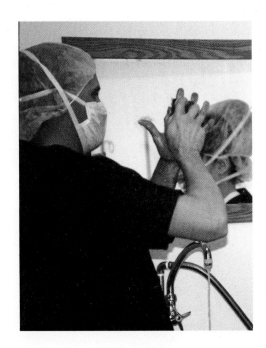

FIG. 8-1 Scrubbing for surgery. Note fingers held above the level of the elbow.

TECHNICAL ACTION	RATIONALE/AMPLIFICATION
	9c. Using the gloved hand, pick up the remaining glove by inserting the fingers into the cuff and pulling it onto the opposite hand.
	9d. Lift the glove cuff over and onto the gown cuff and repeat the process on the other hand.
	9e. Once both gloves are on, smooth any wrinkles from the fingers.
10. Keep hands clasped in front of chest until reaching surgical patient.	**10a.** This will keep you from inadvertently touching something that is not sterile.
	10b. Remove any powder on the outer glove surface by wiping the gloved hands with a damp, sterile gauze before handling any patient tissues.

INSTRUMENT PACK PREPARATION AND STERILIZATION

What instruments are placed in a surgery pack depends on the preferences of the surgeon. Many clinics have several different surgery packs kept sterile and at the ready in order to shorten the time necessary to set up for a surgery. Instruments may be wrapped in muslin or placed in paper and plastic envelopes for sterilization and storage, again depending on the instrument and the surgeon.

Purpose

- Ready source of sterile instruments and gauze sponges
- Contains instruments suitable for most sterile, soft tissue surgeries
- Some packs specialized for orthopedic, ophthalmic, or colic surgeries

Equipment

- 4-8 towel clamps
- 4 curved mosquito hemostats
- 4 straight mosquito hemostats
- 4 curved Kelly hemostats
- 2 straight Kelly hemostats
- 1 Allis tissue forceps
- 1 curved Mayo scissors

- 1 straight Mayo scissors
- 1 stainless steel operating scissors
- 1 curved Metzenbaum scissors
- 1 straight Metzenbaum scissors
- 1 straight Ochsner forceps
- 1 curved Ochsner forceps
- 1 #3 scalpel handle
- 1 #4 scalpel handle
- 1 1 × 2 rat tooth thumb forceps
- 1 Adson or Adson-Brown tissue forceps
- 1 sponge forceps
- 1 grooved director
- 1 stainless steel bowl
- 4 Huck towels
- 20 4 × 4-inch gauze sponges
- Double thickness muslin pack wrap 40 × 40 inches
- Indicator strip
- Indicator tape
- Masking tape

Complications

- Failure to clean instruments completely
- Failure to wrap tightly
- Missing or damaged instruments

Procedure for Surgery Pack Preparation

TECHNICAL ACTION	RATIONALE/AMPLIFICATION
1. Lay pack wrap on table.	1a. Arrange instruments on the towel, like instruments with like, handles closed but not locked (Fig. 8–2A).
2. Arrange the scalpel handles.	2a. Place the handles at the bottom of and perpendicular to the jaws of the instruments.
3. Place indicator strip on top of instruments.	3a. The indicator strip should be positioned in the middle of the pack to ensure complete sterilization of the contents.

TECHNICAL ACTION	RATIONALE/AMPLIFICATION
4. Place gauze sponges in two piles on top of the instruments.	4a. Use 10–20 gauze sponges per pile.
5. Place folded Huck towels on top of the gauze sponges.	5a. Place one towel on top of the other and place the two piles side by side (Fig. 8–2B).
6. Tightly, one corner at a time, wrap up the instruments.	6a. Fold each corner back so that it points in the direction to pull (Fig. 8–3A).
	6b. Last corner is tucked into the folds made by the first three corners (Fig. 8–3B).
7. Secure with masking tape and apply a short strip of autoclave tape on the pack.	—
8. Date and initial the pack using an indelible marker on the masking tape (Fig. 8–4).	—

FIG. 8-2 (A) Instruments arranged on surgical pack wrap. (B) Surgical towels placed on top of instruments and gauze. Note corners folded down.

 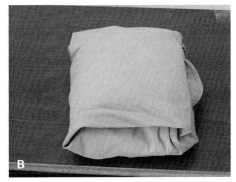

FIG. 8-3 (A) First two corners of pack folded. *Numbers* depict order in which to fold the corners. (B) Pack with final corner tucked into place.

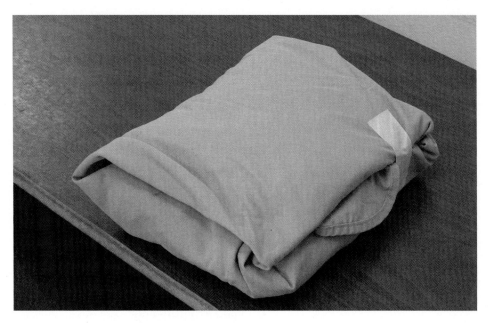

FIG. 8-4 Surgery pack with autoclave tape in place.

STERILIZING INSTRUMENTS

Purpose

- To completely rid an item of all living microorganisms and their spores for use in aseptic procedures (Table 8–1)
- In the case of cold-sterilization, the goal is to disinfect items to decrease the risk of transferring disease or causing infection

Equipment

- Autoclave
- Ethylene oxide and chamber
- Cold-sterilization solutions and water
- Bucket or cold-sterilization tray

Complications

- Inadequate sterilization causing infection, sepsis, or death of the patient
- Exposure to toxic fumes (ethylene oxide)
- Burns, chemical or heat (from steam)

Procedure for Sterilizing Instruments

TABLE 8-1 STERILIZATION METHODS

Steam	Most common means of sterilization in general practice
	Done in a pressure chamber called an autoclave (Fig. 8–5)
	Standard item exposure is 121.5°C at 15 psi for 15 minutes
	Only suitable for items that are heat resistant
	Indicator strip that is sensitive to heat used inside packs, and steam-sensitive tape used on the outside
	Sterilization envelopes generally have indicator tabs for steam sterilization and chemical sterilization
	Periodic confirmation of sterilization obtained by using a strip impregnated with bacterial spores, which is autoclaved in the middle of an instrument pack and then cultured to see if growth occurs
Ethylene Oxide	Also known as gas sterilization
	Requires special ventilation chamber
	Special indicator strips must be used to verify exposure
	Ideal for items damaged by steam sterilization such as electric drills, plastic sleeves, or items containing rubber
	Not commonly found in general large or mixed animal practices
Cold-sterilization Solutions	Often are disinfectants rather than sterilants, with variable efficacy against viruses and spores
	Variety of solutions includes chlorhexidine, povidone-iodine, 70% isopropyl alcohol, chlorine and other iodophors, glutaraldehyde
	Must be mixed and used according to manufacturer's directions
	Commonly used and mixed in a stainless steel bucket for routine field surgeries such as dehorning and castrating
	May be stored in a special sealed glass container with a tray for sterilizing smaller surgical instruments (Fig. 8–6)

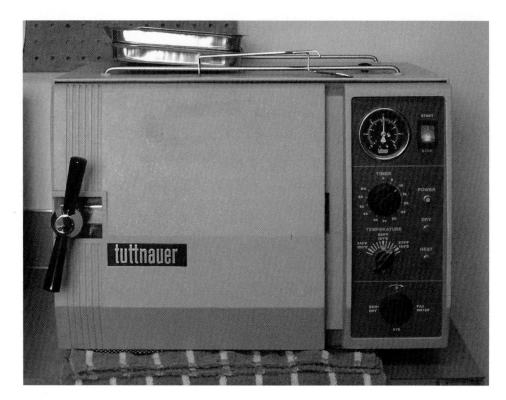

FIG. 8-5 An autoclave.

FIG. 8-6 Cold-sterilization tray filled with cold-sterile solution.

PACKING AND OPERATING AN AUTOCLAVE

Purpose

- Expose an item to enough heat and pressure to render it sterile

Equipment

- Autoclave
- Distilled water
- Perforated metal trays
- Heat-resistant mitts

Complications

- Injury to operator from steam secondary to not closing and locking the door correctly, or opening the door before the steam has vented, or removing items from the chamber with a bare hand while they are still hot
- Burning or singeing the muslin wraps secondary to failing to put water in the autoclave or allowing the wraps to come in contact with the chamber walls
- Damage to instruments secondary to placing heat-susceptible items, such as rubber, plastic, or electrical items, in the autoclave
- Inadequate sterilization secondary to autoclave malfunction such as plugged chamber air vents, heat sensor malfunction
- Inadequate sterilization secondary to overfilling the chamber with items to be sterilized

Procedure for Packing and Operating an Autoclave

TECHNICAL ACTION	RATIONALE/AMPLIFICATION
1. Read manufacturer's operating instructions.	1a. Every autoclave is different. It is important to read and follow manufacturer's instructions.
2. Fill reservoir with distilled water.	2a. This is especially important in areas with hard water. Hard water will leave mineral deposits that plug tubing and vent holes.
3. Place perforated tray(s) in autoclave chamber.	3a. Number of trays depends on items to be sterilized and the size of the autoclave chamber.
4. Place items to be sterilized on the perforated tray(s).	4a. Make sure bulky items do not come in contact with the chamber wall or the tray above,

TECHNICAL ACTION	RATIONALE/AMPLIFICATION
	because they may scorch or burn (Fig. 8–7).
	4b. If doing many flat items, make sure to leave some of the holes in the tray exposed so that steam can circulate evenly.
	4c. It is critical to load the chamber in a way that steam can circulate throughout the chamber and completely around the packages.
5. Fill chamber with water according to manufacturer's instructions.	**5a.** Turn the dial to fill and watch the water flow until it reaches the marker, then turn the dial to "sterilize" to stop the water flow.
	5b. Set the dial according to the manufacturer's instructions.
6. Close autoclave door.	**6a.** Make sure the door is tightly sealed or latched.
	6b. Most autoclaves have a safety feature that prevents the machine from operating until the door is sealed properly.

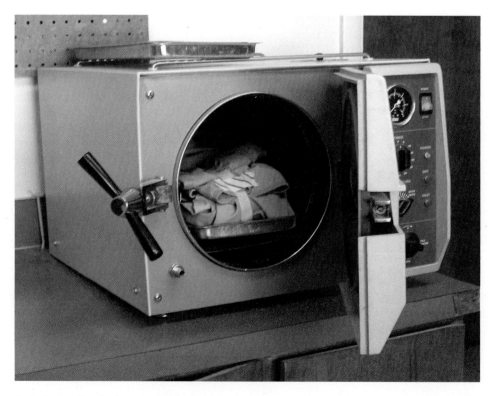

FIG. 8–7 Packed autoclave.

TECHNICAL ACTION	RATIONALE/AMPLIFICATION
7. Set time and temperature.	**7a.** Minimum time for a complete cycle is 30 minutes. This includes time to reach 121°C and 15 psi, the 15 minutes required at 121°C, and the time to return to atmospheric pressure.
	7b. Temperature should be set at 121°C.
8. Start cycle.	**8a.** See manufacturer's operating instructions on how to start.
	8b. You should hear the click of an automatic door lock once the cycle starts.
	8c. Make sure there is nothing occluding the steam vent at the top of the autoclave.
9. Cycle ends.	**9a.** When the cycle is finished, the autoclave will begin to vent steam from the chamber. Some autoclaves have a vent setting to speed this process along.
	9b. When chamber pressure has reached atmospheric pressure, the automatic lock will disengage and allow the door to be opened.
10. Open the autoclave door slowly.	**10a.** There will still be some steam in the chamber, so stand to the side of the machine as you slowly open the door. Failure to do so could result in a steam burn to the face, hands, or arms.
	10b. If using drying function on the autoclave, leave the door cracked open until drying is complete.
11. Remove items from chamber.	**11a.** Always wear heat-resistant mitts.
	11b. Remove entire tray rather than individual items to decrease risk of injury secondary to contact with the sides of the chamber.
	11c. Place items on a table or countertop to cool before use. Once cooled, the items may be used or stored.

PREPARATION OF THE PATIENT FOR SURGERY

For most surgeries, the patient needs to be cleaned to the extent that one can provide a clean, aseptic surgical field. This may involve clipping hair away and scrubbing with **antiseptic** soaps. The extent to which the patient is prepared can depend on what surgery is going to be done, where it is going to be done (in the field versus in a hospital), and how it is going to be done. If time is of the essence (emergency cesarean section for example), the preparation will be shorter and less involved than if it were a planned elective surgery.

THREE-STEP SURGICAL SCRUB

Purpose

- To rid the skin of contaminating and commensal bacteria
- To prepare the skin for an invasive procedure

Equipment

- Surgical scrub soap (povidone-iodine, chlorhexidine)
- Isopropyl alcohol 70%
- Surgical preparation solution (povidone-iodine, chlorhexidine)
- Gauze sponges 4 × 4 inches, pieces of loose cotton, surgical preparation scrub brush (commercially available), or soft sterilized scrub brush
- Bucket of warm water
- Vacuum cleaner (optional)

Complications

- Infection due to inadequate scrubbing
- Skin irritation or dermatitis secondary to products used or excessive force

Procedure for Three-step Surgical Scrub

TECHNICAL ACTION	RATIONALE/AMPLIFICATION
1. Hair removal	1a. Long hair or wool should first be sheared away using a #7 or #10 blade.
	1b. Using a #40 blade and going against the lay of the hair, clip the hair in a neat rectangle at least 2 inches equidistant from the center of the proposed incision site or needle puncture.

TECHNICAL ACTION	RATIONALE/AMPLIFICATION
	1c. Remove loose hairs by vacuuming or rinsing with water.
2. Scrub technique (Fig. 8–8)	**2a.** Wet the skin with water.
	2b. Using gauze sponges, loose cotton, or scrub brush and surgical scrub soap, vigorously scrub the surgical site starting at the proposed incision line and working your way out in concentric circles to the edges of the shaved area.
	2c. *Never* bring your dirty sponge back into the center.
	2d. Rinse soap away, using a gauze sponge or loose cotton soaked in 70% isopropyl alcohol, starting from the center and working your way out. In large preparation areas, this may require more than one sponge or piece of cotton.
	2e. Alternatively, the soap may be rinsed away with copious amounts of warm water containing surgical preparation solution poured over the site.
	2f. Repeat the above two steps 2 more times.
	2g. Spray or wipe surgical preparation solution onto the entire shaved and scrubbed area using the same concentric circles described in Step 2b.

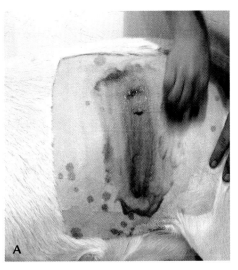 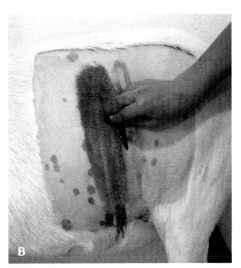

FIG. 8–8 (A) Applying surgical scrub to the flank of a goat. (B) Applying surgical solution to the flank of a goat.

DRAPING THE PATIENT USING ASEPTIC TECHNIQUE

Purpose

- Provide a sterile field in which to operate
- Prevent contaminants on the patient from entering the surgical field
- Act as a barrier to liquids emanating from the patient

Equipment

- Sterile drape(s)

    Fenestrated, muslin (most common, because they are reusable)

    Self-adhesive plastic (more common in equine surgery)

    Paper (disposable but easily torn)

- Sterile Huck towels—optional
- Sterile towel clamps

Complications

- Fenestration too small for proposed incision
- Drapes not properly sterilized
- Drapes tear or fall off, exposing surgical field to contamination
- Patient objects to sight of or securing of drape?

Procedure for Draping the Patient

TECHNICAL ACTION	RATIONALE/AMPLIFICATION
1. Four-towel technique (Fig. 8–9)	1a. Open a sterile four-towel pack on a Mayo stand or other flat surface near the patient.
	1b. Wearing sterile surgical gloves, gown, mask, and cap, pick up one towel.
	1c. Keeping the towel away from you, the patient, and the pack, hold the towel out in the air and allow it to unfold.
	1d. Grasp one corner of a long end in each hand and fold down the long edge away from you approximately 6 centimeters (3 inches).

TECHNICAL ACTION	RATIONALE/AMPLIFICATION
	1e. Carefully lay the towel on the patient parallel to and 2–4 cm (1–2 inches) from the proposed line of incision.
	1f. Following Steps 1b through 1d, apply the second towel perpendicular to the first and 2–4 cm (1–2 inches) from the top of the proposed line of incision.
	1g. Repeat the above steps until all four towels are on the patient, outlining the proposed line of incision.
	1h. Secure the towels by placing a towel clamp on each corner, making sure the clamp goes through both towels and into the patient's skin.
2. Fenestrated drape	**2a.** Some surgeons use a fenestrated drape alone, while others will place the drape over towels described in Steps 1a through 1h.
	2b. Open the sterile drape pack on a Mayo stand or clean, dry surface near the patient.
	2c. Wearing sterile gloves, gown, cap, and mask, carefully pick up the folded drape and carry it to the patient.
	2d. Lay the drape (still folded) on the patient with the fenestration overlying the proposed line of incision.

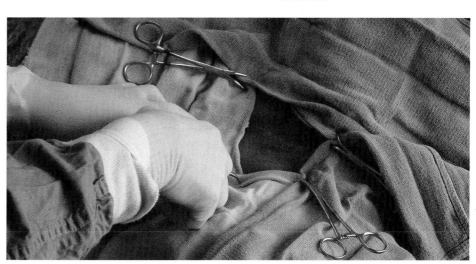

FIG. 8-9 Four-towel technique of draping, applying the final towel clamp.

TECHNICAL ACTION	RATIONALE/AMPLIFICATION
	2e. Grasp the corners of the drape and pull them away from the center to spread it over the patient.
	2f. Secure in position with nonpenetrating towel clamps.
3. Adhesive drape	**3a.** Requires smooth skin to contact, so the patient may need to be shaved with a safety razor before the three-step surgical scrub.
	3b. Using the same aseptic technique described previously, remove the drape from the package and place it on the patient overlying the proposed line of incision.
	3c. Remove the protective backing and tightly adhere the clear plastic to the patient's skin, starting at the proposed line of incision and working outward until the field is covered.
	3d. A larger cloth drape may then be applied to cover the rest of the patient if desired.

CUSHIONING THE RECUMBENT PATIENT

Purpose

- Prevent injury secondary to reduced blood flow to dependent body parts
- Prevent peripheral neuropathy, myositis, or ischemic necrosis by maintaining adequate blood flow to all parts of the body
- Maintain patient comfort

Equipment

- Varies with species, location of surgery, and position of patient
- Straw
- Tarp
- Heavy foam pads
- Blankets or quilting
- Towels or similar-sized quilting

Complications

- Inadequate padding causing pressure necrosis, myositis, or peripheral neuropathy
- Excessive padding occluding respirations
- Hyperthermia secondary to excessive coverage of the patient and heat retention
- Patient injury secondary to getting tangled in the padding
- Dermatitis secondary to fluids pooling between the patient and the padding

Procedure for Cushioning the Recumbent Patient

TECHNICAL ACTION	RATIONALE/AMPLIFICATION
1. Spread a bed of straw on the ground where the surgery will occur.	1a. Use a minimum of one straw bale per 250 kg of patient. 1b. Fluff the straw up with a pitch fork, making a layer of straw at least 30 cm (12 inches) deep.
2. Cover the straw with a large tarp.	2a. Tuck the ends of the tarp up under the straw all around the edges of your surgery bed to keep it from spreading out. 2b. Once the patient has been laid upon the cushion, push the straw around under the tarp to help position the patient.
3. Wrap flakes (sections) of straw in a blanket or tarp to make cushions to go between the legs or under the head.	3a. Patients 250 kg or larger that are placed in lateral recumbency need cushions between their legs to hold the upper leg parallel to the ground so as not to occlude major arterial blood flow. 3b. Padding under the head will decrease the chance of nerve damage, especially important in horses.
4. Wrap a strip of towel or quilting around the legs to go between the rope and the leg.	4a. This will prevent rope burns should the patient struggle against the positioning ropes. 4b. When you remove the rope, the padding should fall off on its own.

SUTURES, NEEDLES, AND SUTURING TECHNIQUES

Sutures are made of a wide variety of materials and come in many sizes (Table 8–2). Size refers to the diameter of the material and is expressed numerically. Increasing whole numbers greater than zero denote larger sizes; therefore size 4 suture is larger (thicker) than size 2 suture. Numbers less than zero are expressed

TABLE 8–2 SUTURE TYPES

SUTURE TYPE	GENERIC NAME	BRAND NAME	QUALITIES
1. Absorbable, monofilament	Polydioxanone (PDS)	PDS II, PDS III	Low tissue reactivity, low memory, long-lasting tensile strength. Use within body and skin closure. Very popular.
	Poliglecaprone 25	Monocryl	Similar to catgut but better knot security, longer lasting. Popular in small animal surgery.
2. Absorbable, pseudomonofilament	Polymerized caprolactam	Vetafil	Tissue reactive compared to nylon, knot slippage may occur. Used in skin closure. Very common in large animal surgery.
	Catgut	Braunamid, Supramid	Available as plain or chromed. Plain gut very reactive and rapidly absorbed. Most use chromed gut. Long memory, tendency to swell and come untied. Loses tensile strength within 5–7 days.
3. Absorbable, braided	Polyglycolic acid	Dexon	Both are soft, low memory. Tendency to drag through tissues and wick moisture. Maintain 67% tensile strength for 7 days.
	Polyglactin 910	Vicryl	
4. Nonabsorbable, monofilament	Nylon	Dermalon, Ethilon	Inert, slides easily through tissues.
	Polypropylene	Prolene	Used in skin primarily. Similar to nylon.
	Stainless steel	—	Inert, very stiff. May be used in skin or within body. Nonreactive.
5. Nonabsorbable, braided	Silk	—	Natural product. Rarely used in large animal surgery.
	Stainless steel		Rarely used.
6. Ligature staples	Stainless steel	Hemoclips	Inert, rapid application, requires special tool to apply.
	Titanium		
7. Skin staples	Stainless steel	Proximate	Inert, rapid application, requires special tools to apply and remove.
8. Tissue glue	Cyanoacrylates	Vet Bond, Nexibond	Rapid setting, no tools needed.

as numbers of zeros, in other words size 000 is called 3/0 or three ott. The more zeros there are, the smaller is the suture diameter. The size and type of suture selected is a function of surgeon preference and experience.

Needles used for suturing also come in a variety of shapes, sizes, and tip shape (Fig. 8-10). Needles that come already attached to the suture material are called swaged-on needles and they do not have an eye through which to thread the suture material. Many large animal veterinarians use needles with eyes and thread their own suture material onto them. In general, cutting point needles are used in the skin because they have sharp edges to slice through the tough skin, whereas taper-point needles are used in the softer tissues that lie beneath the skin.

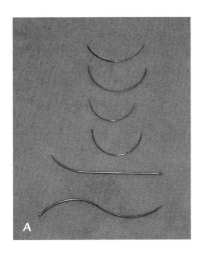
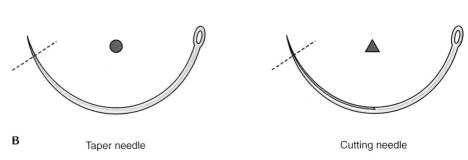

FIG. 8-10 (A) Variety of needles used in large animal surgery. (B) Close-up of the points on a suture needle. Taper point (**left**) and cutting point (**right**).

TYING AND CUTTING A SUTURE KNOT

Purpose

- To secure suture material in place
- Usually at least four throws needed to keep the knot from accidentally coming untied

Equipment

- Suture material (Fig. 8-11)
- Needle
- Needle holder (Fig. 8-12)
- Thumb tissue forceps
- Suture scissors

Complications

- Knot failure
- Suture tied too tight causing tissue necrosis
- Suture breakage
- Suture tearing through skin

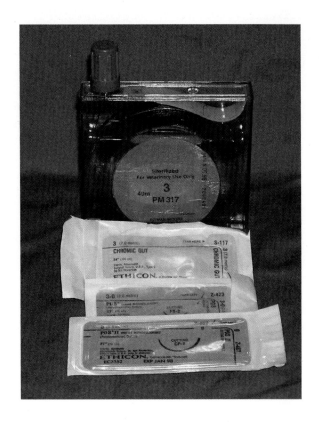

FIG. 8-11 Suture materials in packets and in a cassette.

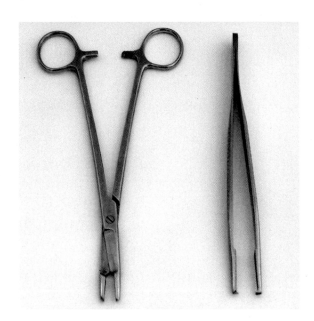

FIG. 8-12 Needle holders and thumb forceps for suturing large animals.

Procedure for Tying and Cutting a Suture Knot

TECHNICAL ACTION	RATIONALE/AMPLIFICATION
1. Grasp long end of suture and needle in the left hand.	1a. This technique is for right-handed operators (Fig. 8–13).
2. Hold needle holder in right hand over the laceration.	—
3. Pass a loop of suture material over and under the jaws of the needle holder (Fig. 8–13B).	3a. For a surgeon's knot, make two loops of suture around the needle holder. This makes the knot less secure, but keeps it from loosening before the second throw can be made.
4. Grasp the short end of suture with the needle holder and pull it thru the loop, crossing your left hand over the wound (Fig. 8–13C).	—
5. Snug wound edges together.	5a. Wound edges should be just touching.
6. Keeping tension on the long end of suture, drop the short end and hold the needle holder back over the middle of the first throw.	6a. Keeping tension on the suture will keep the tie flat.
7. Pass another loop of suture over and under the needle holder.	—
8. Grasp the short end of suture with the needle holder and pull it through the loop, crossing your left hand back over the laceration.	—
9. Snug the knot tight.	9a. Knot should lie flat, directly on the wound edges.

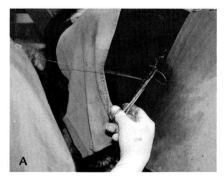 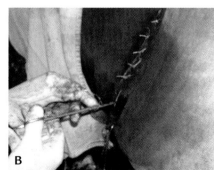 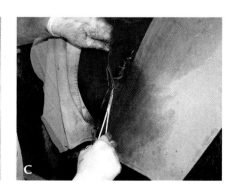

FIG. 8–13 (A) Long end of the suture in the left hand, short end tag in the skin. (B) Suture looped around the needle holders. Surgeon grasping the short end of the suture, getting ready to cross hands. (C) Pulling suture ends apart to establish first throw of the knot.

TECHNICAL ACTION	RATIONALE/AMPLIFICATION
10. Repeat at least 2 more times to secure the knot.	10a. Most surgeons place at least five throws to ensure that the knot does not come undone.
11. Cut ends of suture to approximately 1-cm long.	11a. Suture ends should be longer on cattle for ease of removal.
12. Remove sutures (Fig. 8–14).	12a. Grasp the ends of the suture and pull to lift the knot up off of the skin.
	12b. Slide the hooked portion of the suture scissors under the stitch.
	12c. Cut.
	12d. Pull suture material out and lay it on a piece of gauze or paper towel.
	12e. Repeat until all the sutures are removed.
	12f. Count to verify they are all out.

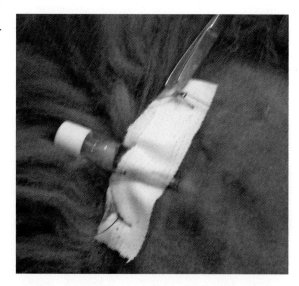

FIG. 8-14 Cutting out a suture by slipping the hook of the suture scissors under the knot.

SUTURE PATTERNS

Purpose

- A variety of patterns to suit the wound or area to be sutured
- Special patterns and stents required for some wounds that are under great tension

- Other factors the surgeon takes into account: The organ to be sutured, the time it will take to suture, the amount of suture material required, and the species being sutured

Equipment

- Suture material
- Needle
- Needle holder
- Thumb tissue forceps
- Suture scissors

Complications

- Dehiscence (failure of suture to hold edges together), which may result in herniation, evisceration, or other untoward consequences
- Tissue necrosis secondary to excessive tension placed on wound edges
- Delayed healing secondary to poor wound edge apposition
- Suture breakage secondary to excessive tension

Procedure for How to Sew Four Suture Patterns

PATTERN	PURPOSE	METHOD
1. Simple interrupted (Fig. 8–15A)	Closure of any tissue type any location Allows precise apposition of wound edges Used to affix butterfly tape on catheters to the skin or to affix other stents in position	1a. Hold the skin edge with the thumb forceps and the needle with the needle holder. 1b. Starting approximately 2 mm ($1/8$ inch) from skin edge, drive the needle through the skin. 1c. Grasp the opposing skin with the thumb forceps. 1d. Drive the needle through the skin, going from inner surface to outer surface, directly across from the starting point. 1e. Pull the thread through until there's just an inch or two left on the other side.

PATTERN	PURPOSE	METHOD
		1f. Tie the two ends together, using the needle holder and placing a square knot. Secure with 3 more throws.
		1g. Cut both ends and repeat until wound is closed, placing the stitches 2–3 mm (⅛–¼ inch) apart.
2. Simple continuous (Fig. 8–15B)	Closure of elastic tissues not under tension Rapid closure with minimal suture material	**2a.** Perform the same steps as simple interrupted, but cut only the short end.
		2b. Place more stitches every 2–3 mm (⅛–¼ inch) to the end of the wound, snugging up the suture as you go.
		2c. Tie a square knot at the end using a loop of suture for the short end. Secure with 3 more throws.
3. Cruciate mattress (Fig. 8–15C)	Closure of tissues under modest tension where maintenance of blood supply to wound edges is important	**3a.** Start as though doing a simple interrupted stitch, but don't tie a knot yet.

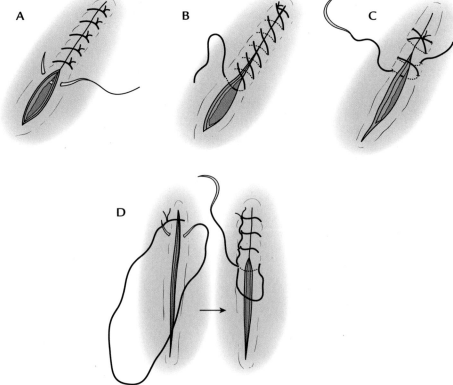

FIG. 8-15 Suture patterns commonly used in large animal surgery. Simple interrupted (**A**), simple continuous (**B**), cruciate mattress (**C**), and Ford-interlocking (**D**).

PATTERN	PURPOSE	METHOD
		3b. After coming out of the skin on the second side, cross back over to a point approximately 2 mm (⅛ inch) away from the first needle insertion.
		3c. Drive the needle through the skin 2 mm (⅛ inch) from the edge and then through the underside of the opposing skin.
		3d. Tie a square knot in the middle overlying the wound edges.
		3e. Final product should look like an X.
		3f. Repeat until wound is closed.
4. Ford-interlocking (Fig. 8–15D)	Modified simple continuous stitch where each loop is locked into place Commonly used to close the skin of cattle following a flank laparotomy, to distribute tension evenly along the whole suture line	**4a.** Start the same as simple continuous pattern.
		4b. After securing the knot, pass the needle through the wound edges the same way as with a simple continuous suture, but then pass the needle through the loop formed by the suture as it comes off the knot.
		4c. Pull suture up snugly.
		4d. Pass needle through wound edges 4–5 mm (¼ inch) down the wound, and again allow needle to pass through the resultant loop of suture material as it exits the skin.
		4e. Pull suture material until snug and repeat until wound is closed.
		4f. To end, hold onto the loop, do not pass the needle through it, and use that loop for to tie with, same as with a simple continuous pattern.

SURGICAL CASTRATION OF THE LARGE ANIMAL

Any male animal not destined for reproduction is castrated, usually at a young age. Sheep, goats, and cattle often are castrated within a few weeks of birth by placing a special, heavy rubber band around the neck of the scrotum. Older calves, lambs, and pigs are castrated surgically in the field. Most horses are castrated between 1 and 2 years of age (as they reach puberty), while llamas generally are not castrated until they are mature (over 2 years old.)

SURGICAL CASTRATION OF CALVES

Purpose

- Decrease aggressive behavior toward each other or handlers
- Prevent reproduction
- Improve carcass quality of food animals
- Remove diseased or damaged testis

Equipment

- Scalpel handle #4 with #20 or #21 blade
- Reimers emasculator (Fig. 8–16A)
- Stainless steel bucket filled with water and disinfectant solution (povidone-iodine, chlorhexidine, or bleach)
- Size 3 chromic gut (optional)
- Stainless steel hemostatic staple (optional) (Fig. 8–16B)

Complications

- Excessive bleeding
- Injury to operator
- Scirrhous cord and peritonitis secondary to infection

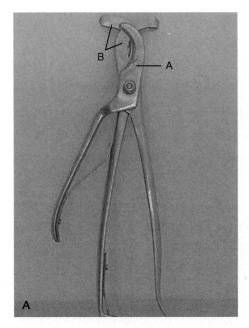 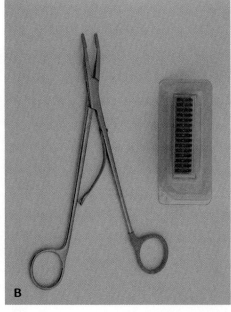

FIG. 8–16 (A) Reimers emasculator. Note separate cutting (**A**) and crushing jaws (**B**). (B) Hemoclips and applicator.

Procedure for Castration of Calves

TECHNICAL ACTION	RATIONALE/AMPLIFICATION
1. Restraint	**1a.** Restrain calf in chute.
	1b. Standing on the right side, and using both hands, push the tail up over the calf's back.
	1c. If a chute is unavailable, calves less than 500 lb may be thrown and restrained in lateral recumbency for castration.
2. Surgical preparation	**2a.** If the calf's perineum and scrotum is covered with feces, the calf should be washed off with a disinfectant solution, especially if it weighs more than 500 lb.
3. Surgery	**3a.** Grasp the calf's scrotum, pulling down on the skin while making a horizontal incision through the skin and vaginal tunic at the widest portion of the scrotum just distal to the testes.
	3b. Transect the entire distal portion of the scrotum and discard (Fig. 8–17A).
	3c. Hold on to the testes and pull firmly ventrad to release traction caused by the cremaster muscle and gubernaculum.
	3d. Firmly strip away fat, fascia, and the cremaster muscle, stripping them proximally.
	3e. Apply emasculator to the spermatic cord with the tightening nut on the tool facing the testes (Fig. 8–17B).

FIG. 8–17 (A) Scrotum cut, testes hanging out of calf. (B) Application of Reimers emasculator. Note position of cutting jaws.

TECHNICAL ACTION	RATIONALE/AMPLIFICATION
	3f. Crush the cord and cut away the testes below the emasculator.
	3g. Wait 60–90 seconds and remove emasculator.
4. Alternative castration	**4a.** Used on calves with large, well-developed testes, bulls, or those with a herd history of bleeding problems.
	4b. After testes are exposed and spermatic cord cleaned, a ligature of heavy absorbable suture or a stainless steel hemostatic staple is placed on the cleaned cord.
	4c. The emasculator is then applied distal to the ligature, the cord crushed, and the testis cut free.
5. Postsurgical care	**5a.** Fly repellent may be applied to prevent infestation with maggots.
	5b. Calves are returned to the group and observed for excess bleeding immediately after castration.
	5c. Owners should note any lack of appetite or lethargy for 5 days after castration.

RECUMBENT SURGICAL CASTRATION OF HORSES

Purpose

- Improve trainability or performance
- Prevent unwanted pregnancy
- Remove diseased or damaged testis

Equipment for Horses and Llamas

All listed under equipment needed for calves plus:

- Flat nylon halter and long soft cotton lead rope
- Clippers with #40 blade
- Surgical scrub and solution
- Isopropyl alcohol 70% (optional)
- Sterile surgical gloves

- Surgical mask and cap and gown (optional)
- Sterile surgery pack (optional)
- Sterile surgery drape or towels (optional)
- Cold-sterilization tray (if not using surgery pack) containing

    Mayo scissors

    Needle holders

    Hemostats

    Scalpel handle and blade

Complications

- Injury to patient, handler, or surgeon
- Infection of surgery site, spermatic cords, or abdomen
- Excessive bleeding
- Excessive swelling secondary to poor drainage or lack of enforced exercise
- Peritonitis secondary to scirrhous cord and infection
- Herniation of intestines through external inguinal ring

Procedure for Recumbent Castration of Horses

TECHNICAL ACTION	RATIONALE/AMPLIFICATION
1. Restraint	1a. Halter horse and restrain for general anesthesia (Chapter 10).
	1b. Lay horse down in left lateral recumbency for right-handed surgeon.
	1c. Place a half hitch around the horse's uppermost fetlock joint on the hind leg, and hold or tie the leg rostrally and dorsally (Fig. 8–18A).
2. Surgical preparation	2a. Clip scrotum and inguinal area with #40 blade and remove loose hairs.
	2b. Perform three-step surgical scrub with ventral scrotum as the center point.
	2c. Surgeon should perform personal preparation at this point.

TECHNICAL ACTION	RATIONALE/AMPLIFICATION
	2d. If they are not in a sterile pouch, place Reimers emasculator in bucket of cold-sterilizing solution.
	2e. Apply sterile drape or towels (optional).
3. Surgery	3a. Grasp lower (right) testis, and incise scrotal skin the length of the testis and 1 cm lateral to the median raphe.
	3b. Continue incision through the scrotal fascia and extrude testis up through the incision.
	3c. Firmly grasp the testis with the left hand while incising into the common tunic.
	3d. Hook your finger into the common tunic as the incision is continued proximally.
	3e. Digitally strip away or cut the common tunic, cremaster muscle, and fascia to clean the vascular portion of the spermatic cord.
	3f. Apply the emasculator with the tightening nut and cutting blade facing the testis (Fig. 8–18B).
	3g. Crush firmly, leaving emasculator in place for 2–3 minutes.
	3h. A ligature may be applied at this time to further prevent bleeding.
	3i. Use the blade on the emasculator to cut off the testis.

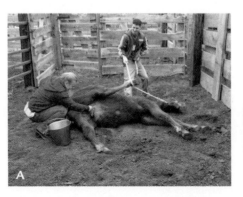 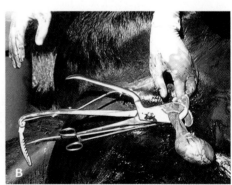

FIG. 8–18   (A) Positioning a horse for castration. Note technique for holding foot out of the way and position of technician scrubbing the scrotal area. (B) Application of Reimers emasculator. Note position of cutting jaws.

TECHNICAL ACTION	RATIONALE/AMPLIFICATION
	3j. Release the testis and repeat procedure on other side.
	3k. Trim away any stray pieces of fat or subcutaneous tissue.
	3l. Stretch the scrotal incisions digitally to a length of 8–10 cm.
	3m. Some surgeons remove the median raphe at this time.
4. Postoperative care	**4a.** Once horse is awake and able to walk, return it to a clean stall or small pen.
	4b. Observe closely for excessive bleeding or intestines protruding from one or both incisions. Notify doctor immediately if either is observed.
	4c. Starting 12–18 hours after surgery, the owner is instructed to force the horse to exercise. First by hand-walking 15–20 minutes 4 times a day, then by light trotting or riding for 5 days.
	4d. Owner to observe horse for severe preputial swelling, loss of appetite, hind leg edema, or difficulty with urination.

SURGICAL CASTRATION OF LLAMAS

Purpose

- Decrease aggressive behavior toward each other or handlers
- Improve trainability
- Prevent unwanted pregnancy

Equipment

- See equipment needed for horse

Complications

- Injury to patient, handler, or surgeon
- Infection of surgery site, cords, or abdomen
- Excessive bleeding
- Scirrhous cord

Procedure for Castration of Llamas

TECHNICAL ACTION	RATIONALE/AMPLIFICATION
1. Restraint	1a. May be done standing or in lateral recumbency.
	1b. Llamas may be held in a standing position in a squeeze chute or against a wall.
	1c. Sedation with xylazine and butorphanol may ease handling (Chapter 10).
2. Surgical preparation	2a. Clip scrotum and inguinal area with #40 blade and remove loose hairs.
	2b. Perform three-step surgical scrub with ventral scrotum as the center point.
	2c. Anesthetize scrotum with 2–5 ml of 2% lidocaine injected along the median raphe.
	2d. Repeat surgical scrub, alcohol, and solution one more time.
	2e. Surgeon should perform personal preparation at this time (see Procedure for Personal Preparation for Surgery).
3. Surgery	3a. Procedure is same as for the horse, except ligature is always applied to the spermatic cord.
	3b. Drapes or towels are not used if llama is castrated while standing.
	3c. Incision may be on midline just rostral to scrotum if llama is being castrated in dorsal or lateral recumbency.
	3d. In this case, the surgical field must be draped or towels applied using aseptic technique, and the incision will be sutured.
4. Postsurgical care	4a. Return llama to clean stall or small pen once he is ambulatory.
	4b. Observe closely for excessive bleeding.
	4c. Keep separate from female llamas.

SURGICAL CASTRATION OF PIGS

Purpose

- Decrease aggressive behavior toward each other or handlers
- Improve carcass quality and feed efficiency

Equipment

- Surgical scrub and solution
- Gauze sponges 3 × 3 inches
- Bucket of water with dipper cup
- Number 3 scalpel handle with #10 blade (pigs less than 20 kg)
- Number 4 scalpel handle with #20 or #21 blade (pigs greater than 20 kg)
- Surgical gloves
- Reimers emasculator for pigs weighing more than 40 kg

Complications

- Hernia of intestines through inguinal rings
- Infection at the castration site
- Peritonitis secondary to infection at the castration site
- Heat stroke if done when ambient temperature above 33°C (90°F)

Procedure for Castration of Pigs

TECHNICAL ACTION	RATIONALE/AMPLIFICATION
1. Restraint	1a. Pigs less than 20 kg (40 lb) may easily be held by assistant in a handstand with belly facing surgeon and hind legs spread apart (Fig 8-19).
	1b. Pigs up to 40–60 kg (80–120 lb) may be held in a squeeze chute with a hog snare on the snout, but are more easily done with some chemical restraint (Chapter 10).
	1c. Pigs greater than 60 kg (120 lb) require general anesthesia and must be done on a cool day to prevent heat stroke.

TECHNICAL ACTION	RATIONALE/AMPLIFICATION
2. Surgical preparation	2a. Scrotum is scrubbed with surgical scrub and gauze sponge, rinsed with water, and both repeated 2 times more.
3. Surgical procedure	3a. With gloved hands, push the testes proximally into the inguinal area, making the skin bulge.
	3b. An incision is made into the skin over each testicle, down through the subcutaneous fascia and fat.
	3c. With blunt dissection, a testis is grasped and pulled free of the scrotal ligament, leaving the common tunic intact.
	3d. Testis and tunic are scraped with the scalpel blade until they tear and testis is torn free from its attachment.
	3e. For pigs over 40 kg (80 lb), the spermatic cord and common tunic will be crushed in toto with an emasculator, and then the testis is cut free.

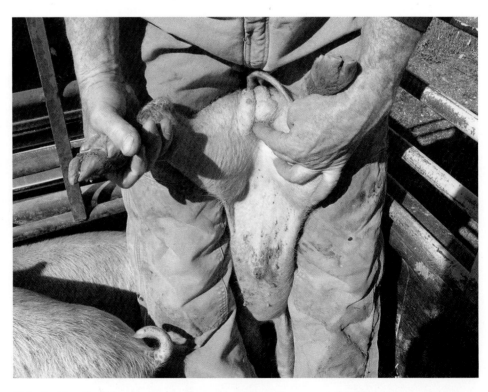

FIG. 8-19 Positioning a pig for castration. Testes pushed into scrotal sac.

TECHNICAL ACTION	RATIONALE/AMPLIFICATION
	3f. In both cases, the cord is allowed to retract into the body, and the process is repeated on the other testis.
	3g. The incisions are left open to provide drainage.
4. Postsurgical care	**4a.** The pigs are released into a clean, dry pen and kept there until the incisions have healed (5 to 7 days).

PREPARATION OF THE PATIENT FOR LAPAROTOMY

A **laparotomy** is a surgical entry into the abdomen. In large animal medicine, this can be done with the animal standing or lying down. Positioning depends on the procedure, the species, and the surgeon's preference. Most horse laparotomies are done at referral hospitals, in part for safety of the patient and personnel.

PREPARATION FOR A FLANK LAPAROTOMY

Purpose

- Treat some obstipation forms of colic
- **Ovariectomy**
- Remove abdominal testis
- Treat left or right displaced abomasum
- Rumenotomy
- Cesarean section in all large species but horses
- Other

Equipment

- Clippers with #40 blade
- Surgical scrub and solution
- Bucket of water
- Lidocaine 2%, 1–2 bottles
- Syringes: 12 ml to 60 ml
- Needles for local or general anesthetic, or both, as well as for antibiotics
- Abdominal pack
- Drape or sterile surgery towels
- Surgery needles: Taper ½ circle, cutting ½ or ¼ circle

- Suture material: Depends on veterinarian's preference, with an example below of what a veterinarian might use (size depends on species of animal)

    Chromic gut size 0–3

    Polydiaxone size 0–2

    Polymerized caprolactam size 0–3

- Sterile surgical gloves and sleeves
- Cap, mask, and gown

Complications

- Accidental intestinal evisceration from the animal falling down during the surgery or excessive abdominal contractions
- Postoperative seroma formation and wound dehiscence
- Herniation of abdominal muscles secondary to inadequate rest postoperatively or infection
- Peritonitis secondary to gross contamination or bowel rupture
- Injury to patient or operator secondary to the animal falling down intraoperatively or inadequate restraint

Procedure for Standing Flank Laparotomy of the Horse

TECHNICAL ACTION	RATIONALE/AMPLIFICATION
1. Restraint	1a. Apply halter and lead rope and place the patient in a stock.
	1b. Administer agents for standing anesthesia or analgesia (Chapter 10).
2. Surgical preparation	2a. Insert a large-bore jugular catheter and administer fluids at the minimum flow of 10 ml/kg/hr. Catheter placement technique is covered in Chapter 6.
	2b. Clip a rectangle extending from the fourteenth rib to the tuber coxae and from the dorsal spinous processes to the level of the stifle (Fig. 8–20).
	2c. Wash or vacuum away hair.
	2d. Perform three-step surgical scrub on the entire clipped area.

CHAPTER 8 Surgical Preparation **341**

TECHNICAL ACTION	RATIONALE/AMPLIFICATION
	2e. Using 2% lidocaine, block the proposed line of incision following the steps to an infiltrative line block (Chapter 10).
	2f. Rescrub, rinse, and apply surgical preparation solution to the prepared field.
	2g. Lay out surgical pack, gloves, sterile sleeves, drapes, suture needles, and other suture materials.
	2h. The surgeon may perform personal preparation for surgery at this time.
3. Surgery	**3a.** Using aseptic technique, a sterile drape or towels are placed surrounding the proposed line of incision.

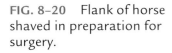

FIG. 8-20 Flank of horse shaved in preparation for surgery.

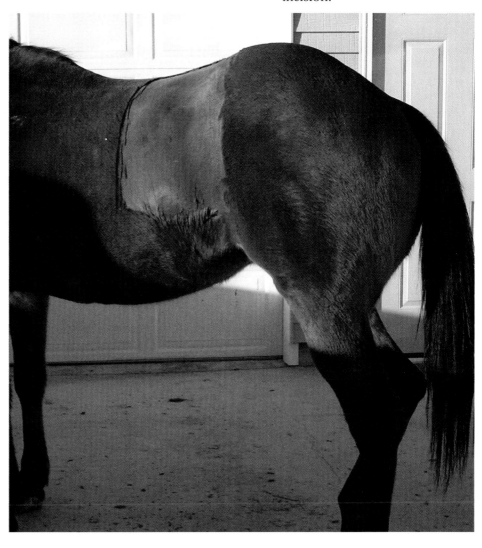

TECHNICAL ACTION	RATIONALE/AMPLIFICATION
	3b. Flank anesthesia is verified by pricking the proposed line of incision with an 18g needle.
	3c. A 15–20 cm (7–10 inch) incision is made through the skin and underlying fascia in the paralumbar fossa, starting level with the tuber coxae.
	3d. The three muscle layers are divided along the direction of their fibers. Large bleeding vessels are ligated or clamped with hemostats.
	3e. The peritoneum is much thicker in the horse than the ruminant. It is grasped with thumb forceps and incised with blunt-blunt Mayo scissors. This is a very painful moment for the horse.
	3f. Following completion of the planned surgery, the abdomen may be lavaged with sterile 0.9% saline or other isotonic fluid.
	3g. The incision is closed in multiple layers, starting with the peritoneum and then closing each muscle layer individually with a heavy absorbable suture material such as size 3 polydioxanone.
	3h. The skin may be closed with nonabsorbable monofilament suture or staples.
4. Postoperative care	**4a.** Antibiotics, antiinflammatory agents, or analgesics may be given before returning the horse to its stall.
	4b. The intravenous fluids are discontinued, leaving the catheter in place, and the horse is led back to a clean stall or pen.
	4c. Vital signs, including bowel sounds, are monitored closely. Intravenous fluids may be restarted; otherwise the jugular catheter should be covered with tape to keep it clean.

TECHNICAL ACTION	RATIONALE/AMPLIFICATION
	4d. Fresh feed and water are offered at the veterinarian's discretion.
	4e. The incision is checked twice daily for swelling or discharge, and the horse is given pain medications or antiinflammatory agents for 5–7 days after surgery.
	4f. Most horses may be discharged within 3–5 days of surgery provided there are no complications and it is eating well.
	4g. The owner is advised to keep the horse confined to a stall or small pen for at least 3 weeks or until the incision has healed completely.
	4h. Sutures or staples may be removed in 14–21 days.

Procedure for Flank Laparotomy of the Ruminant

TECHNICAL ACTION	RATIONALE/AMPLIFICATION
1. Restraint	**1a.** Standing 　**1.** Position patient in a stock or chute. A belly band may be used to prevent the patient from lying or falling down. 　**2.** A xylazine caudal epidural (0.05 mg/kg) may be used to improve analgesia (Chapter 10). **1b.** Recumbent 　**1.** Ropes may be used to pull animal down and secure it in lateral (cattle, sheep, and goats) or sternal recumbency. 　**2.** Alternatively, profound sedation (Chapter 10) may be used to induce the patient to lie down and ropes used to secure the limbs.
2. Surgical preparation	**2a.** Clip hair from flank region in a rectangle extending from the twelfth rib to the tuber coxae, and from the dorsal spinous processes to the level of the stifle (Fig. 8–21A).

TECHNICAL ACTION	RATIONALE/AMPLIFICATION
	2b. Wash or vacuum away hair.
	2c. Perform three-step surgical scrub on the entire clipped area.
	2d. Using 2% lidocaine, block the flank region following the steps to either a simple line block, an inverted L block, or a paravertebral block (Chapter 10).
	2e. Rescrub, rinse, and apply surgical preparation solution to the prepared field.
	2f. Lay out surgical pack, gloves, sterile sleeves, drapes, suture needles, and other suture materials.
	2g. The surgeon may perform personal preparation for surgery at this time.
3. Surgery	**3a.** Using aseptic technique, a sterile drape or towels are placed surrounding the proposed line of incision.
	3b. Flank anesthesia is verified by pricking the proposed line of incision with an 18g needle.
	3c. An incision is made through the skin in the middle of the paralumbar fossa, starting at the level of the tuber coxae and extending into the underlying musculature. Length of the incision varies greatly with the type of surgery (Fig. 8–21B).
	3d. Large bleeding vessels are ligated or clamped with hemostats.

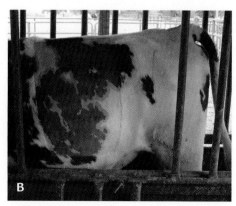

FIG. 8-21 (A) Shaving the flank of a cow. Going against the lie of the hair. (B) Flank of cow shaved in preparation for surgery.

TECHNICAL ACTION	RATIONALE/AMPLIFICATION
	3e. The peritoneum is penetrated carefully with a scalpel or with blunt force with a pair of Kelly hemostats.
	3f. The planned surgery is performed and the body wall is closed with a monofilament or pseudomonofilament absorbable suture material such as chromic catgut.
	3g. The skin is closed with a nonabsorbable monofilament suture such as nylon.
4. Postoperative care	**4a.** Antibiotics may be injected while the animal is still sedated or restrained.
	4b. Fly spray may be applied to the incisional area as well.
	4c. If recumbent, the animal is released from any ropes and allowed to get up when it is ready.
	4d. If in a chute or a stock, the animal is released when it is able to ambulate safely.
	4e. The patient is returned to a clean stall or pen with fresh feed and water.
	4f. The owner is advised to observe the incision daily for swelling or discharge. Patient's appetite and water intake should be monitored for at least 5 days.
	4g. Skin sutures may be removed in 14–21 days.

VENTRAL LAPAROTOMY

Purpose

- Abdominal exploratory surgery
- Cesarean section (horses)
- Displaced abomasum (rare)
- Umbilical hernia repair

Equipment

- Same as flank laparotomy
- 15–30 liters sterile saline (not necessary in hernia repairs)

- Cushions for limbs, body, and head
- Equipment for general anesthesia (Chapter 10)

Complications

- Dehiscence of incision and herniation or evisceration
- Seroma at incision site
- Peritonitis
- Myositis secondary to poor padding technique
- Peripheral neuropathy (facial or limbs) secondary to poor padding
- Death
- Injury to personnel during induction and positioning of patient or during recovery
- Injury to patient (limb fracture, ruptured cruciate ligament, head trauma) during induction or recovery

Procedure for Ventral Laparotomy in Ruminants

TECHNICAL ACTION	RATIONALE/AMPLIFICATION
1. Restraint	1a. Restrain animal with a halter and lead rope.
	1b. General anesthesia or heavy sedation is administered.
	1c. In some hospitals, the patient may then be placed on a cushioned surgery table.
	1d. Once the patient is recumbent, it is rolled onto its back on cushioning, with head outstretched.
	1e. The front and hind legs are secured with ropes to a tie ring or posts or leg holders (if on a surgery table) (Fig. 8–22).
	1f. Place covered straw bales or heavy cushions against the shoulders of the animal to support it in dorsal recumbency.
2. Surgical preparation	2a. Cover the feet with plastic obstetrical sleeves or examination gloves secured with tape.
	2b. Clip a large rectangle from udder to xiphoid extending at least 8 inches from the midline on either side.

TECHNICAL ACTION	RATIONALE/AMPLIFICATION
	2c. Wash or vacuum away hair.
	2d. Perform three-step surgical scrub on the entire clipped area.
	2e. Perform infiltrative line block on the midline along proposed incision site. In the adult sheep block at least 30 cm (12 inches), and in the adult cow block at least 45 cm (18 inches) (Chapter 10).
	2f. Rescrub, rinse, and apply solution to the prepared field.
	2g. Lay out surgical pack, gloves, sterile sleeves, drapes, suture needles, and other suture materials.
	2h. The surgeon may perform steps to personal preparation for surgery at this time.
3. Surgery	3a. Using aseptic technique, a sterile drape or towels are placed surrounding the proposed line of incision.

FIG. 8-22 Calf positioned in dorsal recumbency, having had surgery to repair an umbilical hernia. Pads are removed, and ropes are being readied to be removed.

TECHNICAL ACTION	RATIONALE/AMPLIFICATION
	3b. The incision is usually on the ventral midline, starting at the udder and extending cranially to the umbilicus.
	3c. Ventral midline incision for some abdominal exploratory surgeries or displaced abomasum correction may begin 2–4 cm (1–2 inches) caudal to the umbilicus and extend cranially to within 4 cm (2 inches) of the xiphoid.
	3d. Following completion of the surgical procedure, the abdomen may be lavaged with 10–30 liters of warm saline to remove any contaminants.
	3e. The abdomen is closed with a heavy absorbable suture, such as 3 catgut, in at least two layers.
	3f. The skin is closed with heavy nonabsorbable monofilament suture or a long-lasting suture such as polydioxanone.
4. Postoperative care	**4a.** Antibiotics may be injected while the animal is still sedated or restrained.
	4b. Carefully release the patient from restraining ropes, but leave the halter in place.
	4c. Roll the patient into sternal recumbency, supporting it with straw bales if necessary. Leave head and neck outstretched if it is unable to raise its head on its own.
	4d. Monitor the patient until it is able to stand unaided.
	4e. Remove halter and return the patient to a clean, dry pen or stall containing fresh feed and water.
	4f. Owner is advised to monitor the patient's food and water intake as well as fecal and urine production for at least 5 days.
	4g. Sutures may be removed in 14–21 days.

OBSTETRICAL PROCEDURES

In large animal practice, the female patient may need a variety of reproduction-associated procedures, including assisted vaginal delivery of a fetus, cesarean section, and reduction of vaginal or uterine prolapse. All of these procedures are performed by veterinarians; the role of the technician is to prepare the patient and assist if needed.

ASSISTED VAGINAL DELIVERY OF A FETUS

Purpose

- Overcome structural impediment (such as the pelvic canal being too small for the fetus)
- Prevent damage from physiologic conditions (such as calcium deficiency or hypoglycemia)
- Remove dead fetus in a timely manner
- Provide expert medical help for a very complicated delivery

Equipment

- Bucket of disinfectant solution in water
- Surgical scrub and solution
- Lidocaine 2%
- 3–12 ml syringes
- 18–22 g needles
- Obstetrical sleeves
- Lubricant jelly or powder
- Tail wrap (for horses)
- Obstetrical chains or ropes and handles (Fig. 8–23A)
- Calf jack (cattle) (Fig. 8–23B)
- Towels (for drying the fetus)

Complications

- Injury to dam from pulling a too-large fetus through the pelvic canal
- Injury to operator by unsafe operation of the calf jack or inadequate restraint
- Injury to fetus from having the chains applied incorrectly, pulling too hard, or getting stuck in the pelvic canal

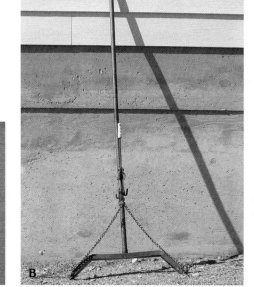

FIG. 8-23 (A) Obstetrical chains and handles. (B) Calf jack assembled. (C) Assisted vaginal delivery of a calf using chains and calf jack.

Procedure for Assisted Vaginal Delivery of Horses

TECHNICAL ACTION	RATIONALE/AMPLIFICATION
1. Place obstetrical chains and handles in bucket of water and dilute disinfectant solution.	1a. Use povidone-iodine or chlorhexidine solution at 1–2 ounces per gallon of warm water.
2. Restrain patient	2a. Most mares need just a halter and lead rope and are held in the pen or stall. If she is recumbent, do not force her to stand.
	2b. The horse may be placed in cross-ties or a stock if admitted to the hospital or if the delivery will be complicated.
3. Apply tail wrap.	3a. See Chapter 6.
4. Wash perineum quickly with soap and water.	4a. Delivery must be very rapid in order for the foal to survive, so little time is spent cleaning unless the fetus is already dead.
5. Set out obstetrical sleeves and lubricant for veterinarian.	—

Procedure for Assisted Vaginal Delivery of Ruminants

TECHNICAL ACTION	RATIONALE/AMPLIFICATION
1. Place obstetrical chains and handles in bucket of water and dilute disinfectant solution (Fig. 8–23).	1a. Most people use povidone-iodine, chlorhexidine, or chlorine bleach diluted to approximately 2 ounces per gallon.
2. Restrain patient.	2a. Cows are placed in headlock, a stock, or tied to a fence.
	2b. Goats and sheep may be haltered and restrained manually.
	2c. Llamas may be haltered and restrained manually or in a chute.
3. Wash and prepare patient for caudal epidural anesthesia.	3a. See Chapter 10 Procedure for Caudal Epidural in the Ruminant. Some veterinarians prefer to do the caudal epidural themselves.
4. With surgical soap or detergent and water, wash manure and debris from the perineum and underside of the tail.	4a. Clipping is not necessary, even if it is a wooly animal.
	4b. Pay particular attention to the vulva and surrounding tissues.
	4c. If part of the fetus is protruding, you may carefully rinse it off, as well. Do not get any antiseptic in the calf's eyes.
5. Assemble calf jack and set it nearby.	5a. This is necessary if the calf is being pulled at the producer's place.
6. Set out obstetrical sleeves and lubricant for veterinarian.	6a. Not all veterinarians wear obstetrical sleeves when doing assisted deliveries for calves.
	6b. Veterinarians who don't wear obstetrical sleeves will want a bucket of dilute disinfectant in which to wash their hands and arms.

TECHNICAL ACTION	RATIONALE/AMPLIFICATION
7. Set out one towel per fetus.	**7a.** Straw may be used to dry the fetus if towels are not available. **7b.** Clear nose and mouth of afterbirth, and tickle nostrils with straw or fingers to encourage the neonate to cough or sneeze. **7c.** Leave some afterbirth on the baby to help the dam bond with newborn. **7d.** See Chapter 7 for further neonatal care instructions.

Procedure for Assisted Vaginal Delivery of Pigs

TECHNICAL ACTION	RATIONALE/AMPLIFICATION
1. Soak pull ropes or pig puller in a cold-sterilizing solution.	
2. Restrain sow.	
3. Wash perineum with small amount soap and water containing either povidone-iodine solution or chlorhexidine solution.	
4. Set out sterile surgery gloves and obstetrical sleeves and lubricant for veterinarian.	
5. Set up heat lamp and have a stack of clean dry towels at the ready.	

CESAREAN SECTION

Definition

- Cesarean section is the removal of a full-term fetus through an abdominal incision in the dam

Purpose

- Primarily done because the dam is unable to expel the fetus vaginally, either due to structural reasons or physiologic reasons
- Done electively on occasion in the horse when the foal is at risk of dying in utero and the mare cannot be induced to start labor

Equipment

- Laparotomy equipment
- Obstetrical chains or ropes
- Towels to dry the fetus

Complications

- Laparotomy complications
- Retained placenta or metritis
- Death of the dam secondary to metabolic problems, shock, or hemorrhage
- Death of the fetus

Comparison of Cesarean Section in Large Animals

Because of variations in anatomy of large animals, there is no single way that cesarean sections are done. Veterinarian preference also affects where the incision will be located and whether the patient will be standing or lying down. Table 8–3 can be used as a guide to how the patient should be prepared for surgery and how many offspring to plan for.

VAGINAL PROLAPSE REPAIR

Definition

- Displacement of engorged vaginal floor, usually during gestation

Purpose

- Allow the calf to descend pelvic canal during **parturition**
- Prevent prolapse or rupture of the bladder

TABLE 8-3 COMPARISON OF CESAREAN SECTION

	CATTLE	SMALL RUMINANTS	HORSES	SWINE
Restraint	Stock or chute	Tied in lateral recumbency	Ventral recumbency in surgical suite under general anesthesia	Tied in lateral recumbency
Surgical preparation	Standing left flank laparotomy most common	Low left flank laparotomy preparation	Ventral laparotomy preparation	Right or left flank laparotomy
Anesthesia	Local or regional	Local, regional, or lumbosacral	General	Lumbosacral epidural or local anesthesia
Number of fetuses	One, occasionally two	Two to three, rarely four	One, rarely two	Seven to fourteen

Equipment

- Water bucket with dilute surgical disinfectant
- Examination gloves
- Obstetrical sleeves
- Supplies for caudal epidural
    - 18g × 1.5-inch needle (cattle)
    - 12-ml syringe
    - Lidocaine 2%
- Surgery (Huck) towel
- Buhner needle
- Umbilical tape ½ inch
- Number 21 scalpel blade
- Utility scissors

Complications

- Inadequate caudal epidural causing straining
- Infection of perivulvar skin
- Ruptured urinary bladder
- Vaginal wall tear
- Failure to remove purse-string suture at parturition resulting in third-degree perineal laceration or dead fetus or both

Procedure for Vaginal Prolapse Repair of the Cow

TECHNICAL ACTION	RATIONALE/AMPLIFICATION
1. Restraint	1a. Secure the cow in a chute or a stock.
	1b. When catching the cow, do not close the gate behind her until her head is safely caught. This will prevent her from rupturing the prolapsed tissue against the back gate (Fig. 8–24).
2. Preparation	2a. Scrub the cow's tail head and administer caudal epidural anesthesia (Chapter 10).
	2b. Secure the tail to the side of the cow away from the prolapse.
	2c. Wearing gloves, wash her perineum, vulva, and protruding vaginal wall with disinfectant in water and a small amount of soap.
	2d. Rinse well to be sure that no soap residue is left on the vaginal tissue.
	2e. Lift prolapsed tissue dorsally to let the bladder drain if need be.

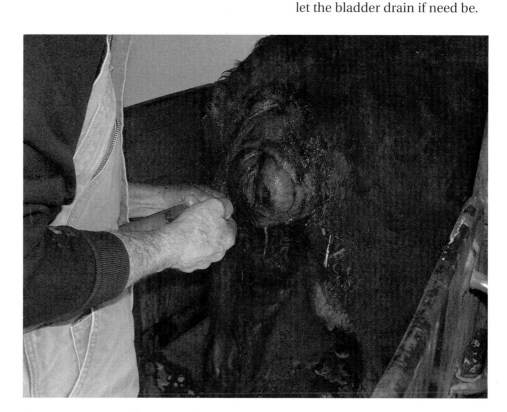

FIG. 8–24 Vaginal prolapse in a cow.

356 SECTION 4 Surgical, Radiographic, and Anesthetic Preparation

TECHNICAL ACTION	RATIONALE/AMPLIFICATION
3. Prolapse reduction	3a. Wearing gloves, wrap a towel completely around the protruding mass of vaginal tissue, with the long end of the towel around the base of the mass.
	3b. Twist the ends of the towel together like a corkscrew so that the towel tightly and completely envelops the mass.
	3c. Continue twisting the towel ends together, squeezing fluid from the mass until it can be replaced easily in the vaginal vault.
	3d. Palpate the vagina and make sure the tissue has been returned to an anatomically correct position.
4. Recurrent prolapse prevention	4a. Put on fresh gloves and wash the vulva and perivulvar tissues again.
	4b. Pull out and cut and soak in disinfectant solution approximately 1 meter of umbilical tape.
	4c. Make a small stab incision midway between the dorsal commissure of the vulva and the anus.
	4d. Make a second stab incision approximately 3 cm (1-$\frac{1}{4}$ inches) below the ventral commissure.
	4e. With one hand in the vagina, insert the Buhner needle deeply into the ventral stab incision and direct the needle to the dorsal stab incision.
	4f. Thread the umbilical tape through the eye of the needle.
	4g. Withdraw the needle back through the ventral incision, taking care to retain one end of the umbilical tape at the dorsal stab incision.
	4h. Remove the tape from the needle and, repeating Step 4e, send the needle up the contralateral (opposite) side of the vulva.

TECHNICAL ACTION	RATIONALE/AMPLIFICATION
	4i. Thread the dorsal end of the umbilical tape through the eye of the needle and withdraw the needle.
	4j. Both ends of the umbilical tape now exit the ventral stab incision.
	4k. Pull on the ends of the tape until only two or three fingers can enter the vagina.
	4l. Tie the ends in either a square knot or a bow (using a bow if the cow is within a week or two of calving).
5. Aftercare	**5a.** The cow must be monitored closely so that the tape can be removed at parturition.

UTERINE PROLAPSE

Definition

- Eversion of the entire uterine body and horns through the vulva

Purpose

- Replace the uterus in the abdominal cavity in correct anatomic position
- Prevent the death of the dam secondary to hemorrhage, hypovolemic shock
- Preserve the dam's ability to reproduce

Equipment

- Large, clean tarp
- Water bucket with dilute surgical disinfectant
- Waterproof coveralls
- Examination gloves
- Obstetrical sleeves
- Supplies for caudal epidural

    18 g × 1.5-inch needle (cattle)

    12 ml syringe

    Lidocaine 2%

- Buhner needle
- Umbilical tape ½ inch
- Number 21 scalpel blade
- Utility scissors
- Lubricant powder

Complications

- Death of the patient secondary to hypovolemic shock or massive hemorrhage
- Injury to operator secondary to inadequate restraint

Procedure for Uterine Prolapse Repair

TECHNICAL ACTION	RATIONALE/AMPLIFICATION
1. Restraint	**1a.** This is an emergency procedure done on location. Most animals will not survive transport to a veterinary hospital.
	1b. Secure the patient with a halter and lead rope, keeping the patient as quiet as possible.
	1c. Horses and small ruminants may require light sedation or analgesia.
	1d. If possible, the patient can be cast or induced to lie down in sternal recumbency.
	1e. A large cow should be positioned (with the aid of ropes) with her hind legs frog-legged out behind her (Fig. 8–25).

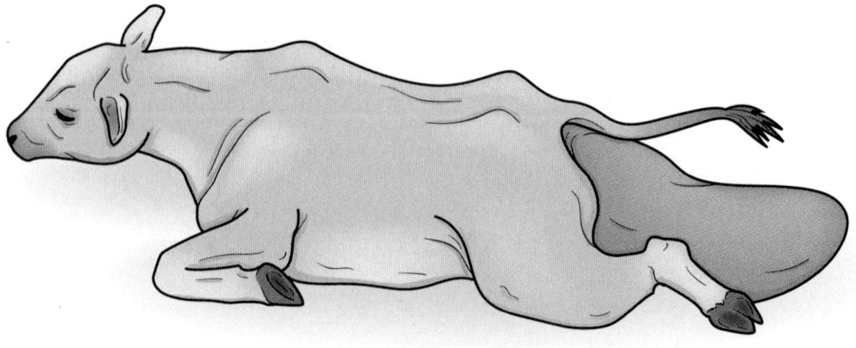

FIG. 8–25 Position of a cow for uterine prolapse repair. Note legs pulled caudally.

CHAPTER 8 Surgical Preparation

TECHNICAL ACTION	RATIONALE/AMPLIFICATION
2. Preparation	2a. For small ruminants and horses, insert an intravenous jugular catheter as soon as possible and start infusing a balanced isotonic electrolyte solution at the rate of at least 40 ml/kg/hr.
	2b. Place a large, clean tarp on the ground under the uterus.
	2c. Prepare for and administer caudal epidural anesthesia (Chapter 10).
	2d. Wash the uterus liberally with water and antiseptic solution.
3. Prolapse reduction	3a. Put on waterproof coveralls and obstetrical sleeves. Some veterinarians prefer to wear surgical gloves, too.
	3b. Ligate any significant arterial bleeding vessels and repair any lacerations.
	3c. Gently remove remaining placental tissues.
	3d. Gather the uterus up in your arms and gently work the swollen tissues back into the vaginal vault. Massage or knead the uterine tissue with your closed hands, not your fingertips, to avoid perforating the friable tissues.
	3e. If the uterus is greatly engorged, lubricant powder may be sprinkled all over it to help shrink the tissues and make them slicker.
	3f. Once the uterus is back inside the cow, palpate to ensure that the uterus is fully restored to normal anatomic position.
	3g. Place a vulvar purse-string suture as described in Procedure for Vaginal Prolapse Repair of the Cow.

TECHNICAL ACTION	RATIONALE/AMPLIFICATION
4. Aftercare	**4a.** Some cows may need intravenous calcium along with oxytocin.
	4b. Small ruminants and horses require large quantities of a balanced isotonic fluid given intravenously in addition to antibiotics, oxytocin, and possibly calcium.
	4c. The purse-string suture generally can be removed in 5–10 days.

MISCELLANEOUS COMMON LARGE ANIMAL PROCEDURES

Routine procedures at the large animal or mixed animal practice include dehorning, eye **enucleation**, rectal prolapse reduction and, if horses are treated, endoscopy. Preparations for these procedures are covered here.

DEHORNING CALVES

Purpose

- Protect other animals in herd from being injured by a horned animal
- Protect handlers from being injured or gored
- Remove a broken or badly damaged horn at risk for infection

Equipment

- Barnes dehorner or hardback saw or wire saw (Fig. 8–26)
- Clippers with #40 blade
- Surgical scrub and solution
- 18g × 1.5-inch needle
- Lidocaine 2%
- Straight Kelly hemostats
- Cotton balls
- Blood-stop powder
- Fly spray

Complications

- Excessive bleeding leading to anemia, weakness, and reduced feed intake
- Sinus infection
- Fly strike or maggot infestation

FIG. 8-26 Barnes dehorner.

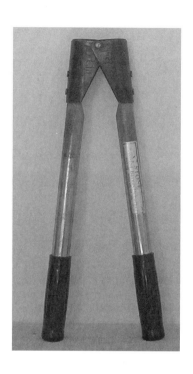

Procedure for Dehorning Calves

TECHNICAL ACTION	RATIONALE/AMPLIFICATION

1. Restraint in chute	1a. Secure head by doing one of the following:
	  1. Place calf's head in head holder
	  2. Place halter on calf's head and tie head to side of chute
	  3. Place nose tongs on calf and pull head up and to the side of chute

2. Surgical preparation	2a. None is required unless the horns are very large or the owner requests anesthesia. 2b. Clip hair away from base of horns in rectangular shape, including temporal ridge. 2c. Perform three-step surgical scrub. 2d. Palpate cornual nerve (ophthalmic branch of fifth cranial nerve) located on the upper third of the temporal ridge (Fig. 8-27A).

TECHNICAL ACTION	RATIONALE/AMPLIFICATION
	2e. Inject 3–5 ml of 2% lidocaine subcutaneously (Fig. 8–27B).
	2f. Wait 10–15 minutes for onset of analgesia.
	2g. Test for anesthesia by pricking the skin at the base of the horn with a 16–18g needle.

 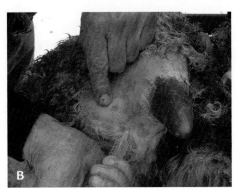

FIG. 8-27 (A) Palpating cornual nerve (B) Injecting lidocaine around cornual nerve.

3. Dehorning procedure	**3a.** Place Barnes dehorner over horn and push onto base of horn, angling the dehorner to match the angle of the calf's skull (Fig. 8–28A).
	3b. If the dehorner does not fit completely over the base of the horn, a hardback saw or wire saw may be used to cut the horn free (Fig. 8–28B).

FIG. 8-28 (A) Application of Barnes dehorner following the angle of the skull. (B) Using a hardback saw to dehorn. (C) Pulling an artery out (arrow).

TECHNICAL ACTION	RATIONALE/AMPLIFICATION
	3c. Pushing down hard onto the base of the horn, rapidly separate the handles of the dehorner to cause the blades to cut through the it.
	3d. Discard horn and locate exposed arteries that are bleeding.
	3e. Sometimes the dehorner does not go all the way through the skin. The remaining tag of skin may be cut or torn, depending on the age of the calf.
	3f. Grasp ends of arteries in tips of hemostat, and wrap the artery around the jaws of the hemostat until it has been pulled free from the calf (Fig. 8–28C).
	3g. Arteries are located at the ventral or 6 o'clock position as well as the 3 o'clock and 9 o'clock positions.
	3h. Pulling the arteries will cause them to retract back under the tight tissues of the head, where tissue compression will stop the bleeding.
	3i. For alternative bleeding control, apply a hot cautery or branding iron to the exposed tissues to cauterize the bleeding vessels.
	3j. Tissues should stop bleeding and turn a belt-leather brown color.
	3k. Put small ball of cotton loosely in any opening into the cornual sinus; this keeps debris from entering cornual sinus.
	3l. Repeat on other horn.
	3m. Sprinkle on blood-stop (coagulant) powder.
	3n. Apply fly spray to the area if dehorning is done during the warmer months.

Special Notes

- Calves may be disbudded by hot iron or caustic paste during the first few weeks of life.
- Use of a local anesthetic block is becoming more common in some areas of North America, so it is included in the procedure.
- Cattle with very thick or large horns will need to be dehorned with a saw.
- Cosmetic dehorning can be done by suturing the skin back over the exposed cornual sinus. This usually requires local and systemic anesthesia.

DEHORNING GOATS

Purpose

- Protect other animals in herd from being injured by a horned animal
- Protect handlers from being injured or gored
- Remove a broken or badly damaged horn at risk for infection

Equipment

- Wire saw
- Clippers with # 40 blade
- Surgical scrub and solution
- 18–20 g × 1.5-inch needle
- Lidocaine 1% or 2%
- Straight Kelly hemostats
- Cotton balls
- Sterile gauze pads 4 × 4 inches
- Blood-stop powder
- Fly spray
- Tetanus toxoid
- Antibiotic of choice

Complications

- Death secondary to shock from pain, sepsis, or tetanus
- Lidocaine toxicosis: Do not exceed 2 mg/kg of lidocaine
- Sinusitis secondary to poor aseptic technique, dirty instruments, or lack of postoperative care

Procedure for Dehorning Goats

TECHNICAL ACTION	RATIONALE/AMPLIFICATION
1. Restraint	1a. Goats have a much lower tolerance for pain and have been known to die from shock associated with dehorning without anesthesia or analgesia. See Chapter 10 for anesthetic regimen suitable for a goat.
	1b. Head is restrained manually once goat is sedated.
2. Surgical preparation	2a. Shave hair away from base of horns down to the zygomatic arch and below the temporal ridge.
	2b. Perform three-step surgical scrub, being careful not to get soap in the goat's eyes.
	2c. Palpate the infratrochlear branch of the fifth cranial nerve where the temporal ridge meets the zygomatic arch. The more mature the goat, the more difficult it is to palpate the nerve because of the thickness of skin (Fig. 8–29).
	2d. Inject 2 ml of 2% lidocaine into the tissues at a depth of a $\frac{1}{2}$ to $\frac{3}{4}$ inch.
	2e. Palpate the lacrimal branch of the fifth cranial nerve as it passes dorsomedial to the eye.
	2f. Inject another 2 ml of 2% lidocaine as close as possible to the zygomatic arch to a depth of $\frac{1}{2}$ inch.
	2g. The nerve is difficult to palpate, so use landmarks given for injection site.
	2h. Fan out your injection to cover nerve as well as possible. This amount is for a goat weighing more than 40 kg (80 lb). Use 1% lidocaine for smaller goats.
	2i. Wait 10–15 minutes for onset of anesthesia.

TECHNICAL ACTION	RATIONALE/AMPLIFICATION
3. Dehorning procedure	3a. Make a small incision into the skin at the base of the horn at the medial-most aspect.
	3b. Place the wire saw into the skin incision and, grasping the ends of the saw, rapidly work the wire back and forth until you have cut all the way through the base of the horn and its attached skin.
	3c. Pull the superficial temporal artery using a hemostat.
	3d. Apply a sterile gauze pad to cover the hole into the frontal sinus.
	3e. Repeat process on other side.
	3f. Bandage head with elastic veterinary wrap in a figure-of-eight around the ears and over the holes.
	3g. Administer penicillin and tetanus toxoid intramuscularly.

FIG. 8-29 Location of cornual nerves in the goat.

Special Notes

- Ideally goats should be disbudded within the first 3 weeks of life.
- Goats have a much thinner skull than calves and can sustain a skull fracture if a Barnes dehorner is used.
- Goats also have a low threshold for pain and a low tolerance for lidocaine, making it difficult to perform a local cornual block. Mature goats, therefore, should be given general anesthesia and excellent postoperative pain control if they are to be dehorned.

BOVINE EYE ENUCLEATION

Definition

Technically, enucleation involves removal of the globe of the eye. In cattle we usually remove the globe and all its support structures.

Purpose

- Preserve the life of the patient by removal of cancerous tissues
- Preserve the function and comfort of the patient when the eyelids can no longer function to keep the eye moist and healthy
- Preserve the function and comfort of the patient when the globe of the eye has been injured irreparably

Equipment

- Clippers with #40 blade
- Surgical scrub, 70% isopropyl alcohol, surgical solution
- Lidocaine 2%, 100-ml bottle
- 18 g × 1.5-inch needle (2–3)
- 20 ml syringes
- 14–16 g 8–10 cm stainless steel needle
- Number 3 polymerized caprolactam suture
- Number 2 or 3 catgut suture
- Sterile surgery pack
- Needle holders

Complications

- Excessive bleeding
- Accidental injection of lidocaine into the cerebral cavity causing seizure or death or both
- Injury to operator or handler after the cow is released from the chute

Procedure for Bovine Eye Enucleation

TECHNICAL ACTION	RATIONALE/AMPLIFICATION
1. Restraint	1a. Restrain the cow in a squeeze chute.
	1b. Place a halter on the cow and secure its head to one side of the chute, or secure the head in a head holder attached to the chute.
2. Surgical preparation	2a. Clip the hair away from the eye, creating a 15 × 15 cm (6 × 6 inch) square.
	2b. Trim long eyelashes to 1 cm (½ inch).
	2c. Perform a three-step surgical scrub, using the eyelids as the center of your preparation area.
	2d. Perform an infiltrative ring block completely around the eye, approximately 2 cm from the eyelid margins.
	2e. The surgeon will perform a four-point retrobulbar block using 4–5 ml of lidocaine per site. This is accomplished by injecting through the eyelids at the 12 o'clock and 6 o'clock positions and into the medial and lateral canthus (Fig. 8–30).
	2f. The surgeon will then perform an intraorbital block using the 8–10 cm (4–6 inch) needle and injecting 15–20 ml of lidocaine over the foramen orbitorotundum deep within the eye socket.
	2g. The area is scrubbed once more while the surgeon scrubs and dons sterile surgical gloves.
3. Surgery	3a. The surgeon may drape the area using a four-towel technique.
	3b. A towel clamp is used to close the eyelids, and a transpalpebral incision is made 1 cm from the lid margins, completely encircling the orbit.

TECHNICAL ACTION	RATIONALE/AMPLIFICATION
	3c. Blunt and sharp dissection is used to free the musculature from the bony orbit 360 degrees around the globe of the eye.
	3d. The optic stalk is clamped with right-angle forceps or similar tool, and a ligature of catgut is placed around the stalk deep to the forceps.
	3e. The stalk is cut with scissors, and the entire globe, musculature, and retrobulbar fascia are removed *en masse*.
	3f. The lids are sutured shut using polymerized caprolactam or other nonabsorbable suture.
	3g. The cavity will fill with blood that will clot once the pressure has built against the sealed lids.

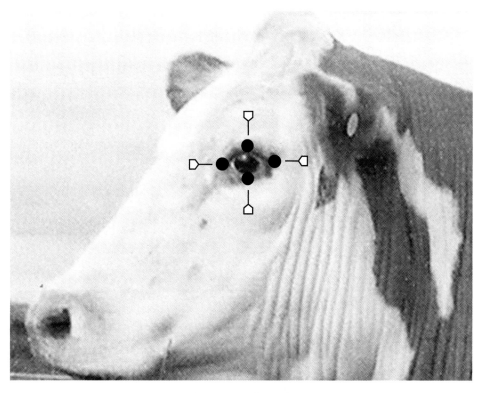

FIG. 8-30 Locations to inject lidocaine for a four-corner ocular block in preparation for eye enucleation.

TECHNICAL ACTION	RATIONALE/AMPLIFICATION
	3h. Some surgeons inject an antibiotic into the socket after closure.
4. Postoperative care	**4a.** The patient is released to a clean, small pen.
	4b. Be sure to herd the cow from its visual side and proceed slowly to keep it from slamming into the fence on its blind side.
	4c. The owner is advised to watch for excessive swelling or signs of infection.
	4d. The sutures may be removed in 10 days.

RECTAL PROLAPSE REDUCTION

Definition

A rectal prolapse is the eversion and protrusion of the rectum from the anus.

Purpose

- Return the rectal tissues to an anatomically correct position
- Relieve pain and distress associated with a rectal prolapse
- Prevent future rectal prolapse

Equipment

- Bucket of water with antiseptic (surgical) solution
- Surgical scrub
- Supplies for caudal epidural
    - 18g × 1.5-inch needle (cattle)
    - 12 ml syringe
    - Lidocaine 2%
- Surgery (Huck) towel
- Buhner needle
- Umbilical tape ½ inch
- Utility scissors
- Obstetrical sleeve

Complications

- Torn rectum acquired while restraining patient
- Hind limb ataxia secondary to caudal epidural

Procedure for Rectal Prolapse Reduction

TECHNICAL ACTION	RATIONALE/AMPLIFICATION
1. Restraint	1a. Place animal in squeeze chute or a stock.
	1b. When catching the animal, do not close the gate behind it until its head is safely caught. This will prevent it from rupturing the prolapsed tissue against the back gate.
2. Preparation	2a. Scrub the tail head and administer caudal epidural anesthesia (Chapter 10).
	2b. Wash debris and necrotic tissue from the prolapsed area, using copious amounts of water and antiseptic.
3. Prolapse reduction	3a. Wrap the prolapsed tissue in the Huck towel so that one long end is tight against the anus.
	3b. Twist the ends of the towel together like a corkscrew so that the towel tightly and completely envelops the mass.
	3c. Continue twisting the towel ends together, squeezing fluid from the mass of rectal tissue until it can be easily pushed back inside the body.
	3d. With a gloved arm, verify that the rectum is fully returned to normal anatomic position.
4. Anal purse-string suture placement	4a. Wash the perineum again with antiseptic solution.
	4b. Secure tail to side of patient or chute.
	4c. Pull out and cut and soak in disinfectant solution approximately 1 meter of umbilical tape.

TECHNICAL ACTION	RATIONALE/AMPLIFICATION
	4d. Starting 2–3 cm (1-¼ inches) dorsal to the anus at the 1 o'clock position, pass the Buhner needle into the perineal skin and direct it laterally to the 11 o'clock position (Fig. 8–31A).
	4e. Pass umbilical tape through the eye of the needle and withdraw tape and needle, pulling a piece of umbilical tape through the skin and through the entry site.
	4f. Thread the needle and put it into the exit wound created by your first needle pass. Direct the needle ventrally to the 5 o'clock position and exit the skin (Fig. 8–31B).
	4g. Grasp the umbilical tape, pulling it out of the eye of the needle and withdraw the needle.
	4h. Repeat on contralateral side, going 11 o'clock to 7 o'clock.
	4i. Repeat on ventral side, going 7 o'clock to 5 o'clock.
	4j. Pull on both ends of tape until the anus will admit only two to three fingers (Fig. 8–31C).
	4k. Tie the ends together in a bow. Trim excess umbilical tape.
	4l. Purse-string suture may be removed by the owner in 7–10 days.

  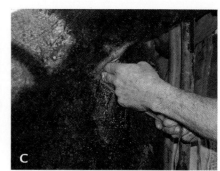

FIG. 8–31 (A) Positioning of Buhner needle and umbilical tape for the first step in applying an anal purse-string suture. (C) The 1 o'clock to 5 o'clock insertion of the umbilical tape. (C) Adjusting the tension on the umbilical tape before tying off.

PREPARATION FOR ENDOSCOPY IN THE HORSE

Definition

- The use of a flexible fiberoptic tube to examine and sample tissues from internal structures for biopsy

Purpose

- Flexible fiberoptic tube passed through the ventral nasal meatus of a horse and into the nasopharynx for diagnostic purposes
- A standard 1-meter-long scope used to examine or sample for biopsy the pharynx, larynx, and guttural pouches
- A 3-meter scope used to examine or sample for biopsy the esophagus, stomach, and even into the duodenum
- Alternative sites for diagnostic endoscopy: The vagina and uterus, the urethra, and rectum (in the case of rectal tear)

Equipment

- Flexible 1 cm × 1–3 meter endoscope
- Light source
- Air pump
- Water pump
- Lubricant gel
- Biopsy forceps

Complications

- Damage to the fiberoptics within the scope
- Nasal sinus hemorrhage
- Esophageal laceration
- Operator or handler injury

Procedure for Preparation for Endoscopy in the Horse

TECHNICAL ACTION	RATIONALE/AMPLIFICATION
1. Preparation of nursing foal for gastric endoscopy	1a. Prevent from nursing 2–4 hours before endoscopy.
2. Preparation of foal on solids for gastric endoscopy	2a. Keep from eating solid feed for 8–12 hours but may nurse up to 2–4 hours before endoscopy.

TECHNICAL ACTION	RATIONALE/AMPLIFICATION
3. Preparation of adult horse for gastric endoscopy	3a. Place a muzzle on the horse and remove all feed and bedding from the stall 8–12 hours before scheduled endoscopy. 3b. Withhold water 2–4 hours before endoscopy. 3c. A nasogastric tube may be inserted just before passage of the endoscope, to siphon off gastric juices.
4. Rectal endoscopy preparation	4a. Rectum is cleared of feces manually (adult) or by repeated enemas (foal). 4b. The tail is wrapped to keep stray hairs out of the way. 4c. Just before endoscopy, the rectum may be rinsed with 2% lidocaine to reduce discomfort and straining.
5. Restraint for any endoscopy	5a. Sucking foals may be restrained manually in lateral recumbency or standing. 5b. Older foals may require light sedation. 5c. Adult horses are placed in a stock or cross-ties. 5d. Sedation is highly recommended for patient safety, operator safety, and for preservation of the endoscope. 5e. For short procedures on a calm horse or when sedation is not advisable, a nose twitch may be applied to keep the horse quiet.
6. Endoscope set-up	6a. The light source (and video monitor) are set up close to the patient and in line-of-site of the operator (Fig. 8–32). 6b. The flexible endoscope is attached to the light source. 6c. The endoscope can be broken easily if it is bent at a sharp angle or if the end is hit against a hard surface.

TECHNICAL ACTION	RATIONALE/AMPLIFICATION
	6d. A water or air pump is hooked up to the air port on the scope.
	6e. Biopsy forceps and other endoscopy tools are set out at the ready.
7. Endoscopy of the nose, pharynx, esophagus, and stomach	**7a.** The scope is passed into the ventral medial meatus of the nares.
	7b. The nasal passages and nasopharynx are observed as the scope passes.
	7c. If the operator wishes to pass the tube into the esophagus, a small amount of water will be dribbled onto the rima glottis (opening into the larynx) to cause the patient to swallow.
	7d. Examination of the stomach is facilitated by pumping air into the stomach and using a nasogastric tube to siphon off any gastric juices.

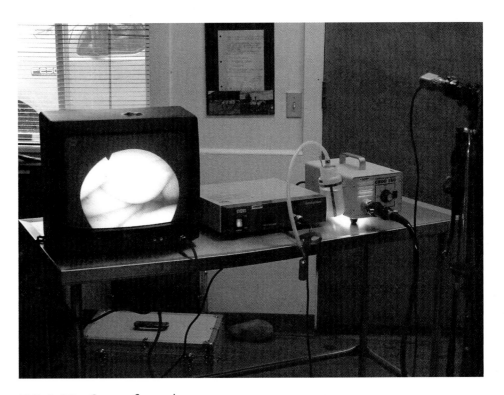

FIG. 8–32 Set-up for endoscopy.

REVIEW QUESTIONS

1. What are the roles of the surgical scrub assistant?
2. What are the roles of the circulating assistant?
3. Why is it important to keep your fingers elevated above your elbows when scrubbing for surgery?
4. Is size 3-0 suture material thicker or thinner than size 3 suture material?
5. Why is it important that packages placed in an autoclave do not touch the walls or each other?
6. What temperature and pressure must the autoclave be set at to sterilize something in 15 minutes?
7. Which side of the abdomen is prepared for a cesarean section in a ruminant?
8. Why is pain control important when dehorning a goat?
9. In what species is anesthesia regularly used when performing a castration?
10. List the complications associated with castration of the horse.
11. Ventral laparotomy most often is done in which species?
12. List four reasons to do a flank laparotomy.
13. Which is an emergency, a vaginal prolapse or a uterine prolapse?
14. What equipment is necessary for assisted vaginal delivery of a calf?
15. How long should a horse fast before undergoing gastric endoscopy?

BIBLIOGRAPHY

Auer, J. (1992). *Equine surgery,* Philidelphia W. B. Saunders.

Capucille, D. J., Poore, M. H., & Rogers, G. M. (2002). Castration in cattle: Techniques and animal welfare issues. *Compend Contin Educ Pract Vet, 24* 9, S66–S73.

Fowler, M. (1998). *Medicine and surgery of South American camelids,* (2nd ed.). Ames: Iowa State Press.

Knecht, C. D., Algernon, R. A., Williams, D. J., & Johnson, J. H. (1987). *Fundamental techniques in veterinary surgery,* (3rd ed.). Philadelphia: W. B. Saunders.

Loesch, D. A., & Rodgerson, D. H. (2003). Surgical approaches to ovariectomy in mares. *Compend Contin Educ Pract Vet, 25* 11, 862–871.

May, K. A., & Moll, H. D. (2002). Recognition and management of equine castration complications. *Compend Contin Educ Pract Vet, 24* 2, 150–162.

McCurrin, D. (1998). *Clinical textbook for veterinary technicians,* (4th ed.). Philidelphia: W. B. Saunders.

Murray, M. J. Endoscopy of the gastrointestinal tract: Current approach. In P. Chuit, A. Kuffer, & S. Montavon (Eds.), 8me Congrs de mdecine et chirurgie quine - 8. Kongress fr Pferdemedezin und -chirurgie. [8th Congress on Equine Medicine and Surgery. 2003 - Geneva, Switzerland.] Ithaca NY: IVIS. Retrieved January 31, 2006, from //www.ivis.org: Search term: P0713.1203.

Nordlund, G. (1994). Rumenocentesis: A technique for the diagnosis of sub acute rumen acidosis in dairy herds. *Bov Pract, 28,* 109–112.

Oehme, F. W., & Prier, J. E. (1974). *Textbook of large animal surgery,* Baltimore: Williams & Wilkins.

Parish, S. M., Tyler, J. W., & Ginsky, J. V. (1995). Left oblique celiotomy approach for cesarean section in standing cows. *JAVMA, 207,* 751–752.

Pavletic, M., Monnet, E., MacPhail, C., Crowe, D., Hendrikson, D., & Trout, N. (2002). Suturing and stapling. *DVM Best Pract, Oct.*

Pratt, P. (1998), *Principles and practice of veterinary technology,* (4th ed.). St. Louis: C. V. Mosby.

Risco, C. A., & Reynolds, J. P. (1988). Uterine prolapse in dairy cattle. *Compend Contin Educ Pract Vet, 10,* 1135–1142.

Turner, A. S., & McIlwraith, C. W. (1982). *Techniques in large animal surgery,* Philadelphia: Lea & Febiger.

CHAPTER 9

Selected Lower Limb Radiographic Procedures

There is something about the outside of a horse that is good for the inside of a man.
WINSTON CHURCHILL

KEY TERMS

artifact
cannon
carpus
cassette tunnel
caudal
coffin bone
cranial
DLPM
DMPL
dorsal
dorsopalmar
dosimetry badge

fetlock
lateral
medial
navicular bone
navicular box
oblique view
palmar
pastern
plantar
radiograph
tarsus
ventral

OBJECTIVES

- Identify and describe the function of basic radiographic equipment used in equine lower limb studies.
- Identify reasons for radiographic hoof and leg preparation.
- Discuss the views required to provide diagnostic quality films for the equine lower limb.
- Compare and contrast the standard views required for each radiographic series.
- Describe techniques used to prepare the limb for ultrasonographic examination.

378

BASIC RADIOGRAPHIC EQUIPMENT

Appropriate equipment is essential to obtaining a diagnostic quality **radiograph**. Table 9–1 lists some of the basic equipment and supplies required to obtain these views.

TABLE 9–1 RADIOGRAPHIC EQUIPMENT AND SUPPLIES

EQUIPMENT	FUNCTION
Lead apron, gloves, thyroid shield, goggles (Fig. 9–1)	Protect personnel from scatter radiation
Dosimetry badge	Measure amount of radiation exposure to personnel
Cassettes	Hold x-ray films
	Typically contain screens that intensify x-ray energy and decrease amount of x-ray exposure needed
Handle-type cassette holders	Assist with positioning of cassettes and decrease personnel exposure to radiation
Positioning blocks	Lift foot off ground to improve positioning and film quality
	Blocks with slots provide dual purposes as positioner and cassette holder
Cassette tunnel	Protective case for cassette capable of bearing horse's weight
Navicular box	Specialized combination of cassette holder and positioning block, which facilitates filming navicular series
Film marker	Identification markers used to label film as right or left, owner name, date, animal identification, and hospital
	Various types available including photoflash markers, lead tape, or lead letter identification clips
Thin wire or paperclip	Taped to hoof
	Used to assist in determining angulation between hoof wall and P3 (coffin)
Radiograph view box	Use to examine films
Bright light	Intensified light source that is used to examine high-density areas of the film
Radiograph tripod	Supports portable radiographic units
Radiography machine	Portable and stationary units available

FIG. 9-1 Proper radiation protection attire and equipment used when taking radiographs.

BASIC RADIOGRAPHIC SAFETY

The guaranteed, most reliable method of protecting oneself from the harmful effects of ionizing radiation associated with radiography is simple. Just don't take radiographs. Fortunately, such a ridiculous, radical approach is neither feasible nor justified. Radiographic equipment is very safe, so long as it is used in a safe manner. Wise technicians always retain a healthy respect for the dangers associated with radiographic production. This respect is manifested through the use of protective equipment, dosimetry (personal monitoring devices), and good technique, thus enabling technicians to safely use one of the most powerful diagnostic tools ever invented. In addition to the basic safety information provided here, all technicians and veterinarians are strongly encouraged to periodically review in-depth explanations regarding radiographic production and safety guidelines.

SAFETY GUIDELINES

- No one under 18 years of age should take radiographs. It is unsafe and illegal in the United States. All technicians should be familiar with individual state guidelines, because radiation safety protocols vary with each state.

- Minimize the time individuals are exposed to radiation. This can be accomplished by:

    1. Rotating personnel. (If you are not necessary for the procedure exit the area.)

2. Minimizing the number of retakes. (This is accomplished by using good technique.)

3. Using fast screens and film.

- Increase the distance between the technician and the primary beam using collimation and positioning aids such as cassette holders. No part of the technician should ever be in the primary beam.

- Always use personal monitoring devises (dosimeters) to monitor the amount of radiation exposure. Film badges remain the most common method of monitoring personnel exposure.

- Always employ direct body shielding with lead aprons, gloves, goggles, and thyroid shields.

- Routinely inspect and service radiographic equipment (annual calibration by service technician) and ancillary equipment such as gloves, gowns, cassettes, and processing chemicals.

- Maintain appropriate radiography logs.

HOOF PREPARATION FOR RADIOGRAPHS

Correct preparation of the hoof is essential for producing diagnostic quality radiographs. Brushing, removing shoes, and packing hooves are standard preparatory procedures designed to minimize artifact. Although preparation may prove time consuming, it is well worth the effort.

Purpose

- Minimize artifact and increase diagnostic quality of radiographs

Complications

- Damage to hoof wall during improper shoe removal

Equipment

- Hoof pick
- Rasp or old set of nippers
- Pullers
- Hoof knife
- Radiolucent packing material (Play-Doh)
- Paper towel
- Hard brush

Procedure for Hoof Preparation

TECHNICAL ACTION	RATIONALE/AMPLIFICATION
1. Clean foot using hoof pick. Place foot on hoof stand.	1a. Occasionally a hoof knife is used to remove overgrown sole to decrease artifact. 1b. If a stand is not available, the hoof can remain on the ground.
2. Use smooth side of rasp to file off clinches. Alternatively, an old pair of nippers can be used to cut the clinches if shoes are old or loose.	2a. Some owners will not permit shoe removal. Obtain owner consent before removal. 2b. Clinches are the ends of the nail in the hoof wall that have been bent over or clinched to secure the shoe. 2c. Clinches must be removed or the hoof wall will be damaged during shoe removal.
3. Holding hoof in a farrier stance, use pullers to grasp the heel portion of the shoe. Apply pressure and begin prying the shoe from the hoof. Alternate between the medial and lateral sides of the shoe. Gradually work toward the toe.	3a. Alternating sides minimizes hoof wall damage. 3b. It is advisable to remove the shoe on the contralateral side as well. This will prevent unevenness and soreness for the horse until the farrier is contacted. 3c. Always place removed shoes in bag and return to owner.
4. Once the shoe is removed, use a stiff or wire brush to clean sole and hoof wall.	4a. Some individuals recommend trimming the hoof at this point to minimize artifact. 4b. A scrub bucket with water may be necessary to clean the hoof adequately.
5. Pack foot with radiolucent material such as Play-Doh. Pay particular attention to the sulci and clefts.	5a. Play-Doh prevents air pocket artifacts. 5b. Clefts and sulci are depressions around the frog of the hoof.
6. Place paper towel over sole to prevent contamination of prepared hoof.	—

LOWER LIMB RADIOGRAPHS

Most large animal radiographs involve the lower limb. Their use as a diagnostic tool is essentially unparalleled. The importance of attaining views appropriate for each diagnostic workup can not be overstressed.

Purpose

- Obtain diagnostic quality radiographic films

Complications

- Personnel trauma
- Damage to radiographic equipment
- Nondiagnostic films because patient is uncooperative

Equipment

- Hoof preparation materials (see above)
- Cassettes loaded with film (12 × 14 inches is standard)
- Tranquilizer
- Positioning block
- Cassette tunnel
- Lead apron, gloves, goggles, thyroid shield
- Dosimetry badge
- Brush
- Lead markers, identification tape
- Handle-type cassette holder
- Coffin bone–specific equipment: Wire and tape
- Navicular bone–specific equipment: Navicular box
- Metacarpal-specific equipment: 7 × 17-inch cassettes

Procedure

- Procedures variable based on the radiographic series selected

Procedure Used for All Distal Limb Radiographs

TECHNICAL ACTION	RATIONALE/AMPLIFICATION
1. Halter horse and be sure that it is standing on a flat surface.	1a. Ideally the horse should be standing on cement.
	1b. Two assistants should be available, one to restrain the animal and one to hold the cassettes.
2. Sedate horse if necessary.	2a. Horses often spook during radiographic procedures. Tranquilizers can be used to calm horses and minimize movement.
3. Prepare hoof if filming P2 (short pastern), P3 (coffin), or navicular bones. See steps listed in hoof preparation.	3a. Proper hoof and leg preparation minimizes artifact and increases diagnostic quality of films.
	3b. Brush limb if evaluating any structure proximal to the coffin (P3) bone.
4. Properly mark and label radiograph.	4a. Proper identification of films assists in diagnosis and meets legal requirements. Remember, for legal purposes the identification must be within the film emulsion.
	4b. Oblique views should always be marked as dorsomedial–palmar lateral (DMPL) or dorsolateral–palmar medial (DLPM). This is vitally important when evaluating symmetric joints such as the fetlock.
	4c. Photo imprinters make film identification much easier. Information that should be on each film includes: horse and owner name, age, breed, sex, area of interest, date, clinic name and address, and veterinarian name.
	4d. Radiographic logs should be maintained within all hospitals.

Understanding directional terms is an important aspect of equine radiology (Fig. 9–2). All individuals involved in obtaining these radiographs should be familiar with basic anatomic directional terms. As a general rule, remember that the name of the radiographic view reflects where the beam comes from and goes to.

In addition to understanding directional terms, technicians must recognize both anatomically correct and lay terminology associated with the lower limb structures in the horse. See the appendices for a review of these terms.

CHAPTER 9 Selected Lower Limb Radiographic Procedures

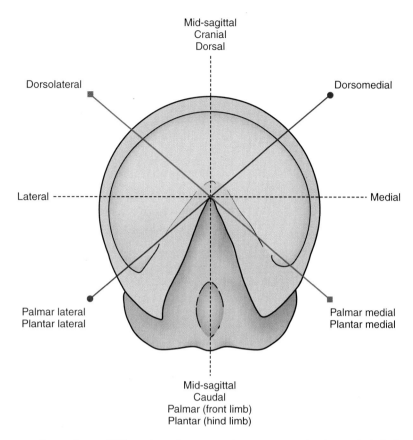

FIG. 9-2 Location of Directional nomenclature or to the horse's left foot.

Radiographic Views of Coffin

TABLE 9-2 RADIOGRAPHIC VIEWS OF COFFIN

VIEW	PURPOSE
Standard Views	
Lateral	Evaluates hoof wall
45-Degree dorsopalmar	P3 (coffin) rotation
Optional Views	
Horizontal 0-degree dorsopalmar	Evaluates extensor process, P2–P3 joint, hoof wall, hoof-ground surface relationship
DMPL oblique	Assess nondisplaced fracture, P3 wing
DLPM oblique	Assess nondisplaced fractures, wing P3

Procedure for Lateral View of Coffin

TECHNICAL ACTION	RATIONALE/AMPLIFICATION
1. Complete Steps 1–4 listed under procedure for all distal limb radiographs.	1a. The coffin bone is also called the distal or third phalanx.
2. Prepare hoof as described in hoof preparation.	—
3. Place foot on positioning block so that hoof is as close as to the medial edge as possible.	3a. Placing hoof on edge minimizes the film-to-object distance, thereby limiting artifactual magnification.
	3b. Elevating the opposite forelimb or holding hoof off ground helps to minimize motion.
	3c. Fig. 9–3.
4. Place cassette on medial aspect of limb as close to the block as possible.	4a. The cassette can be placed in the cassette slot depending on the type of positioning block used
5. Center beam on coronary band. Expose film.	5a. The coronary band is essentially the skin-hoof interface.
	5b. The entire hoof should be on the film with P3 centered.

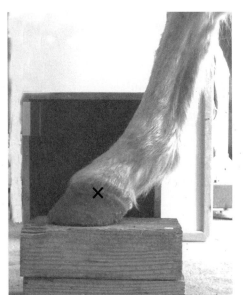
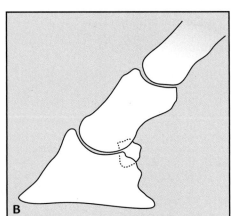

FIG. 9–3 (A) Positioning of the horse and cassette for a lateral view of the coffin bone. X shows where to aim the beam. (B) Drawing of resultant radiograph.

TECHNICAL ACTION	RATIONALE/AMPLIFICATION
6. Tape a thin-gauge wire to the **dorsal-cranial** surface of the hoof wall.	**6a.** Wire is used to facilitate examination for P3 rotation. The wire provides a radiodense approximation of the hoof wall.
	6b. This view would be used to diagnose laminitis with rotation.
7. Repeat Steps 5–7 above.	—

Procedure for 45-degree Dorsopalmar View of Coffin

TECHNICAL ACTION	RATIONALE/AMPLIFICATION
1. Complete Steps 1–4 listed under procedure for all distal limb radiographs.	—
2. Prepare hoof as described in hoof preparation.	—
3. Place cassette in cassette tunnel and set on ground directly cranial to hoof.	**3a.** Tunnels provide protection to cassettes.
	3b. Fig. 9–4.
4. Place foot on center of tunnel.	**4a.** Elevating or lifting the opposite forefoot often helps to minimize motion by the horse.
5. Standing cranial to the leg, center the beam at a 45-degree angle (to the ground), 1 inch proximal to the coronary band. Expose film.	**5a.** The coronary band is essentially at the hoof and skin interface.
	5b. Quality films taken at the correct angles will show an equal size of both wings of P3.

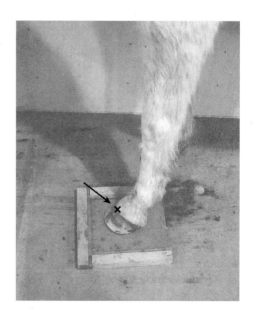

FIG. 9–4 Positioning of horse and cassette for a 45-degree dorsopalmar view of the coffin bone. *Arrow* shows direction of the beam.

Procedure for Horizontal Zero-degree Dorsopalmar View of Coffin

TECHNICAL ACTION	RATIONALE/AMPLIFICATION
1. Complete Steps 1–4 listed under procedure for all distal limb radiographs.	—
2. Prepare hoof as described in hoof preparation.	—
3. Place hoof on positioning block so that heel is as close to the caudal edge as possible.	3a. Placing hoof on edge minimizes the film-object distance, thereby limiting artifactual magnification. 3b. Elevating the opposite forelimb or holding hoof off ground helps to minimize motion.
4. Put cassette in holder and place cassette caudal to limb as close to the block as possible.	4a. The cassette can be placed in the cassette slot, depending on the type of positioning block used. 4b. Fig. 9–5.
5. Standing cranial to the limb, center the beam parallel to ground, aiming at the coronary band. Expose film.	5a. This view causes significant distortion of P3 and should therefore be used only as an additional view. 5b. View should include the entire hoof.

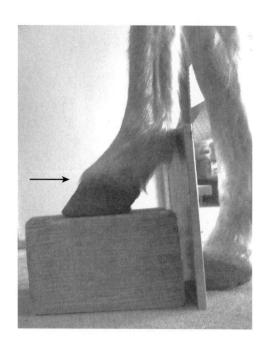

FIG. 9–5 Positioning of horse and cassette for horizontal zero-degree dorsopalmar view of the coffin bone. *Arrow* shows direction of the beam.

CHAPTER 9 Selected Lower Limb Radiographic Procedures

Procedure for Oblique Views of Coffin

TECHNICAL ACTION	RATIONALE/AMPLIFICATION
1. Complete Steps 1–4 listed under procedure for all distal limb radiographs.	—
2. Prepare hoof as described in hoof preparation.	—
3. Place cassette in cassette tunnel and set on ground directly cranial to hoof.	3a. Tunnels provide protection to cassettes.
4. Place foot on center of tunnel.	4a. Elevating or lifting the opposite forefoot often helps to minimize motion of the horse.
5. DMPL: Standing cranial and medial to the leg, angle the beam 45 degrees from the ground and direct it medially (45 degrees medial to a true dorsopalmar view), just distal to the coronary band. Expose film.	5a. This view is used to visualize the medial wing of P3. 5b. Fig. 9–6.
6. DLPM: Standing cranial and lateral to the limb, angle the beam 45 degrees from the ground and direct it laterally (45 degrees lateral to a true dorsopalmar), just distal to the coronary band. Expose film.	6a. This view is used to visualize the lateral wing of P3. 6b. See Fig. 9–6.

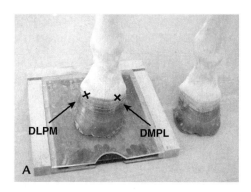
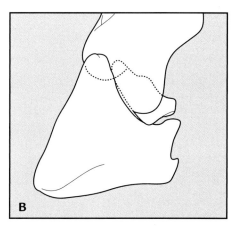

FIG. 9–6 (A) Positioning of horse and cassette for doromedial-palmarlateral (DMPL) and dorsolateral-palmarmedial (DLPM) obliques of the coffin bone. X shows target of the beam for each shot. (B) Drawing of resultant radiograph.

Radiographic Views of Navicular

TABLE 9–3 RADIOGRAPHIC VIEWS FOR NAVICULAR

VIEW	PURPOSE
Standard Views	
Palmar proximal (skyline, flexor)	Evaluates navicular medullary cavity, cortex, and flexor surface
Lateral	Evaluates changes in navicular shape associated with chronic degeneration
	Commonly used to assess for navicular disease, fractures, and street nail disease
65-Degree dorsopalmar (upright pedal)	Evaluates navicular bone, specifically distal border
Optional Views	
Oblique views	
65-Degree dorsoproximolateral	Evaluate wings of navicular without superimposition of adjacent bones
65-Degree dorsoproximomedial	Typically used to assess for fractures

Procedure for Palmar-proximal or Skyline View of Navicular

TECHNICAL ACTION	RATIONALE/AMPLIFICATION
1. Complete Steps 1–4 for all distal limb radiographs.	1a. The navicular is also called the distal sesamoid.
	1b. Be certain to prepare foot properly using steps listed in hoof preparation.
	1c. It is highly recommended that this view be performed first if the horse has received a nerve block. The horse is much more likely to stand properly for this view if the block's anesthetic properties are still in effect.
	1d. Additional names for this view include flexor and caudal tangential.
2. Prepare hoof as described in hoof preparation.	—

TECHNICAL ACTION	RATIONALE/AMPLIFICATION
3. Place film in cassette tunnel and place caudal to hoof. Place hoof on cassette.	3a. Place the cassette as far caudally (with the fetlock still in the extended position) as the horse will allow. This will assist in placing P2 in a more vertical position, thus permitting better visualization of the navicular. 3b. Fig. 9–7.
4. Stand caudal to limb. The radiography unit must be directly caudal to the hoof. Angle 65 degrees from the ground and aim directly between the bulbs of the heel.	4a. Use extreme caution when obtaining this view. Damage to equipment and personnel can easily occur. (The operator is underneath the belly during the procedure.)

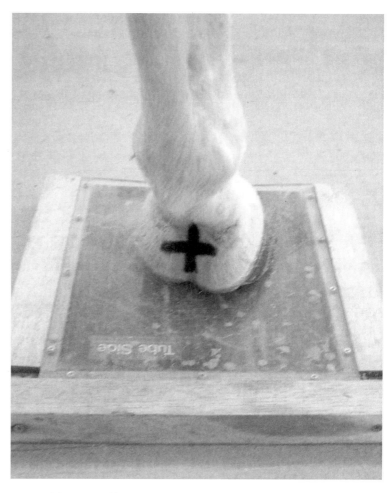

FIG. 9–7 Positioning of horse and cassette for palmarproximal (skyline) view of the navicular bone. Aim beam at the marker.

Procedure for Lateral View of Navicular

TECHNICAL ACTION	RATIONALE/AMPLIFICATION
1. Complete Steps 1–4 for all distal limb radiographs.	—
2. Prepare hoof as described in hoof preparation.	—
3. Complete steps 5–7 listed under lateral radiograph for coffin (P3).	3a. Do not complete Step 8 of lateral P3 radiographs. 3b. Fig. 9–8.

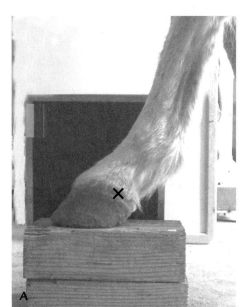
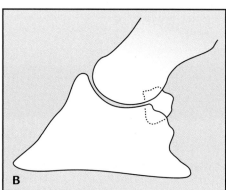

FIG. 9-8 (A) Positioning of horse and cassette for lateral view of the navicular bone. *X* shows target of the beam. (B) Drawing of resultant radiograph.

Procedure for 65-degree Dorsopalmar or Upright Pedal View of Navicular

TECHNICAL ACTION	RATIONALE/AMPLIFICATION
1. Complete Steps 1–4 for all distal limb radiographs.	—
2. Prepare hoof as described in hoof preparation.	2a. Be certain to prepare foot properly.
3. Two methods are available to obtain this view. The technician can select either method one (Step 4) or method two (Step 5) to make this film.	3a. This view is also called the upright pedal, because it projects the navicular bone dorsal to the P2–P3 joint.

CHAPTER 9 Selected Lower Limb Radiographic Procedures 393

TECHNICAL ACTION	RATIONALE/AMPLIFICATION
	3b. Obtaining the view is easier using the navicular box (method one), but a cassette tunnel (method two) can also be used.
4. Method one: Place hoof in groove of navicular box and cassette in the slot caudally. Keeping beam parallel to the floor, aim beam at the coronary band. Expose film.	**4a.** The groove of the box holds the foot in a flexed position. This effectively maintains the dorsal hoof wall perpendicular to the ground and parallel to the cassette.
	4b. Most of the horse's weight must be on the opposite forelimb.
5. Method two: Place cassette in tunnel and set directly cranial to the hoof. Place hoof on cassette. Standing cranial to the leg, direct the beam at a 65-degree angle to the ground, 1 inch proximal to the coronary band. Expose film.	**5a.** It is very important that the beam be directly midsagittal. This will minimize distortion of the navicular bone.
	5b. Fig. 9–9.

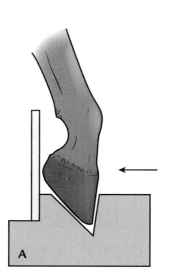 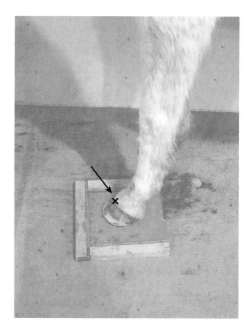

FIG. 9-9 (A) Positioning of horse and navicular box for the upright pedal using method one. Positioning of horse and cassette for 65-degree dorsopalmar (upright pedal) view of the navicular bone using method two. *Arrow* shows direction of the beam; *X* shows the target.

Procedure for Oblique Views of Navicular

TECHNICAL ACTION	RATIONALE/AMPLIFICATION
1. Complete Steps 1–4 for all distal limb radiographs.	—
2. Prepare hoof as described in hoof preparation.	2a. Be certain to prepare foot properly.
3. Place cassette in tunnel and set on ground cranial to hoof. Place hoof on cassette.	3a. Tunnels provide protection to the cassette.
4. 65-Degree dorsoproximolateral view is done as follows: Stand cranial and lateral to the hoof. Angle the beam 65 degrees from the ground and direct it toward the coronary band, 25 degrees lateral to midline. Expose film.	4a. Fig. 9–10.
5. 65-Degree dorsoproximomedial view is done as follows: Stand cranial and medial to the limb. Angle the beam 65 degrees from the ground and direct it toward the coronary band, 25 degrees medial to midline. Expose film.	5. Fig. 9–10.

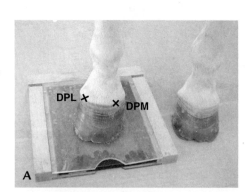
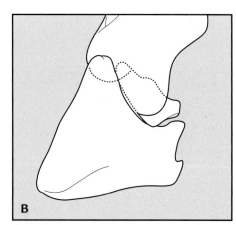

FIG. 9-10 (A) Positioning of horse and cassette for 65-degree dorsoproximal-lateral (DPL) and 65-degree dorsoproximal-medial views of the navicular bone. *Arrows* show direction of the beam and *X* shows the target for each exposure. (B) Drawing of resultant radiograph.

Radiographic Views of Pastern

TABLE 9-4 RADIOGRAPHIC VIEWS FOR PASTERN

VIEW	PURPOSE
Standard Views	
Dorsopalmar	Evaluate P1–P2 articular surfaces, joint space width, subchondral bone, and presence of any periarticular bone
Lateral	Evaluates P1, P2, and general hoof axis
Optional Views	
Oblique views	
DMPL	Permits evaluation of palmar medial and dorsolateral aspects of pastern
DLPM	Permits evaluation of palmar lateral and dorsomedial aspects of pastern

Procedure for Dorsopalmar View of Pastern

TECHNICAL ACTION	RATIONALE/AMPLIFICATION
1. Complete Steps 1–4 for all distal limb radiographs.	1a. Hoof preparation is not necessary.
2. Brush leg, removing all visible dirt and debris.	2a. Dirt causes radiographic artifact and decreases diagnostic quality of the films.
3. Place hoof on positioning block so that heel is as close to the caudal edge as possible.	3a. All attempts should be made to keep legs perpendicular to the ground in a normal weight-bearing position.
	3b. Placing hoof on edge minimizes the film-object distance, thereby limiting artifactual magnification.
	3c. Elevating the opposite forelimb on additional block helps to maintain normal weight bearing.
	3d. Lifting opposite forefoot can help to minimize motion.
4. Place cassette caudal to pastern. The cassette must be held parallel to the pastern.	4a. Alternatively, the cassette can be placed in positioning box cassette groove.
	4b. Fig. 9-11.

TECHNICAL ACTION	RATIONALE/AMPLIFICATION
5. Standing cranial to the leg, direct the beam at a 45-degree angle to the ground, aiming midpastern. Expose film.	5a. Midpastern is the proximal interphalangeal joint.
	5b. The view should include both P1 and P2 and usually P3.
	5c. It is critical that the beam always be perpendicular to the pastern and cassette. Thus the actual angle of the beam can vary from 30–45 degrees, depending on the horse's conformation.

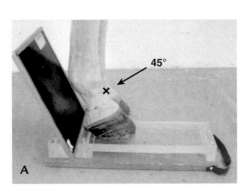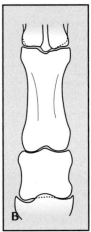

FIG. 9-11 (A) Positioning of horse and cassette for dorsopalmar view of the pastern bones. *Arrow* shows direction of the beam. (B) Drawing of resultant radiograph.

Procedure for Lateral View of Pastern

TECHNICAL ACTION	RATIONALE/AMPLIFICATION
1. Complete Steps 1–4 for all distal limb radiographs.	1a. Hoof preparation is not necessary.
2. Brush leg, removing all visible dirt and debris.	2a. Dirt causes radiographic artifact and decreases diagnostic quality of the film.
3. Place foot on positioning block so that hoof is as close as to the medial edge as possible.	3a. Placing hoof on edge minimizes the film-object distance, thereby limiting artifactual magnification.
	3b. Elevating the opposite forelimb or holding hoof off ground helps to minimize motion.

CHAPTER 9 Selected Lower Limb Radiographic Procedures

TECHNICAL ACTION	RATIONALE/AMPLIFICATION
4. Put cassette in holder and place medial to pastern. The cassette must be held perpendicular to the floor.	4a. Fig. 9–12.
5. Direct beam parallel to the ground and aim midpastern. Expose film.	5a. Midpastern is the proximal interphalangeal joint. 5b. View should include P1, P2, and P3.

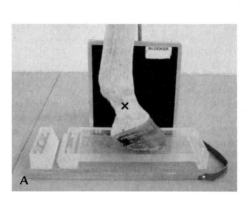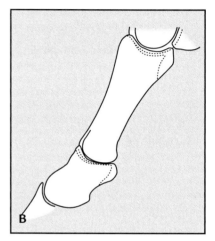

FIG. 9–12 (A) Positioning of horse and cassette for lateromedial view of the pastern bones. *X* shows the target of the beam. (B) Drawing of resultant radiograph.

Procedure for Oblique Views of Pastern

TECHNICAL ACTION	RATIONALE/AMPLIFICATION
1. Complete Steps 1–4 for all distal limb radiographs.	1a. Hoof preparation is not necessary.
2. Brush leg, removing all visible dirt and debris.	2a. Dirt causes radiographic artifact and decreases diagnostic quality of the films.
3. Place hoof on positioning block.	3a. All attempts should be made to keep legs perpendicular to the ground in a normal weight-bearing position. 3b. Elevating the opposite forelimb on additional block helps to maintain normal weight bearing. 3c. Lifting opposite forefoot can help to minimize motion.

TECHNICAL ACTION	RATIONALE/AMPLIFICATION
4. DMPL: Stand cranial and medial to leg. Place cassette caudal and lateral to pastern. Direct beam at a 45-degree angle to the ground, 25 degrees medial to midline. Expose film.	**4a.** The pastern or fetlock is symmetric. Be certain to mark the film DMPL. **4b.** The cassette must be parallel to the pastern. **4c.** Directing 45 degrees to the ground makes the beam perpendicular to the pastern. **4d.** Fig. 9–13.
5. DLPM: Stand cranial and lateral to leg. Place cassette caudal and medial to pastern. Direct beam at a 45-degree angle to the ground, 25 degrees lateral to midline. Expose film.	**5a.** The pastern or fetlock is symmetric. Be certain the mark the film DLPM. **5b.** The cassette must be parallel to the pastern. **5c.** Directing 45 degrees to the ground makes the beam perpendicular to the pastern. **5d.** Fig. 9–14.

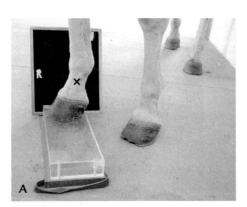 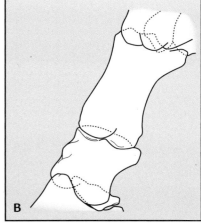

FIG. 9–13 (A) Positioning of horse and cassette for dorsomedial–palmar lateral (DMPL) **oblique view** of the pastern bones. *X* shows target of the beam. (B) Drawing of resultant radiograph.

 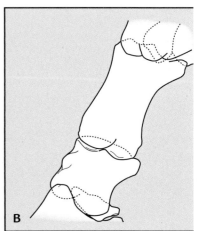

FIG. 9–14 (A) Positioning of horse and cassette for dorsolateral–palmar medial (DLPM) oblique view of the pastern bones. *X* shows target of the beam. (B) Drawing of resultant radiograph.

Radiographic Views of Fetlock

TABLE 9-5 RADIOGRAPHIC VIEWS FOR FETLOCK

VIEW	PURPOSE
Required	
Dorsopalmar	Shows articular surfaces, joint space width, subchondral bone, and presence of any periarticular bone
Extended lateral	Shows all information listed above
	Additionally valuable for examining the palmar surfaces of metacarpals or metatarpals and sesamoid bones
Flexed lateral	Especially valuable for evaluation of articular surfaces
Oblique views	Provide ability to specifically evaluate medial and lateral aspects of bones
DMPL	Highlights aspect dorsolateral and medialpalmar
DLPM	Highlights aspect dorsomedial and lateralpalmar
Optional	
Flexor surface or skyline	Evaluation of sesamoids or bony changes secondary to suspensory tears

Procedure for Dorsopalmar View of Fetlock

TECHNICAL ACTION	RATIONALE/AMPLIFICATION
1. Complete Steps 1–4 listed under all limb radiographs.	1a. Hoof preparation is not necessary.
2. Brush leg to remove all dirt and debris.	2a. Brushing minimizes artifact.
3. Place hoof on ground, ensuring that limb is directly under body and hoof is bearing full weight.	3a. Severely injured animals may not be willing to bear weight. Do not force weight bearing in these animals.
	3b. If needed, the opposite forelimb can be elevated to minimize motion while filming.
	3c. If the opposite limb is elevated, the joint space can appear altered artifactually due to abnormal weight bearing. Typically this would cause a narrowing of the joint space.

TECHNICAL ACTION	RATIONALE/AMPLIFICATION
4. Put cassette in holder and place directly caudal to fetlock.	4a. The cassette should be in contact with the heel surface.
	4b. Keeping the cassette as close as possible to the fetlock will minimize distortion.
	4c. Fig. 9–15.
5. Stand cranial to the hoof. Center the beam directly midsagittal on the fetlock. Ensure that beam is perpendicular to the hoof axis and cassette. Expose film.	5a. Fetlock is synonymous with metacarpophalangeal joint (MC3–P1).
	5b. To maintain beam perpendicular to the hoof axis or cassette, the beam will need to be angled slightly downward. It will not be perpendicular to the floor. The degree of downward angulation is dependent on the horse's conformation.
	5c. Field of view should include fetlock joint, distal metacarpal, and proximal phalanx.

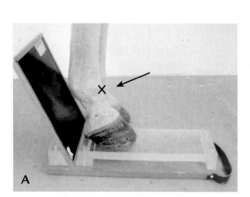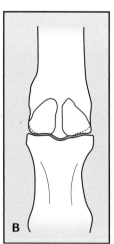

FIG. 9-15 (A) Positioning of horse and cassette for dorsopalmar view of the fetlock joint. *Arrow* shows direction of the beam and *X* shows the target. (B) Drawing of resultant radiograph.

Procedure for Extended Lateral View of Fetlock

TECHNICAL ACTION	RATIONALE/AMPLIFICATION
1. Complete Steps 1–4 listed under all limb radiographs.	1a. Hoof preparation is not necessary.
2. Brush leg to remove all dirt and debris.	2a. Brushing minimizes artifact.
3. Place hoof on ground, ensuring that limb is directly under body and hoof is fully weight bearing.	3a. Severely injured animals may not be willing to bear weight. Do not force weight bearing in these animals.
	3b. If needed, the opposite forelimb can be elevated to minimize motion while filming.
	3c. If the opposite limb is elevated, the joint space can be altered artifactually because of abnormal weight bearing. Typically this would cause a narrowing of the joint space.
4. Put cassette in holder and place on medial aspect of fetlock. Cassette should be held perpendicular to the floor.	4a. Cassette should be in contact with fetlock.
	4b. Fig. 9–16.
5. Stand lateral to the limb. Aim beam directly at fetlock joint parallel to the floor. Expose film.	5a. A true lateral view will show metacarpal condyles, visible joint space, and superimposed sesamoids. Most lateral films taken lack correct technique and obliterate the joint space.

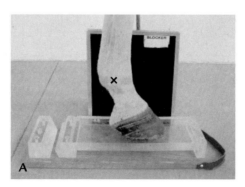
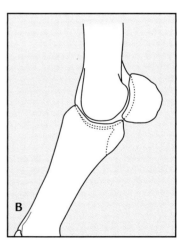

FIG. 9–16 (A) Positioning of horse and cassette for extended lateral view of the fetlock joint. X shows the target of the beam. (B) Drawing of resultant radiograph.

Procedure for Flexed Lateral View of Fetlock

TECHNICAL ACTION	RATIONALE/AMPLIFICATION
1. Complete Steps 1–4 listed under all limb radiographs.	1a. Hoof preparation is not necessary.
2. Brush leg to remove all dirt and debris.	2a. Brushing minimizes artifact.
3. Assistant should stand caudal to limb and hold hoof off the ground at the level of the opposite **carpus** (knee). Grasp the toe of the hoof and flex the fetlock as much as possible.	3a. Do not allow the hoof to roll medially in your hand as you flex the fetlock. 3b. Do not attempt to pull the entire limb laterally. The limb should remain in a conformationally neutral position under the body.
4. Put cassette in holder and place medial to fetlock. The cassette must be held perpendicular to the ground.	4a. Fig. 9–17.
5. Stand lateral to the limb. Aim directly at the fetlock, maintaining the beam parallel to the ground. Ensure that the beam is collimated properly. Expose film.	5a. Collimation is critical for avoiding direct exposure of the assistant's hand, which is holding the hoof. 5b. View should include fetlock joint, sesamoids, proximal phalanx, and distal metacarpal.

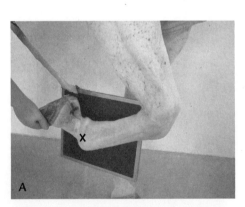 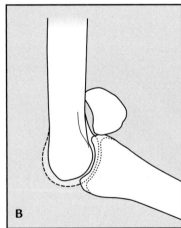

FIG. 9-17 (A) Positioning of horse and cassette for flexed lateral view of the fetlock joint. *X* shows the target of the beam. (B) Drawing of resultant radiograph.

Procedure for Oblique Views of Fetlock

TECHNICAL ACTION	RATIONALE/AMPLIFICATION
1. Complete Steps 1–4 listed under all limb radiographs.	1a. Hoof preparation is not necessary.
2. Brush leg to remove all dirt and debris.	2a. Brushing minimizes artifact.
3. Place hoof on ground, ensuring that limb is directly under body and hoof is bearing full weight.	3a. Severely injured animals may not be willing to bear weight. Do not force weight bearing in these animals.
	3b. If needed, the opposite forelimb can be elevated to minimize motion while filming.
	3c. If the opposite limb is elevated, the joint space can be altered artifactually because of abnormal weight bearing. Typically this would cause a narrowing of the joint space.
4. DMPL view: Put cassette in holder and place on the palmar lateral aspect of the fetlock. The cassette should be touching the hoof and perpendicular to the floor.	4a. The DMPL is also called the lateral oblique view. The name reflects the position of the cassette.
	4b. Because of joint symmetry, correct markers are critical. This cassette should be marked DMPL.
	4c. Fig. 9–18.
5. Stand craniomedial to the limb. Direct the beam medially 45 degrees from a true dorsopalmar (midsagittal) position and center on fetlock joint. Maintain beam parallel to floor and perpendicular to cassette. Expose film.	5a. Angle can vary from 30 to 45 degrees, depending on oblique preference. Forty-five degrees is considered the standard view angle.

FIG. 9-18 (A) Positioning of horse and cassette for dorsomedial–palmar lateral (DMPL) oblique view of the fetlock joint. *X* shows the target of the beam. (B) Drawing of resultant radiograph.

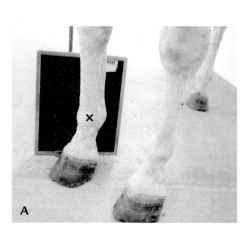
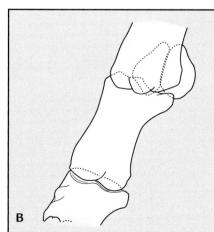

404 SECTION 4 Surgical, Radiographic, and Anesthetic Preparation

TECHNICAL ACTION	RATIONALE/AMPLIFICATION
6. DLPM view: Put cassette in holder and place on the palmar medial aspect of the fetlock. The cassette should be touching the hoof and perpendicular to the floor.	6a. The DLPM is also called the medial oblique. The name reflects the position of the cassette.
	6b. Because of joint symmetry, correct markers are critical. This cassette should be marked DLPM.
	6c. Fig. 9–19.
7. Stand craniolateral to limb. Direct beam 45 degrees lateral from a true dorsopalmar (midsagittal) and center on fetlock joint. Maintain beam parallel to floor and perpendicular to cassette. Expose film.	7a. Angle can vary from 30 to 45 degrees, depending on oblique preference. Forty-five degrees is considered the standard view angle.

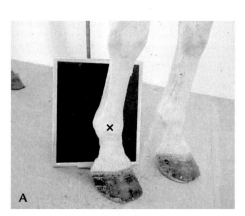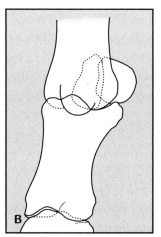

FIG. 9-19 (A) Positioning of horse and cassette for dorsolateral–palmar medial (DLPM) oblique view of the fetlock joint. *X* shows the target of the beam. (B) Drawing of resultant radiograph.

Procedure for Flexor Surface or Skyline View of Fetlock

TECHNICAL ACTION	RATIONALE/AMPLIFICATION
1. Complete Steps 1–4 listed under all limb radiographs.	1a. Hoof preparation is not necessary.
2. Brush leg to remove all dirt and debris.	2a. Brushing minimizes artifact.
3. Place cassette in cassette tunnel and set on ground caudal to hoof. Place hoof on center of cassette.	3a. The cassette tunnel should be placed as far caudally as the horse will permit.
	3b. Fig. 9–20.

TECHNICAL ACTION	RATIONALE/AMPLIFICATION
4. Stand caudal to limb. Position radiography unit directly dorsal and caudal to fetlock. Direct beam as close to perpendicular to the ground as possible. Aim between the sesamoids. Expose film.	4a. This view is also called the proximopalmar distopalmar view of the proximal sesamoids. 4b. Because of the location of the radiography unit, this view is especially apt to result in injury to equipment or personnel.

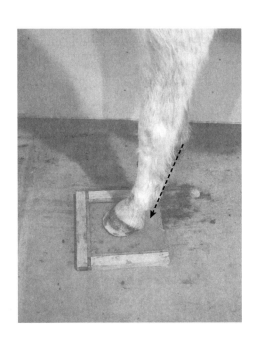

FIG. 9-20 Positioning for flexor surface (skyline) view of the fetlock joint. Direct beam between the sesamoid bones (*arrow*).

Radiographic Views of Metacarpals and Metatarsals

TABLE 9-6 RADIOGRAPHIC VIEWS FOR METACARPALS AND METATARSALS

VIEW	PURPOSE
Standard Views	
Dorsopalmar	Typically used to evaluate sesamoids, metacarpals 3 and 4, and the carpometacarpal articulation
Lateral	See above
Oblique views	
DMPL DLPM	Both oblique views are especially valuable for assessing splint bones (metacarpals [MC] 2 and 4) DMPL shows medial splint bone (MC2) DLPM shows lateral splint bone (MC4)

Procedure for Dorsopalmar View of Metacarpals and Metatarsals

TECHNICAL ACTION	RATIONALE/AMPLIFICATION
1. Complete Steps 1–4 listed under all limb radiographs.	1a. Hoof preparation is not necessary.
2. Brush leg to remove all dirt and debris.	2a. Brushing minimizes artifact.
3. Place hoof on ground in normal weight-bearing position.	3a. The horse should be standing squarely, with weight evenly distributed on all four limbs.
4. Put cassette in holder and place caudal to the metacarpals (**cannon**). Maintain cassette in contact with the cannon, parallel to the leg and perpendicular to the ground.	4a. Longer cassettes typically are used for this series (7 × 17-inch cassettes are recommended). The increased length permits visualization of the cannon in its entirety.
	4b. Fig. 9–21.
5. Stand cranial to the limb. Direct the beam midsagittally, parallel to the ground. Aim toward the center portion of the cannon. Expose film.	5a. View should include all of metacarpal bones 2, 3, 4, and a portion of the carpal and metacarpophalangeal joints.

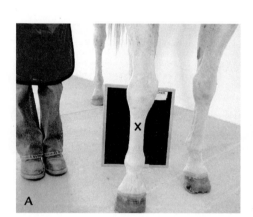 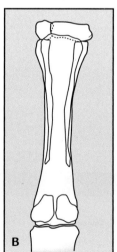

FIG. 9–21 (A) Positioning of horse and cassette for dorsopalmar view of metacarpal bones. *X* shows target of the beam. A dorsoplantar of the metatarsals is positioned in similar fashion. (B) Drawing of resultant radiograph.

Procedure for Lateral View of Metacarpals and Metatarsals

TECHNICAL ACTION	RATIONALE/AMPLIFICATION
1. Complete Steps 1–4 listed under all limb radiographs.	1a. Hoof preparation is not necessary.
2. Brush leg to remove all dirt and debris.	2a. Brushing minimizes artifact.
3. Place hoof on ground in normal weight-bearing position.	3a. The horse should be standing squarely, with weight evenly distributed on all four limbs.
4. Put cassette in holder and place on the medial aspect of the metacarpus (cannon). Cassette should be parallel to the leg and perpendicular to the ground.	4a. The cassette should contact the leg. 4b. Longer cassettes should be used (7 × 17 inches). 4c. Fig. 9–22.
5. Stand lateral to the leg. Direct the beam parallel to the ground, aiming midcannon (halfway between fetlock and knee). Expose film.	5a. A true lateral image will superimpose the second and fourth metacarpal bones (splint bones). 5b. View should include all of metacarpal bones 2, 3, and 4 and a portion of the knee and fetlock.

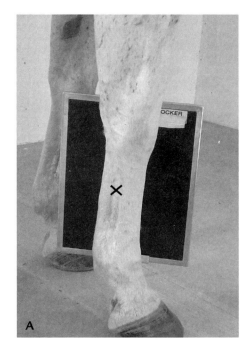 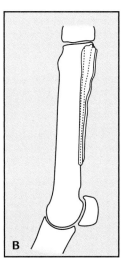

FIG. 9–22 (A) Positioning of horse and cassette for lateral view of metacarpal bones. *X* shows target of the beam. A lateral of the metatarsals is positioned in similar fashion. (B) Drawing of resultant radiograph.

Procedure for Oblique Views of Metacarpals and Metatarsals

TECHNICAL ACTION	RATIONALE/AMPLIFICATION
1. Complete Steps 1–4 listed under all limb radiographs.	1a. Hoof preparation is not necessary.
2. Brush leg to remove all dirt and debris.	2a. Brushing minimizes artifact.
3. Place hoof on ground in normal weight-bearing position.	3a. The horse should be standing squarely, with weight evenly distributed on all four limbs.
4. DLPM: Put cassette in holder and place on palmar medial aspect of leg. Cassette should be parallel to leg and perpendicular to ground.	4a. This view is also called the medial oblique. The name reflects the position of the cassette.
	4b. Longer cassettes are advised (7 × 17 inches).
	4c. Cassette should be in contact with leg.
	4d. Marking the cassette DLPM is critical.
	4e. Fig. 9–23.
5. Stand craniolateral to limb. Maintain the beam parallel to the ground. Direct beam 45 degrees lateral to a true dorsopalmar (midsagittal) view. Center beam in mid-metacarpal area. Expose film.	5a. Mid-metacarpal is halfway between the knee and fetlock.
	5b. This view highlights the metacarpal 4 (lateral splint) bone.
	5c. View should include metacarpal bones 2, 3, and 4 in entirety and a portion of the knee and fetlock.

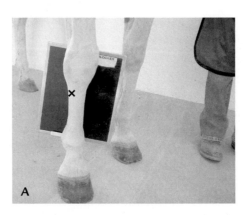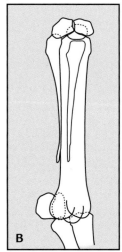

FIG. 9–23 (A) Positioning of horse and cassette for dorsolateral–palmar medial (DLPM) oblique view of the metacarpal bones. *X* shows target of the beam. A dorsolateral-palmarmedial (DLPM) oblique view of the metatarsals is positioned in the same fashion. (B) Drawing of resultant radiograph.

TECHNICAL ACTION	RATIONALE/AMPLIFICATION
6. DMPL: Put cassette in holder and place on palmar lateral aspect of leg. Cassette should be parallel to the leg and perpendicular to the ground.	6a. This view is also called the lateral oblique. The name reflects the position of the cassette. 6b. Longer cassettes are advised (7 × 17 inches). 6c. Cassette should be in contact with leg. 6d. Marking the cassette DMPL is critical. 6e. Fig. 9–24.
7. Stand craniomedial to the limb. Maintain the beam parallel to the ground. Direct the beam 45 degrees medial to a true dorsopalmar (midsagittal) view. Center beam in mid-metacarpal area. Expose film.	7a. Mid-metacarpal is halfway between the knee and fetlock. 7b. This view highlights the metacarpal 2 (medial splint) bone. 7c. View should include metacarpal bones 2, 3, and 4 in entirety and a portion of the knee and fetlock.

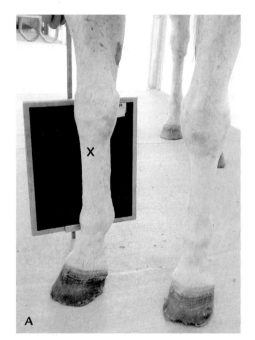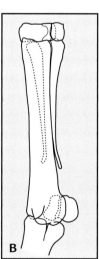

FIG. 9–24 (A) Positioning of horse and cassette for dorsomedial–palmar lateral (DMPL) oblique view of the metacarpal bones. X shows the target of the beam. (B) Drawing of resultant radiograph.

Radiographic Views of Carpus

TABLE 9-7 RADIOGRAPHIC VIEWS FOR CARPUS

VIEW	PURPOSE
Standard Views	
Dorsopalmar	Shows radiocarpal, intercarpal, and carpometacarpal joints
	Distinguishes size, shape, and density of carpal bones
	Shows distal end on radial physis
Extended lateral	Shows conformation, palmar aspect of ulnar, third, fourth, and accessory carpal bones
	Permits examination of suspensory ligament attachment on proximal palmar surface of third metacarpal
Flexed lateral	Shows articular surfaces of carpal bones and distal radius
Oblique views	
DLPM	Shows dorsomedial surface of radial and third carpal bones, palmar lateral aspect of ulnar and fourth carpals
	Distinguishes accessory carpal bone
DMPL	Shows dorsolateral aspect of intermediate and third carpals, palmar medial aspect of radial and second carpal bones
Optional Views	
Skyline proximal carpal bones	Evaluates dorsal surface of radial, intermediate, and ulnar carpal bones
Skyline distal carpal bones	Evaluates dorsal surface of second, third, and fourth carpal bones

Procedure for Dorsopalmar View of Carpus

TECHNICAL ACTION	RATIONALE/AMPLIFICATION
1. Complete Steps 1–4 listed under all limb radiographs.	1a. Hoof preparation is not necessary.
2. Brush leg to remove all dirt and debris.	2a. Brushing minimizes artifact.
3. Place hoof on ground in normal weight-bearing position.	3a. The horse should be standing squarely, with weight evenly distributed on all four limbs.
4. Put cassette in holder and place caudal to knee. Cassette should be parallel to the leg and perpendicular to the ground.	4a. Cassette should contact the limb.
	4b. The opposite forelimb can be elevated to minimize motion.
	4c. Fig. 9–25.

CHAPTER 9 Selected Lower Limb Radiographic Procedures

TECHNICAL ACTION	RATIONALE/AMPLIFICATION
5. Stand cranial to the limb. Direct the beam midsagittally, parallel to the ground. Center on the midcarpal region. Expose film.	5a. To determine a true midsagittal view, the operator can draw an imaginary line from the middle of the hoof to the radius. The beam is then centered on that line.
	5b. View should include distal radius, carpal bones, and proximal metacarpal bones.
	5c. An accurate dorsopalmar projection will show the radial and intermediate carpal joint space, with no other bones superimposed.

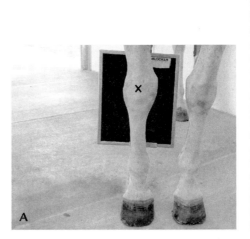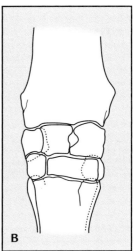

FIG. 9-25 (A) Positioning of horse and cassette for dorsopalmar view of the carpus. *X* shows target of the beam. (B) Drawing of resultant radiograph.

Procedure for Extended and Flexed Lateral Views of Carpus

TECHNICAL ACTION	RATIONALE/AMPLIFICATION
1. Complete Steps 1–4 listed under all limb radiographs.	1a. Hoof preparation is not necessary.
2. Brush leg to remove all dirt and debris.	2a. Brushing minimizes artifact.
3. Place cassette on medial aspect of carpus. Cassette should firmly contact the leg.	3a. A cassette holder is not used for this view.
	3b. Fig. 9-26.

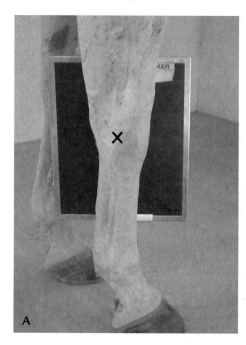
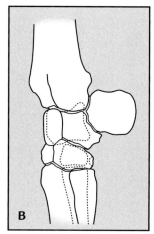

FIG. 9-26 (A) Positioning of horse and cassette for extended lateral view of the carpus. X shows the target of the beam. (B) Drawing of resultant radiograph.

TECHNICAL ACTION	RATIONALE/AMPLIFICATION
4. Extended lateral view: Stand lateral to the limb. Direct the beam parallel to the ground and aim just dorsal to the accessory carpal bone. Ensure collimation of the beam. Expose film.	4a. Collimation is critical for minimizing exposure to personnel. 4b. View should include the distal radius, carpal bones, and proximal metacarpal bones.
5. Flexed lateral view: Holding the hoof or pastern, flex the carpus approximately 60 degrees. Keep the cannon (metacarpal bones) parallel to the ground and the radius perpendicular to the ground.	5a. Sixty degrees is approximately three-fourths of full carpal flexion. 5b. Hold the hoof at the same height as the opposite carpus. 5c. Do not overflex the carpus. Do not pull the limb medially or laterally. Maintain in a neutral anatomic plane.
6. Place cassette on medial aspect of carpus. Cassette should firmly contact the leg.	6a. A cassette holder is not often used for this view. 6b. Fig. 9-27.
7. Stand lateral to the limb. Direct the beam parallel to ground and aim just dorsal to the accessory carpal bone. Ensure collimation of beam. Expose film.	7a. Collimation is critical for minimizing exposure to personnel. 7b. View should include the distal radius, carpal bones, and proximal metacarpal bones.

CHAPTER 9 Selected Lower Limb Radiographic Procedures

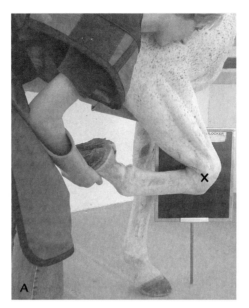
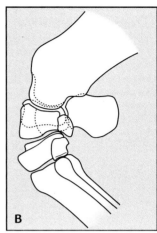

FIG. 9–27 (A) Positioning of horse and cassette for flexed lateral of the carpus. X shows target of the beam. (B) Drawing of resultant radiograph.

Procedure for Oblique Views of Carpus

TECHNICAL ACTION	RATIONALE/AMPLIFICATION
1. Complete Steps 1–4 listed under all limb radiographs.	1a. Hoof preparation is not necessary.
2. Brush leg to remove all dirt and debris.	2a. Brushing minimizes artifact.
3. Place hoof on ground in normal weight-bearing position.	3a. The horse should be standing squarely, with weight evenly distributed on all four limbs.
4. DLPM: Put cassette in holder and place on palmar medial aspect of carpus. Cassette should contact the limb and be held perpendicular to the ground.	4a. This view is also called the medial oblique. The name reflects the position of the cassette. View can also be called the palmar medial–dorsolateral oblique.
	4b. Fig. 9–28.
5. Stand cranial and lateral to the limb. Direct the beam 45 degrees lateral to a true dorsopalmar projection (midsagittal). Maintain beam parallel to ground and perpendicular to cassette. Center on middle of carpus. Expose film.	5a. The oblique angles can vary from 45 to 60 degrees off of the midsagittal plane.

FIG. 9-28 Positioning of horse and cassette for dorsolateral–palmar medial (DLPM) oblique view of the carpus. *X* shows target of the beam.

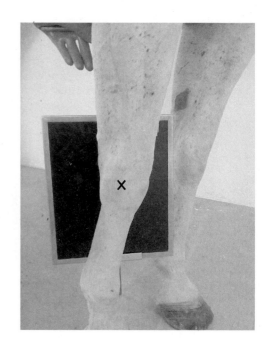

TECHNICAL ACTION	RATIONALE/AMPLIFICATION
6. DMPL: Put cassette in holder and place on palmar lateral aspect of carpus. Cassette should contact limb and be held perpendicular to the ground.	**6a.** Alternatively the cassette may be placed on the dorsomedial aspect of the carpus. This allows the operator to stay on the lateral side of the leg and still obtain a DMPL projection. The purpose of this is to minimize the number of times the operator must move the radiography tube.
	6b. Fig. 9–29.
7. Stand cranial and medial to the limb. Direct beam 45 degrees medial to a true dorsopalmar projection (midsagittal). Maintain beam parallel to ground and perpendicular to cassette. Center on middle of carpus. Expose film.	**7a.** Alternatively, if the cassette is placed on the dorsomedial aspect, the operator will remain caudal and lateral to the limb. The beam is then directed 30–40 degrees lateral to a midsagittal plane.
	7b. This view is also called the lateral oblique. The name reflects the position of the cassette. View can also be called palmar lateral–dorsomedial oblique.
	7c. The oblique angles can vary from 45 to 60 degrees off of the midsagittal plane (on the dorsopalmar projection).

CHAPTER 9 Selected Lower Limb Radiographic Procedures 415

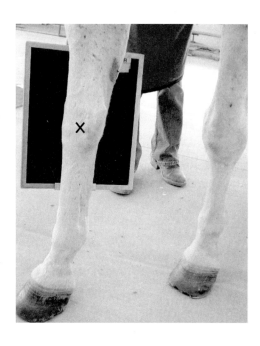

FIG. 9-29 Positioning of horse and cassette for dorsomedial–palmar lateral (DMPL) oblique view of the carpus. *X* shows target of the beam.

Procedure for Skyline Views of Carpus

TECHNICAL ACTION	RATIONALE/AMPLIFICATION
1. Complete Steps 1–4 listed under all limb radiographs.	1a. Hoof preparation is not necessary.
2. Brush leg to remove all dirt and debris.	2a. Brushing minimizes artifact.
3. Holding fetlock or cannon, flex carpus, and push carpal joint cranially. Maintain the metacarpus parallel to the ground.	3a. The flexed carpus should be cranial to the opposite leg.
4. Hold cassette and place **ventral** to the flexed carpus. Keep cassette parallel to the floor and in contact with the carpus and proximal metacarpal bones.	4a. Typically the cassette is not completely perpendicular to the floor on a skyline view, and this causes some elongation distortion of the carpus.
5. Proximal carpal row: Stand cranial to the limb. Position the beam dorsal to the knee (almost a 90-degree angle to the knee). Aim at center of carpus. Expose film.	5a. This view is also known as the 90-degree dorsoproximal-dorsodistal flexed oblique.
	5b. This view is also called the proximal carpal tangential.
	5c. Fig. 9–30.
6. Distal carpal row: Stand cranial to the limb. Position the beam cranial and dorsal to the flexed carpus. Direct the beam 30 degrees dorsal to the cassette. Aim at center of the carpus. Expose film.	6a. This view is also known as the 30-degree dorsoproximal-dorsodistal flexed oblique.
	6b. This view is also called the distal carpal tangential.
	6c. Fig. 9–31.

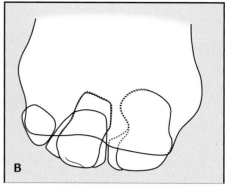

FIG. 9-30 (A) Positioning of horse, cassette, and assistant for skyline view of proximal row of carpal bones. *Arrow* shows direction of the beam. (B) Drawing of resultant radiograph.

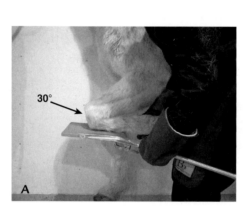
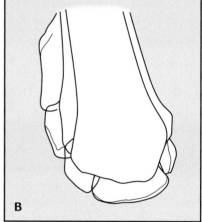

FIG. 9-31 (A) Positioning of horse, cassette, and assistant for view of distal row of carpal bones. *Arrow* shows direction of the beam. (B) Drawing of resultant radiograph.

Special Note

To minimize moving radiographic equipment during a carpal series, the following sequence often is used: **dorsopalmar**, **medial** oblique, extended lateral, lateral oblique, flexed lateral, skyline proximal, and skyline distal.

Radiographic Views of Tarsus

TABLE 9-8 RADIOGRAPHIC VIEWS FOR TARSUS

VIEW	PURPOSE
Required	
Dorsoplantar	Shows intertarsal, tarsometatarsal joints
	Excellent evaluation of medial aspect of joint
Lateral	General tarsal evaluation
Oblique views	
Dorsolateral–plantar medial (DLPM)	Shows dorsomedial aspect of tarsal bones
	Provides very good visualization of fourth metatarsal, talus, trochlear ridge, and intermediate ridge of distal tibia
Plantarlateral–dorso medial (PLDM)	Specifically good for evaluating lateral trochlear ridge of talus, second metatarsal, and central and third tarsal bones
Optional	
Flexed lateral	Shows tarsocrural joint and dorsal area of calcaneus
Flexed dorsoplantar	Shows calcaneus

Procedure for Dorsoplantar View of Tarsus

TECHNICAL ACTION	RATIONALE/AMPLIFICATION
1. Complete Steps 1–4 listed under all limb radiographs.	1a. Hoof preparation is not necessary.
2. Brush leg to remove all dirt and debris.	2a. Brushing minimizes artifact.
3. Stand horse with equal weight bearing on all four limbs. Place hoof on ground, with a lateral rotation such that horse appears toed out.	3a. Leg will appear as a cow-hocked conformation.
	3b. Placement permits the radiography equipment to be placed more lateral, away from the underbelly of the horse.
	3c. Elevating the forelimb on the same side can help to minimize motion.
4. Put cassette in the holder and place on the plantar aspect of the hock.	4a. Cassette should contact the hock.
	4b. *Never* stand directly behind the horse. Always stand to the side of the hind limbs.
	4c. Fig. 9–32.

TECHNICAL ACTION	RATIONALE/AMPLIFICATION
5. Direct the beam midsagittally, parallel to the ground and perpendicular to the cassette. Center on the central tarsal bone. Expose film.	5a. View should include the entire tarsus, distal tibia, and proximal metatarsal. 5b. Some individuals advocate taking a second dorsopalmar view, angling the beam 10 degrees downward from parallel.

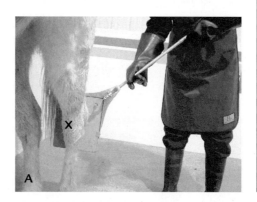 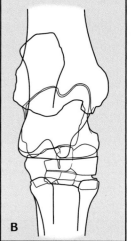

FIG. 9-32   (A) Positioning of horse, cassette, and assistant for dorsoplantar view of the tarsal joint. X shows the target of the beam. (B) Drawing of resultant radiograph.

Procedure for Lateral and Flexed Lateral Views of Tarsus

TECHNICAL ACTION	RATIONALE/AMPLIFICATION
1. Complete Steps 1–4 listed under all limb radiographs.	1a. Hoof preparation is not necessary.
2. Brush leg to remove all dirt and debris.	2a. Brushing minimizes artifact.
3. Stand horse with equal weight bearing on all four limbs.	3a. Do not rotate the limb laterally. Toe must point forward (cranially).
4. Extended lateral view: Put cassette in holder and place on medial aspect of hock. Maintain cassette parallel to the leg and perpendicular to the ground.	4a. Cassette should contact limb. 4b. Individual holding cassette should be cranial to the limb. 4c. Fig. 9-33.
5. Stand lateral to the limb. Direct the beam parallel to the floor and center the beam 4 inches distal to the point of the hock. Expose film.	5a. The most common error of this view is centering the beam too high. 5b. A slight 3–5 degree downward angle can be used to enhance the quality of this film. The angulation is used to compensate for the intertarsal joint space angulation.

CHAPTER 9 Selected Lower Limb Radiographic Procedures

TECHNICAL ACTION	RATIONALE/AMPLIFICATION
6. Flexed lateral view: The flexed lateral is an optional view and can be taken easily at this time. Grasp hoof, flexing both fetlock and hock. Hold so that the cannon is not parallel to the ground but angled approximately 30 degrees downward from parallel. Cassette is placed on medial aspect of tarsus.	6a. Do not use a cassette holder for this view. 6b. Do not abduct the limb laterally. 6c. Fig. 9–34.
7. Stand lateral to the limb. Direct the beam parallel to the ground and center on the hock. Expose film.	7a. The assistant holding the cassette should use reasonable caution during this procedure.

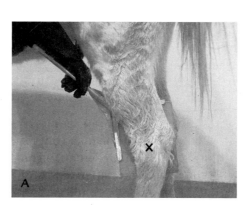 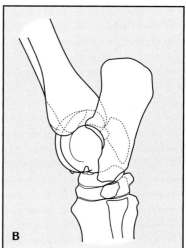

FIG. 9–33 (A) Positioning of horse and cassette for extended lateral view of the tarsal joint. X marks the target of the beam. (B) Drawing of resultant radiograph.

 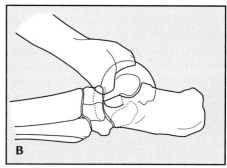

FIG. 9–34 (A) Positioning of horse, cassette, and assistant for flexed lateral view of the tarsal joint. X marks the target of the beam. (B) Drawing of resultant radiograph.

Procedure for Oblique Views of Tarsus

TECHNICAL ACTION	RATIONALE/AMPLIFICATION
1. Complete Steps 1–4 listed under all limb radiographs.	1a. Hoof preparation is not necessary.
2. Brush leg to remove all dirt and debris.	2a. Brushing minimizes artifact.
3. Stand horse with equal weight bearing on all four limbs.	3a. Do not laterally rotate the limb. Toe must point forward (cranially).
4. Dorsolateral–plantar medial (DLPM): Put cassette in holder and place on medial plantar aspect of hock. Maintain cassette perpendicular to floor.	4a. Cassette should contact limb 4b. Fig. 9–35.
5. Stand cranial and lateral to limb. Direct beam parallel to ground, 45 degrees lateral to a true dorsopalmar (midsagittal). Center beam on hock. Expose film.	5a. This view is also called the medial oblique. The name reflects cassette placement. 5b. A slight 3–5 degree downward angle can be used to enhance the quality of this film. The angulation is used to compensate for the intertarsal joint space angulation

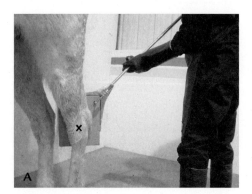 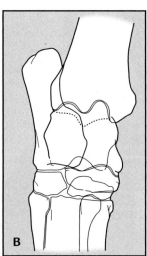

FIG. 9–35   (A) Positioning of the horse and cassette for a dorsolateral–plantar medial (DLPM) oblique view of the tarsal joint. *X* marks the target of the beam. (B) Drawing of resultant radiograph.

CHAPTER 9 Selected Lower Limb Radiographic Procedures

TECHNICAL ACTION	RATIONALE/AMPLIFICATION
6. Plantarlateral–dorsomedial (PLDM): Put cassette in holder and place on dorsomedial aspect of hock. Maintain cassette perpendicular to floor.	6a. Cassette should contact limb. 6b. The cassette is placed on the dorsomedial aspect of the limb instead of the plantar lateral aspect, to minimize the potential for damage to equipment and personnel. 6c. Assistants holding cassettes on the hind limbs should always stand cranial to the limbs. 6d. Fig. 9–36.
7. Stand lateral and caudal to the limb. Direct the beam parallel to the ground, 50–60 degrees lateral to a true midsagittal aspect. Center the beam on the central tarsal area. Expose film.	7a. This view is also called the lateral oblique. 7b. Beam can be angled 3–5 degrees upward. 7c. View should include distal tibia, tarsal bones, and proximal metatarsal bones. 7d. This view can also be obtained by placing the cassette on the plantar lateral aspect of the hock. Beam would then be directed from a craniomedial position. This view is not recommended, because it places equipment and personnel in danger.

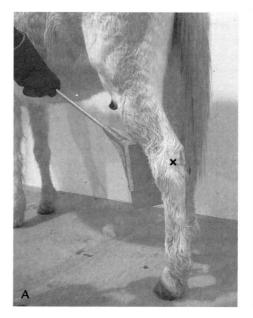 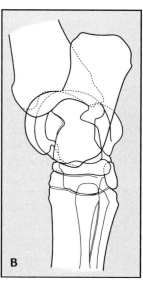

FIG. 9–36 (A) Positioning of the horse, cassette, and assistant for a plantar lateral–dorsomedial (PLDM) oblique view of the tarsal joint. X shows the target of the beam. (B) Drawing of resultant radiograph.

Procedure for Flexed Dorsoplantar View of Tarsus

TECHNICAL ACTION	RATIONALE/AMPLIFICATION
1. Complete Steps 1–4 listed under all limb radiographs.	1a. Hoof preparation is not necessary.
2. Brush leg to remove all dirt and debris.	2a. Brushing minimizes artifact.
3. Grasp cannon and flex hock. Metatarsal bones (cannon) should remain parallel to ground. Push entire limb caudally.	3a. Keeping the cannon parallel to the ground permits full flexion of the hock.
4. Place cassette ventral to the flexed hock, parallel to the ground. Cassette should firmly contact the hock.	4a. A cassette holder typically is not used. 4b. Fig. 9-37.
5. Position radiography tube dorsal to the flexed hock. Direct beam almost perpendicular to the plate. Use a 10-degree cranial angulation. Center 3 inches cranial to flexed point of hock.	5a. Operator should always stand lateral to hind limbs when positioning the radiography tube. 5b. The point of the hock is formed by the calcaneus. 5c. This film is difficult to obtain if using heavy radiography units.

FIG. 9-37 Positioning of the horse, cassette, and assistant for a flexed dorsoplantar view of the calcaneus. *Arrow* shows direction of the beam and *X* marks the target.

PREPARATION OF THE LIMB FOR ULTRASONOGRAPHIC EXAMINATION

Ultrasonographic examination is an essential component of many workups for lameness. Appropriate preparation of the limb minimizes artifact and facilitates accurate diagnosis.

Purpose

- Prepare limb for diagnostic quality ultrasonographic examination

Complications

- Injury to personnel

Equipment

- Brush
- Clippers with #40 blade
- Shaving razor
- Shaving gel
- Isopropyl alcohol 70%
- Ultrasonography gel

Procedure for Ultrasonographic Preparation

TECHNICAL ACTION	RATIONALE/AMPLIFICATION
1. Halter horse and stand on flat surface.	1a. Ideally the horse should be standing on cement.
2. Sedate horse if necessary.	2a. Horses can spook during ultrasonographic examination.
	2b. Tranquilizers can be used to calm horses and minimize movement.
3. Brush limb.	3a. Mud and dirt rapidly dull clipper blades.
	3b. Excessively muddy limbs can be washed and toweled dry.
4. Use a #40 clipper blade and clip area to be examined. Shave in the direction of hair growth.	4a. Equine limb ultrasonography most commonly involves examination of tendons.
	4b. The clip should be very neat, with squared edges.

TECHNICAL ACTION	RATIONALE/AMPLIFICATION
	4c. Always inform the owner that the horse will be clipped. Unauthorized removal of hair on competition horses should never be performed.
	4d. Should the owner refuse to permit clipping, an alcohol-gel slurry can be made and applied to the area of interest. Mix 1 part alcohol with 3 parts gel. The examination will be of lesser quality, and repeated attempts to perform this type of examination will shorten the life of the ultrasound probe.
5. Apply shaving gel and shave clipped area using standard razor.	**5a.** A disposable razor is commonly used.
	5b. Shaving the area minimizes wear or damage to the ultrasonographic probe heads.
6. Wipe area with alcohol. Apply ultrasonographic gel and begin examination.	—

REVIEW QUESTIONS

1. State the function of a dosimetry badge.
2. Describe the procedure used to remove a shoe.
3. Name the substance used to remove air pocket artifacts on hoof radiographs.
4. Diagram and label all the bones in the front limb of the horse.
5. Explain the reason for using a thin wire during a lateral coffin (P3) radiograph.
6. Identify the standard and optional views used during a navicular series.
7. Describe techniques that can be used to encourage a horse to maintain a limb in weight-bearing position.
8. Diagram the radiographic beam direction for DLPM and DMPL views of the fetlock.
9. State an alternative name for a DLPM fetlock radiograph.
10. Identify the proper cassette size for use in a metacarpal series.
11. Describe the technique used to determine the true midsagittal position on the carpus.

12. Name the view that is used to assess the calcaneus.
13. Compare the two methods of preparing for ultrasonographic examination. State which method increases the diagnostic quality and lengthens probe life.

BIBLIOGRAPHY

Han, C., & Hurd, C. (1999). *Practical diagnostic imaging for the veterinary technician* (2nd ed.). St. Louis: C. V. Mosby.

Lavin, L. (1999). *Radiography in veterinary technology* (2nd ed.). Philadelphia: W. B. Saunders.

Morgan, J. (1993). *Techniques of veterinary radiography* (5th ed.). Ames: Iowa State University Press.

Thrall, D. (1986). *Textbook of veterinary diagnostic radiology.* Philadelphia: W. B. Saunders.

CHAPTER 10
Anesthesia

The cautious seldom err.
CONFUCIUS

KEY TERMS

analgesia
anesthesia
apnea
aspiration
ataxia
bradycardia
cyanosis
dyspnea
dysrhythmia

fasciculations
hyperthermia
hypothermia
malignant
nystagmus
palpebral
regurgitation
tachycardia
tachypnea

OBJECTIVES

- Identify the roles a veterinary technician may have in performing routine anesthetic procedures of the horse, cow, llama, goat, and pig.
- Discuss use and maintenance of anesthesia equipment.
- Describe general standards for application and maintenance of **anesthesia**.
- Identify routine anesthetic protocols in the horse, cattle, llama, goat, sheep, and pig.
- Discuss complications associated with anesthesia in the various animal species.

VETERINARY ANESTHESIA IN LARGE ANIMAL PRACTICE

Unlike that of small animal practice, anesthesia in large animal practice runs the gamut from local to general anesthesia. Procedures may be performed out in the field or barn or in a sterile surgery room with a full battery of anesthesia equipment. Monitoring may range from a simple stethoscope and watch with a second hand to blood pressure monitors and pulse oximeters. Regardless of the circumstance, the technician often performs the duties of veterinary anesthetist—administering and monitoring anesthesia under veterinary supervision—and is responsible for the safety of both patient and doctor.

Roles of the Veterinary Anesthetist

- Calculate ordered drug doses
- Draw up and understand how to administer sedatives, analgesics, or anesthetics
- Maintain drug logs in compliance with state and federal regulations
- Use and maintain monitoring equipment
- Use and maintain anesthesia equipment
- Monitor anesthesia, maintaining a safe anesthetic plane for patient and surgeon

PREANESTHETIC PERIOD

Although most large animal surgeries are done with restraint devices and local anesthetics, many times the animal will be at least sedated. Furthermore, general anesthesia in large animals has all the associated risks found in small animals plus the difficulty of maintaining circulation in such a large body mass. It is, therefore, important to be attentive to the preanesthetic period and ensure that the patient is prepared correctly for whatever anesthetic regimen the doctor has planned.

MINIMUM DATA BASE

Purpose

- Ensure that the correct patient is presented for anesthesia
- Ensure that paperwork has been completed and that the owner is aware of the risks
- Ensure that a physical examination is done before anesthesia is administered

ALL SPECIES	REASONING
1. Patient identification or signalment	1a. Make sure the correct patient has been presented for surgery or anesthesia, verifying breed, sex, and patient name, or identification tag.
	1b. Verify procedure or surgery to be performed and what body part will be worked on.
2. Patient history	2a. Pay special attention to previous anesthetic events and any difficulties.
	2b. Chapter 4 contains techniques on gathering patient history.
3. Physical examination	3a. Findings may determine what anesthetic regimen is selected.
	3b. Important to identify any pathology that may affect response to general anesthesia.
	3c. Chapter 4 contains techniques for physical examination.
4. Owner consent	4a. Most states require a legal anesthetic consent form to be signed by the owner (Fig. 10–1).
	4b. Information regarding the effects of anesthetic drugs and possible consequences should be included on the form.

PREPARATION OF HORSES FOR GENERAL ANESTHESIA OR HEAVY SEDATION

Purpose

- Ensure a positive outcome by attempting to prevent complications, especially those leading to anesthesia-related death

Equipment

- Syringe with which to rinse mouth
- Water

ANIMAL MEDICAL CLINIC

OWNER'S NAME: _____   CLIENT ID: _____

CLIENT ADDRESS: _____   CLIENT PHONE: _____

ANIMAL NAME: Horse   DATE OF BIRTH: _____

ANESTHESIA CONSENT FORM

Today _____ will undergo a procedure requiring anesthesia. I understand that even under the best of circumstances, there is a risk associated with undergoing anesthesia. Risks include, but are not limited to, low blood pressure, irregular heart or respiratory rate, sudden death. I consent to the anesthesia and authorize any life-saving measures deemed necessary by the doctor. I understand that additional expense may be incurred and that results are not guaranteed.

In order to best assess the current state of _____'s health, we recommend the following laboratory work:

    1) Pre-Anesthetic Blood Profile   Y_____   N_____   39.00

    2) General Health Blood Profile   Y_____   N_____   67.50

    3) Complete Blood Count (CBC)   Y_____   N_____   34.00

    4) Urinalysis   Y_____   N_____   22.50

I consent to the following tests: _____

I decline the recommended tests: _____

I assume full financial responsibility for _____ and understand that payment is due in full at time of discharge.

Owner's Signature: _____   Date: _____

FIG. 10-1 Sample anesthesia release form.

- Items for jugular catheter placement (Chapter 6)
- Items for blood sampling (Chapter 5)

Complications

- Injury to patient or handler
- Anesthetic emergency secondary to inadequate preparation
- Dehydration
- Hypoglycemia
- Hyper/Hypothermia

Procedure for Preparing Horses for General Anesthesia

TECHNICAL ACTION	RATIONALE/AMPLIFICATION
1. Food withdrawal	1a. Controversial. May lead to ileus and colon impaction or rupture, so some do not withhold feed for more than a couple hours.
	1b. Adult horses should have feed withheld for 12–24 hours.
	1c. Foals may be allowed to suckle up to 2 hours before anesthesia.
	1d. Withhold fermentable feed such as grain or pelleted feeds for at least 24 hours.
2. Water withdrawal	2a. Controversial (see above). The risk of regurgitation and aspiration is negligible.
3. Jugular catheter	3a. Size and length of catheter depends on the size of the horse, length of procedure, and whether catheter will be left in after recovery.
	3b. Chapter 6 covers catheter placement technique.
	3c. Use of lidocaine topically or subcutaneously will ease placement.
4. Blood work	4a. Whether to do blood work and which to do is determined by the veterinarian.
	4b. Complete blood count may detect anemia, septicemia, or inflammatory process.
	4c. Blood chemistry may detect renal or hepatic problems or hidden muscular damage that influence type of anesthetic used.

TECHNICAL ACTION	RATIONALE/AMPLIFICATION
5. Rinse mouth with water	5a. Especially important when horse will be intubated, to prevent feed material from being carried into the trachea.
	5b. Advisable whenever heavy sedation is used.
	5c. Most horses are cooperative with the procedure.
6. Estimate weight	6a. May be done with weight tape around girth or on livestock scale (Fig. 10–2).

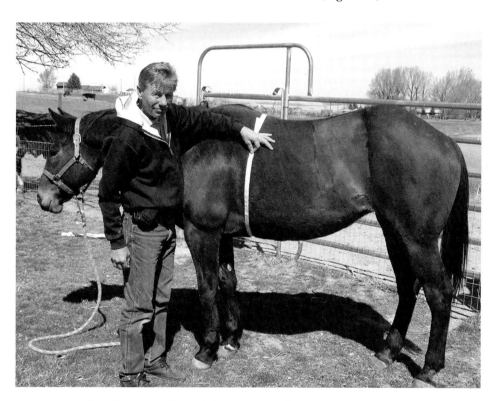

FIG. 10–2   Application of a weight tape to a horse.

PREPARATION OF RUMINANTS FOR GENERAL ANESTHESIA OR HEAVY SEDATION

Purpose

- Ensure a positive outcome by attempting to prevent complications, especially those leading to anesthesia-related death

Equipment

- Items for jugular catheter placement (Chapter 6)
- Items for blood sampling (Chapter 5)

Complications

- Injury to patient or handler
- Anesthetic emergency secondary to inadequate preparation
- Dehydration
- Hypoglycemia
- Bloat

Procedure for Preparing Ruminants for Anesthesia

TECHNICAL ACTION	RATIONALE/AMPLIFICATION
1. Food withdrawal in general	1a. Done to prevent aspiration of food into the lungs while under anesthesia, rumen tympany, and bradycardia.
	1b. Not recommended for suckling animals, because hypoglycemia may result. Nursing animals may continue to nurse up to 2 hours before anesthesia.
2. Calves 2–4 months old eating solid feed	2a. Withhold feed 4–8 hours, sucking permissible.
	2b. Younger calves need shorter fasting because of lack of a rumen.
3. Small ruminants such as goats and sheep	3a. Withhold feed 12–18 hours.
	3b. Use caution with ewes or does in advanced pregnancy. They do not tolerate prolonged feed withdrawal.
4. Adult cattle (over the age of 6 months)	4a. Withhold feed 24–36 hours.
	4b. Once the rumen is fully functional, prolonged feed withdrawal is required to allow microbial fermentation to subside.
	4c. Large bulls may require 48 hours of feed withdrawal.
5. Water withdrawal	5a. Done to prevent regurgitation and subsequent aspiration of fluid into lungs and to decrease incidence of bloat.
	5b. In extreme heat, water withdrawal times are shortened so as to prevent dehydration.

TECHNICAL ACTION	RATIONALE/AMPLIFICATION
6. Calves, lambs, kids older than 3 months	6a. Withhold water 4–8 hours.
7. Adult animals	7a. Withhold water 12–18 hours.
8. Jugular catheter 18g for sheep and goats 16–18g for calves 12–14g for adult cattle	8a. Chapter 6 covers placement technique. 8b. Use of lidocaine topically or subcutaneously will ease placement.
9. Blood work	9a. As ordered by veterinarian. Chapter 5 includes blood collection techniques. 9b. Complete blood count may detect anemia secondary to blood parasites or other organic disease. Packed cell volume should be at least 30%. 9c. Blood chemistry may detect renal or hepatic problems that influence type of anesthetic used.
10. Estimate of body weight	10a. Ruminants under 300 lb and halter or collar broken may be weighed on a small animal scale. 10b. Large ruminants may be weighed in a vehicle on a livestock scale. 10c. If a scale is not available, a weight-calculation tape may be used.

PREPARATION OF LLAMAS FOR GENERAL ANESTHESIA OR HEAVY SEDATION

Purpose

- Ensure a positive outcome by attempting to prevent complications, especially those leading to anesthesia-related death

Equipment

- Items for placement of jugular catheter (See Chapter 6)
- Items for blood sampling (See Chapter 5)

Complications

- Injury to patient or handler
- Anesthetic emergency secondary to inadequate preparation

- Dehydration
- Hypoglycemia
- Bloat

Procedure for Preparing Llamas for Anesthesia

TECHNICAL ACTION	RATIONALE/AMPLIFICATION
1. Food withdrawal	1a. Done to prevent bloating, which would compromise respirations.
	1b. Also helps to prevent regurgitation and subsequent risk of aspiration.
	1c. Adult llamas should have feed withheld 24 hours, longer if the rumen is very full or a prolonged general anesthesia is expected.
	1d. Crias less than 3 months old are allowed to suckle up to time of anesthesia.
	1e. Crias eating solid feed may be made to fast for up to 12 hours, but still allowed to suckle.
2. Water withdrawal	2a. Withhold water for 12 hours for adults.
	2b. Decreases bloating, regurgitation, and aspiration.
	2c. Withhold water for a shorter period if the ambient temperature is higher than 90°F.
3. Jugular catheter placement	3a. Chapter 6 shows how to place a jugular catheter.
	3b. Use of lidocaine topically or subcutaneously will ease placement.
4. Rinse mouth with water	4a. May require sedation to complete. Often the llama will regurgitate immediately.
	4b. Especially important when general anesthesia will necessitate intubation, because rinsing the mouth will prevent feedstuffs from being carried into the trachea by the tube.
5. Estimate of weight	5a. Small llamas and alpacas should be weighed on scale to prevent drug overdose.

PREPARATION OF PIGS FOR GENERAL ANESTHESIA OR HEAVY SEDATION

Purpose

- Ensure a positive outcome by attempting to prevent complications, especially those leading to anesthesia-related death

Equipment

- Syringe with which to rinse mouth
- Water
- Items for jugular catheter placement (Chapter 6)
- Items for blood sampling (Chapter 5)

Complications

- Injury to patient or handler from inadequate restraint
- Anesthetic emergency secondary to inadequate preparation
- Dehydration
- Hyperthermia
- Hypoglycemia
- Vomiting

Procedure for Preparing Pigs for Anesthesia

TECHNICAL ACTION	RATIONALE/AMPLIFICATION
1. Food withdrawal	**1a.** Done to decrease the chance of aspiration of feed secondary to vomiting during recovery. **1b.** Withhold feed in pigs 35 pounds or more for 12 hours. **1c.** Sucking pigs may nurse up to 2 hours before anesthesia.
2. Water withdrawal	**2a.** Done to prevent aspiration of gastric juices. **2b.** Withhold water 6 hours in the adult pig. **2c.** When ambient temperature exceeds 85°F, shorten the time water is withheld.

TECHNICAL ACTION	RATIONALE/AMPLIFICATION
3. Estimate of body weight	3a. Small pigs may be weighed easily on a pediatric or small animal scale; larger pigs require walk-on livestock scale.
4. Ear vein catheter 22g × 1 inch for small pigs Up to 18g × 1.5 inch for larger pigs	4a. Requires excellent restraint; chute may be necessary in pigs over 150 lb. Pigs over 200 lb may require chemical restraint. 4b. Chapter 6 covers placement techniques. 4c. Topical lidocaine will ease placement in awake pigs. 4d. Application of 70% isopropyl alcohol will make ear vein easier to see.
5. Rinse mouth with water	5a. To remove food particles prior to intubation.

STAGES OF ANESTHESIA

General anesthesia is divided into stages and planes describing its depth. Understanding these stages and planes will help you determine the depth of anesthesia by monitoring vital signs and reflexes. Parameters to monitor and their meaning are covered in the section, "Parameters to Monitor in All Patients Under Anesthesia". They are influenced greatly by the type of anesthetic used.

DESCRIPTION OF STAGES OF ANESTHESIA

TABLE 10-1 DESCRIPTION OF STAGES OF ANESTHESIA

1. Stage 1 immediately after administration of a general anesthetic for all species.	1a. Patient is still conscious but disoriented. 1b. Heart rate and respiratory rate may be increased. 1c. The animal may exhibit signs of anxiety or fear if not pretreated with a sedative or tranquilizer.
2. Stage II begins with loss of consciousness, often known as the excitatory phase.	2a. All reflexes are present and possibly exaggerated. 2b. Motor neuron inhibition is lost, causing thrashing limb and head movements. 2c. Signs are similar in all species, but a greater danger to patient and personnel exists when anesthetizing horses and food animals. 2d. Breathing may be irregular, or the patient may appear to hold its breath. 2e. There may be catecholamine release causing cardiac arrhythmias or arrest, especially in horses and pigs. 2f. It is important that excellent premedication be used to avoid clinical signs of this stage in all large animal species.

TABLE 10-1 DESCRIPTION OF STAGES OF ANESTHESIA—cont'd

3. Stage III divided into four planes	3a.	Surgical anesthesia.
4. Stage III plane 1 light anesthesia for horse and ruminants	4a.	Involuntary limb movement stops.
	4b.	Respirations become even and regular, and ear twitch is still present.
	4c.	Pupillary light reflex is diminished and pupil may be constricted.
	4d.	Nystagmus may be present.
	4e.	The gag reflex is suppressed, allowing intubation.
	4f.	Same occurs in ruminants as does in the horse, but the globe of the eye begins to roll ventromedial.
	4g.	Corneal reflex is present.
5. Stage III plane 2 medium surgical anesthesia		
Horses	5a.	Palpebral, corneal reflex is strong. Ear flick may be present.
Ruminants	5b.	Globe of eye rolled ventral, only sclera visible.
	5c.	Palpebral reflex present but sluggish.
Horses and ruminants	5d.	Respirations regular at 6–10, heart rate is over 40.
	5e.	Unresponsive to most noxious stimuli.
6. Stage III plane 3 deep anesthesia		
Horses	6a.	Corneal reflex diminished or absent as is palpebral reflex.
	6b.	Tear secretions are diminished.
	6c.	Pupils are dilated.
	6d.	Blood pressure is falling. Respirations are shallow or irregular.
Ruminants	6e.	Eye globe headed back toward central, with pupil fixed and dilated.
	6f.	Corneal and palpebral reflexes absent.
	6g.	Respirations shallow and less than 5 breaths per minute. Heart rate dropping, blood pressure falling, capillary refill time exceeds 3–4 seconds.
7. Stage III plane 4		IN DANGER—TOO DEEP. PATIENT IS DYING!!!
Horses	7a.	Mean arterial blood pressure <60 mmHg, heart rate irregular or slow, respirations erratic.
	7b.	Pupil fixed and dilated, no corneal or palpebral reflexes.
Ruminants	7c.	Profound muscle relaxation.
	7d.	Same as horse.
8. Stage IV moribund	8a.	Death is imminent.

COLLEGE OF SOUTHERN IDAHO VETERINARY TECHNOLOGY
LA ANESTHESIA RECORD

Animal Name____________________

Owner Name____________________

Date	Surgeon	Student Assistant
Pre-Op Diagnosis		
Proposed surgery/Procedure		

BodyWt	Temp	HR	RR	PCV	TP	MM	CRT

Anesthetist	Student Anesthetist

Pre-Anesthetic Drugs				Anesthesia Induction			
Drug	Dose (mg)	Route	Time	Drug	Dose (mg)	Route	Time

Fluids: Type_______________ Total amount_______

ASA Status
1 2 3 4 5

Airway maintenance
- ❑ Mask
- ❑ Endotracheal tube
  Size______

Symbols	Comments
A Start Anesthesia O Start Operation A End Anesthesia *f* End Operation • Pulse rate _ Resp rate x Blood pressure	Extubation time________ Post-op temp______ Quality of recovery:

Respiratory System
- ❑ Mechanical
- ❑ Nonrebreathing
- ❑ Rebreathing
- ❑ Closed
- ❑ Semi-closed
- ❑ Liters/min_____

	15	30	45	_:00	15	30	45	_:00	15
Anesthetic %									
L M D									
140									
130									
120									
100									
90									
80									
70									
60									
50									
40									
30									
20									
10									
5									
	15	30	45	_:00	15	30	45	_:00	30

FIG. 10–3 Sample of an anesthesia record.

PARAMETERS TO MONITOR IN ALL PATIENTS UNDER ANESTHESIA

Purpose

By monitoring vital signs, the anesthetist can maintain the patient at a safe level of anesthesia. Monitoring allows the anesthetist to adjust the patient's anesthesia level throughout surgery to ensure a rapid and safe recovery (Fig. 10–3).

Equipment

- Watch with a second hand
- Stethoscope
- Thermometer
- Pen and paper with clipboard

Complications

- Prolonged recovery from anesthesia
- Premature recovery from anesthesia
- Anesthesia-related death

Procedure for Monitoring Horses

PARAMETER	INFORMATION DERIVED
1. Heart rate and rhythm	1a. Auscultate with stethoscope to side of ventral chest. Best to palpate pulse simultaneously (Fig. 10–4A).
	1b. Normal heart rate 35–50 beats per minute.
	1c. Tachycardia is heart rate greater than 60. May be pain induced, drug induced (ketamine), or a sign that the horse is waking up.
	1d. Bradycardia is heart rate less than 30 bpm. May be drug induced (alpha-2 agonists) or sign of excessively deep anesthetic plane.
	1e. Irregular heart rhythm may be associated with pain, excessively deep anesthesia, or impending death.

TECHNICAL ACTION	RATIONALE/AMPLIFICATION
	1f. Determine type of **dysrhythmia** with electrocardiogram.
	1g. Blood pressure along with physical signs (mucous membrane color, respiratory rate) will help to identify potential negative consequences.
2. Pulse rate is same as the heart rate	**2a.** Pulses are palpable at the following locations: Submandibular, facial, carpal, lateral metatarsal, abaxial, and digital arteries (Fig. 10–4B).
	2b. A pulse cannot be palpated when systolic blood pressure falls below 60 mmHg.
	2c. Thready (difficult to palpate) pulse may be caused by alpha-2 agonists, low blood pressure, or dangerously deep anesthetic plane.
	2d. Bounding pulses are palpated easily or even seen. May be caused by high blood pressure, pain, or light anesthetic plane. Also seen when there is a large difference between diastolic and systolic pressures.
	2e. Pulses that don't match the heart rate indicate significant dysrhythmia and reduced cardiac output. Condition needs immediate diagnosis and management. Use electrocardiogram and blood pressure monitor.

FIG. 10–4 (A) Auscultating the heart of an anesthetized horse. (B) Palpating the facial artery pulse in an anesthetized horse.

PARAMETER	INFORMATION DERIVED
3. Respiratory rate and rhythm	3a. Acquire by watching the chest rise and fall, watching the rebreathing bag inflate and deflate, or by watching the nostrils flare.
	3b. Rate and rhythm are influenced by the anesthetic drugs used and by the position of the patient.
	3c. Horses in dorsal recumbency tend to have shallow respirations. Mucous membrane color, pulse oximetry, and capnometry indicate effectiveness of ventilation.
	3d. **Apnea** is lack of spontaneous respirations. Horses may breath-hold up to 2 minutes without negative consequences.
	3e. Provide ventilatory support if apnea persists beyond 2 minutes by: 1. Placing an endotracheal tube 2. Squeezing the reservoir bag or blowing into the endotracheal tube 3. Providing oxygen via a nasotracheal or endotracheal tube
	3f. Shallow respirations may indicate poor ventilation. See above to diagnose and remedy.
	3g. A deep regular rhythm at 6–12 breaths per minute on inhalant anesthetics is generally indicative of a good plane of anesthesia.
4. Mucous membrane color	4a. Examine color of oral mucous membranes.
	4b. Indicates degree of perfusion and oxygenation. May be influenced by anesthetic agents used and whether the horse is receiving supplemental oxygen.

PARAMETER	INFORMATION DERIVED
	4c. Normal color is pale pink to pink, but may be influenced by anesthetic drugs used or by presence of organic disease.

	4d. Pale, grey, or white may indicate poor perfusion secondary to:
	  1. Poor cardiac output
	  2. Vasoconstriction
	  3. Low blood pressure
	  4. Too deep anesthetic plane

	4e. Cyanosis, or blue color, indicates poor blood oxygenation secondary to:
	  1. Poor ventilation secondary to inadequate respiration rate or depth
	  2. Pulmonary edema
	  3. Pneumothorax
	  4. Airway obstruction

5. Capillary refill time (CRT)	**5a.** Obtained by pressing on the gums until blanched, releasing the pressure, and counting how many seconds until color returns.
	5b. Good indicator of perfusion, hence cardiac output.
	5c. Normal CRT is 1–2 seconds

	5d. Prolonged CRT (longer than 3 seconds) indicates poor perfusion. May be caused by:
	  1. Alpha-2 receptor agonists causing peripheral vasoconstriction without dire consequences
	  2. Anesthetic plane too deep

	5e. An animal with prolonged CRT needs attention.
6. Eye position	**6a.** Globe position generally remains unchanged in horses.
	6b. Globe drifting ventromedially is indicative of deepening anesthetic plane.

PARAMETER	INFORMATION DERIVED
7. Palpebral reflex	7a. Elicited by stroking or tickling the eyelashes or a touch to the medial canthus.
	7b. Palpebral reflex must always be present during equine anesthesia, although it may be difficult to assess when lids are bruised (as in colic surgeries).
	7c. A strong palpebral reflex indicates a light plane of anesthesia.
	7d. A sluggish palpebral reflex indicates a surgical plane of anesthesia that may be approaching deep anesthesia.
	7e. Absent palpebral reflex indicates that anesthesia is getting too deep and the horse needs immediate attention.
8. Corneal reflex	8a. Obtained by dropping sterile saline or artificial tear into eye and seeing global withdrawal. Gloved finger touched gently to eyeball may also be used.
	8b. Corneal reflex must always be present. Heavy lacrimation is common.
	8c. A sluggish corneal reflex and a dry cornea indicate the horse is getting into a dangerously deep level of anesthesia.
	8d. If corneal reflex is absent, anesthesia is dangerously deep and the horse may die soon.
9. Ear twitch	9a. Obtained by lightly stroking the inner ear hairs with a finger.
	9b. Reliability depends on sensitivity of the horse and skill of operator.
	9c. Presence of an ear twitch suggests a light plane of anesthesia.
10. Temperature	10a. Obtained by use of a rectal thermometer and should be checked every 30 minutes.
	10b. Normal temperature 36°–38°C (98°–101°F).

PARAMETER	INFORMATION DERIVED
	10c. Temperature of less than 36°C (97°F) indicates **hypothermia**, which causes decreased cardiac function and may lead to prolonged anesthetic recovery secondary to reduced drug clearance and major organ failure.
	10d. Hypothermia is a more common problem in foals and during prolonged abdominal surgeries.
	10e. Hypothermia may indicate too deep an anesthetic plane.
	10f. Temperature higher than 39°C (102°F) indicates hyperthermia, which can cause major organ damage and failure, cerebral edema, and death.
	10g. Hyperthermia most often occurs when animal is severely stressed before or during induction (rough induction) or when horse is induced outside in the sun on a hot day.
	10h. **Malignant** hyperthermia (extreme rise in temperature, muscle **fasciculations**, and seizure) may occur secondary to some inhalant anesthetics (halothane), although it is very rare.

	10i. Immediate steps to cool horse should be taken including:
	  1. Cool intravenous fluids
	  2. Cool packs or cool water to ears, inguinal area
	  3. Cool water enema
	  4. Ventilate

Procedure for Monitoring Ruminants

PARAMETER	INFORMATION DERIVED
1. Heart rate and rhythm	**1a.** Auscultate with stethoscope on thorax at level of elbow just behind forelimb. Best to palpate pulse simultaneously. **1b.** Arteries commonly palpated are external mandibular, radial artery at the carpal canal, and auricular artery. **1c.** Normal heart rate is 70–100 bpm in the adult cow and 90–130 in calves. **1d.** Tachycardia, or a heart rate greater than 100 bpm, in the adult cow may be pain induced, drug induced (e.g., ketamine), or a sign that the animal is waking up. **1e.** Bradycardia, or a heart rate less than 50 bpm, may be drug induced (e.g., alpha-2 agonists, opioids), or a sign of too deep a plane of anesthesia. **1f.** Irregular heart rhythm may be associated with pain, excessively deep anesthesia, or impending death. **1g.** Diagnose dysrhythmia with electrocardiogram. **1h.** Blood pressure along with physical signs (mucous membrane color, respiratory rate) will identify potential negative consequences.
2. Respiratory rate and rhythm	**2a.** Observe chest rising and falling. Listen for stridor (noise) and watch for rumen bloat. May also watch reservoir bag if using gas anesthesia. **2b.** Rate and rhythm may be influenced by anesthetic used and position of animal. Ruminants in dorsal and lateral recumbency have a tendency to bloat, making respirations shallow. Mucous membrane color and pulse oximetry will help you decide if ventilation or oxygenation is adequate.

PARAMETER	INFORMATION DERIVED
	2c. Normal rate is 20–30 respirations per minute in adult cattle and usually 20–40 respirations per minute in calves.

	2d. Apnea, or no spontaneous respirations, for longer than 2 minutes is an emergency. Supply ventilatory support by:
	   1. Relieving rumen tympany (if present) by placement of orogastric tube and possibly trochar insertion
	   2. Endotracheal tube placement if one isn't present already
	   3. Mechanical respiration by squeezing reservoir bag or blowing into endotracheal tube
	   4. Administering oxygen via nasotracheal tube or endotracheal tube

	2e. Ruminants tend to have rapid, shallow respirations (**tachypnea**) while under anesthesia because of their anatomy, so regularity of respiration is a better indicator of anesthetic depth than depth of respiration.
3. Mucous membrane color	**3a.** Examine oral mucous membranes, vulva, or inside of prepuce for pale pink to pink color.
	3b. Mucous membrane color indicates degree of perfusion and oxygenation. May be influenced by anesthetics used and whether the patient is receiving ventilatory support.

	3c. Pale grey or white mucous membranes indicate poor perfusion secondary to:
	   1. Poor cardiac output
	   2. Vasoconstriction
	   3. Low blood pressure
	   4. Too deep anesthetic plane
	   5. Rumen bloat or tympany

PARAMETER	INFORMATION DERIVED
	3d. Cyanotic or blue mucous membranes indicate poor tissue oxygenation secondary to: 1. Poor ventilation secondary to anesthetic drugs or depth of anesthesia (patients not receiving supplemental oxygen) 2. Pulmonary edema 3. Aspiration of rumen contents 4. Airway obstruction **3e.** If mucous membrane color is abnormal, determine the cause and rectify.
4. Capillary refill time (CRT)	**4a.** Obtained by pressing on the gums until blanched, releasing the pressure, and counting how many seconds until color returns. **4b.** Good indicator of perfusion, hence cardiac output. **4c.** Normal is 2–3 seconds. **4d.** Prolonged CRT (longer than 4 seconds) indicates poor perfusion, possibly low blood pressure. May be from too deep an anesthetic plane. Animal needs immediate attention.
5. Eye position	**5a.** Influenced by type of anesthetic used; monitoring eye position is most effective when the patient is on inhalant anesthesia. **5b.** Normally globe is central, pupils not dilated. **5c.** Globe rotating ventromedially occurs as the animal enters a good plane of anesthesia. **5d.** If the globe then rotates back up and the pupils are fixed and dilated, the anesthesia is too deep and the animal needs immediate attention.
6. Palpebral reflex	**6a.** Elicited by stroking or tickling the eyelashes or a touch to the medial canthus. **6b.** Poor indicator of anesthetic depth in ruminants, because it generally is absent with minimal anesthesia.

PARAMETER	INFORMATION DERIVED
7. Corneal reflex	**7a.** Obtained by dropping sterile saline or artificial tear into eye and seeing global withdrawal. Gloved finger touched gently to eyeball may also be used.
	7b. Loss of reflex is an indicator of extreme anesthetic depth, as is a lack of tear secretions.
8. Ear twitch	**8a.** Not a reliable indicator of anesthetic depth in ruminants.
9. Temperature	**9a.** Obtained by use of a rectal thermometer and should be checked every 30 minutes.
	9b. Normal temperature ranges from 38°–39.5°C (100°–103°F).
	9c. Hypothermia, or temperature less than 37.5°C (99°F), is a more common problem in very young ruminants and prolonged surgeries.
	9d. Hypothermia causes decreased cardiac function and may lead to prolonged anesthetic recovery and major organ failure. May indicate too deep an anesthetic plane.
	9e. Hyperthermia, or temperature greater than 39.8°C (103.5°F), most often occurs when animal is severely stressed before or during induction (rough induction) or when anesthesia is induced outside in the sun on a hot day.
	9f. Hyperthermia can cause major organ damage and failure, cerebral edema, and death.
	9g. Immediate steps to cool ruminant should be taken including: 1. Orogastric intubation with instillation of cool water 2. Cool packs or cool water to ears, inguinal area 3. Cool water enema 4. Cool intravenous fluids

Procedure for Monitoring Pigs

PARAMETER	INFORMATION DERIVED
1. Heart rate and rhythm	1a. Auscultate with stethoscope to side of ventral chest. Best to palpate pulse simultaneously.
	1b. The pulse is best found at caudal auricular artery.
	1c. The saphenous or radial artery may be used in pigs under 50 kg (110 lb) (Fig. 10–5).
	1d. Normal heart rate in the adult pig is 80–130 bpm.
	1e. Irregular heart rate or a pulse rate that doesn't correspond with the heart rate should be analyzed with an electrocardiogram.
2. Respiratory rate and rhythm	2a. Normal respiratory rate in the pig is 10–25 breaths per minute.
	2b. Respiratory rate and rhythm may be influenced by anesthetic drugs.
	2c. When not intubated, special attention should be paid to keeping head and neck extended so as not to obstruct airway.
3. Mucous membrane color	3a. Examine oral mucous membranes, vulva, or inside of prepuce.
	3b. Indicates degree of perfusion and oxygenation. May be influenced by anesthetics used and whether the patient is receiving ventilatory support.

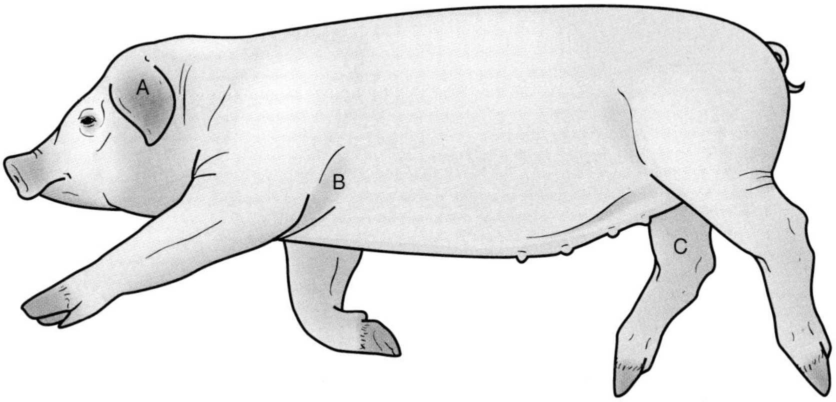

FIG. 10–5 Sites to auscult heart rate (**B**) and to palpate pulse (**A** and **C**).

PARAMETER	INFORMATION DERIVED
	3c. Pale grey or white mucous membranes indicate poor perfusion secondary to: 　**1.** Poor cardiac output 　**2.** Vasoconstriction 　**3.** Low blood pressure 　**4.** Too deep anesthetic plane **3d.** Cyanotic or blue mucous membranes indicate poor tissue oxygenation secondary to: 　**1.** Poor ventilation secondary to anesthetic drugs or depth of anesthesia (patients not receiving supplemental oxygen) 　**2.** Pulmonary edema 　**3.** Aspiration of gastric contents 　**4.** Airway obstruction **3e.** Brick red mucous membranes may be seen with hyperthermia. **3f.** If mucous membrane color is abnormal, determine the cause and rectify.
4. Capillary refill time (CRT)	**4a.** Obtained by pressing on the gums until blanched, releasing the pressure, and counting how many seconds until color returns. **4b.** Good indicator of perfusion, hence cardiac output. **4c.** Normal CRT is 1–2 seconds, prolonged is more than 3 seconds. **4d.** Prolonged CRT indicates poor perfusion, vasoconstriction. May be associated with low blood pressure. May be from too deep an anesthetic plane. Patient needs immediate attention.
5. Eye reflexes (corneal and palpebral)	**5a.** Corneal and palpebral reflex assessment is of no value in the pig.

PARAMETER	INFORMATION DERIVED
6. Temperature	**6a.** Normal temperature in the adult pig is 37.8°–39°C (100–102°F) and in the piglet it is 39°–40°C (102°–104°F)
	6b. A known genetic link to malignant hyperthermia in pigs on inhalant anesthetics is related to porcine stress syndrome.
	6c. Signs of malignant hyperthermia include:
	1. Temperature up to 43°C (108°F)
	2. Tail twitching
	3. Flushed skin
	4. Tachycardia
	5. Muscle fasciculations
	6. Seizure
	7. Death
	6d. Treatment includes:
	1. Cessation of inhalant anesthetic
	2. Cool packs or cool water to ears, inguinal area
	3. Cool water enema
	4. Cool intravenous fluids
	5. Ventilation
	6. Dantrolene
	6e. Baby pigs are highly susceptible to hypothermia (temperature less than 37.8°C [100°F]). Care must be taken to keep them warm during surgery.

MONITORING EQUIPMENT

Monitoring equipment can be as simple as a watch with a second hand and a stethoscope, or it can be as complex as electrocardiograms and pulse oximetry. The degree of monitoring often depends on the species anesthetized and the facilities available. Clearly, out in the field, electronic devices are not available to assist monitoring, nor would hooking them up be feasible or necessary when the expected anesthetic period is short (less than 30 minutes). During the last 30 years, however, technology has brought equine surgery, especially, to a much more sophisticated level, involving sterile surgical suites, inhalant anesthetics, and equipment to monitor a variety of parameters. These advances have led to survival after more complex surgical procedures and have allowed surgery to be performed successfully on patients that in the past would have died on the surgery table. Because most other species of livestock do not carry the same economic or psychologic value, anesthesia for these species generally is done in the field and monitored without the benefit of modern technology. This, by no means, makes the monitoring less important or effective relative to the types of surgeries being done and, therefore, should be undertaken with the same diligence and attention to detail as seen in referral hospitals.

BLOOD PRESSURE MONITORS

Blood pressure monitors measure the pressure generated by the circulation of blood through the arteries. Blood pressure readings tell us if major organs (liver, kidneys, heart, brain, and gastrointestinal tract) are being perfused adequately. It is therefore a measure of cardiac output. Systolic pressure is the pressure generated when the ventricles squeeze blood into the arteries, and diastolic blood pressure is the pressure that remains while the heart is filling with blood. Blood pressure readings are recorded as systolic pressure over diastolic pressure (e.g., 110/60). Depending on the equipment used, systolic, systolic over diastolic, or mean arterial blood pressure may be monitored.

Purpose

Blood pressure readings are most important in horses as a gauge of anesthetic depth. Unlike most small animal species, as anesthesia deepens, blood pressure readings will begin to fall before a change in heart rate or rhythm is seen. By the time heart rate or rhythm is disturbed in the horse, anesthetic depth may have reached critical, if not fatal, levels.

DOPPLER BLOOD PRESSURE MONITOR

Principle

- Indirect measurement
- Uses ultrasonic sound waves bounced off the arterial wall to listen to the pulse
- Cuff and sphygmomanometer used to read the systolic pressure

Sites

- Tail (horses)
- Forelimb or hind limb of any species

Equipment

- Doppler machine (Fig. 10–6)
- Cuff sized so that its width is 40% of the diameter of the limb or tail
- Ultrasonography crystal
- Ultrasonography gel
- 1-inch tape

CHAPTER 10 Anesthesia

FIG. 10-6 Doppler blood pressure machine, crystal, and cuff.

Procedure for Doppler Blood Pressure Monitoring

TECHNICAL ACTION	RATIONALE/AMPLIFICATION
1. Select appropriate size cuff for tail or limb.	1a. Select a cuff width that is 40% of the thickness of limb or tail.
2. Wrap the cuff snugly around the limb or base of tail.	2a. If the cuff is too loose, it will not inflate correctly and may yield inaccurate readings.
3. Palpate the artery distal to cuff.	3a. If not palpable, you may use the ultrasonography crystal to listen for the artery.
	3b. Apply ultrasonography gel to probe.
	3c. Place concave side of the probe over the artery just distal to cuff.
	3d. Turn monitor on and listen for a whooshing sound in time with the pulse rate.
4. Secure probe with tape.	4a. Use 1–2 inch nonelastic tape.
5. Squeeze bulb to inflate cuff until sound disappears.	5a. Generally need to go up to about 180–200 mmHg.
6. Slowly let air out of the cuff while watching the manometer (gauge).	6a. Record the pressure at which the sound first returns. This is the animal's systolic blood pressure.

TECHNICAL ACTION	RATIONALE/AMPLIFICATION
7. Release the rest of the air from the bulb.	7a. Repeat every 5 minutes during anesthesia.
8. Findings: Systolic blood pressure	8a. Blood pressure requirements are relatively stable across species, so the following limits may be used for any large animal.
80	Severe hypotension: Needs attention immediately. Decrease anesthesia level and increase fluids.
80–90	Low: Adjust anesthetic level, increase fluids.
100–150	Normal.
150	Hypertension: Needs attention immediately if surgery is to continue. May be indicative of pain or lightening of anesthesia plane. Check fluid administration; may be excessive.

OSCILLOMETRIC BLOOD PRESSURE MONITORING

Principle

- Indirect measurement
- Machinery reads the vibrations of the artery
- Can determine systolic and diastolic pressures, and calibrates or reads mean blood pressure depending on the machinery

Sites

- Same as for Doppler

Equipment

- Machine (Fig. 10–7)
- Cuff sizing same as for Doppler

FIG. 10-7 Oscillometric blood pressure machine.

Procedure for Oscillometric Blood Pressure Monitoring

TECHNICAL ACTION	RATIONALE/AMPLIFICATION
1. Palpate the artery you wish to monitor.	1a. The tail is used most commonly in horses. 1b. Can use forelimb (dorsal metatarsal artery) or hind limb more easily in small ruminants and pigs.
2. Place the cuff snugly around the tail or limb so that the sensor overlies the palpable artery.	2a. Cuff measurement is same as with Doppler but even more critical, because the sensor is built into the cuff.
3. Turn on machine and check to see that machine's reading of pulse rate matches manual reading.	3a. This will help you calibrate the machine and ensure that you are detecting the animal's artery.
4. Set time between automatic inflations (alarms may also be set to alert you to low or high blood pressure or pulse rate).	4a. Generally set for every 5 minutes. 4b. The alarms may sometimes be annoying or may risk waking the animal if it is in a light plane of anesthesia. It is best to monitor the findings yourself.
5. Push start and record results.	—
6. Findings.	6a. Same as for Doppler.

TECHNICAL ACTION	RATIONALE/AMPLIFICATION
7. Notes.	**7a.** Difficult to maintain large variety of cuffs needed in practices in which many species are anesthetized.
	7b. Difficult to gain accurate readings on swine because of short limbs and thick skin.
	7c. Older models may not tolerate slow heart rates of adult horses because the alarms will go off coninually and the accuracy of the readings will be affected.

DIRECT BLOOD PRESSURE MONITORING

Principle

- Direct measurement
- Invasive technique measures mean arterial pressure and diastolic and systolic pressure within the artery itself when an intraarterial catheter is hooked up to a pressure transducer that converts the pressure into an electrical signal
- Alternatively, a simple aneroid manometer may be used and mean arterial pressure alone obtained

Sites

- Facial and dorsal metatarsal arteries used most commonly in the horse
- Auricular artery in ruminants
- Dorsal pedal artery in pigs

Equipment

- 18–20g × 1.5–2 inch over-the-needle Teflon catheter
- 20–22g × 1 inch in small foals, miniature horses, llamas, and other small ruminants
- Extension set with 3-way stopcock
- 12 cc syringe containing 2.0 units heparin/ml saline
- Nonpliable tubing to attach to the pressure transducer or manometer
- 1-inch white tape
- Razor to shave insertion site
- Number 15 scalpel blade or 16g needle
- Sterile preparation items (scrub, alcohol, solution)

Procedure for Direct Blood Pressure Monitoring

TECHNICAL ACTION	RATIONALE/AMPLIFICATION
1. Prepare site for surgery.	1a. Use a human razor to shave the area.
	1b. Perform the three-step surgical scrub described in Chapter 8.
2. Insert catheter and secure.	2a. Technique is similar to intravenous catheterization except that more force is necessary to penetrate the wall of the artery, and the arterial wall is thicker than that of a vein.
	2b. Gently advance catheter and stylet into the lumen of the artery before sliding catheter off of stylet.
	2c. Secure catheter with superglue and tape.
3. Attach tubing, stopcock, and flush line with heparin solution.	3a. Be sure to fill tubing with heparinized saline and set stopcock so that no blood enters the manometer.
	3b. Heparinized saline supplies the fluid column against which the pressure is measured. In other words, it is a sterile buffer between blood and manometer, in addition to preventing thrombus formation in the artery.
4. Attach manometer.	4a. Attach manometer only after saline column is established and you have decided which direction to turn stopcock.
5. Align the manometer so that it is level with the aortic root. In dorsal recumbency it should be level with the point of the shoulder; in lateral recumbency it should be level with the patient's midline.	5a. This is important to get an accurate reading (Fig. 10–8).
	5b. If manometer is placed too high, the blood pressure will appear lower than it is.
	5c. If manometer is placed too low, the blood pressure will appear to be higher than it really is.
6. Flush catheter and acquire a reading every 5 minutes.	6a. Flushing will prevent thrombus formation, which would cause an inaccurate pressure reading.

TECHNICAL ACTION	RATIONALE/AMPLIFICATION
7. Findings (adults)	**7a.** Same as indirect findings (Doppler or oscillometric).
Mean Arterial Pressure	
60 mmHg	Hypotension: Needs attention immediately. Decrease anesthetic level and increase fluids.
75–100 mmHg	Normal.
>110 mmHg	Hypertension: Needs attention immediately if surgery is to continue. May be indicative of pain or lightening of anesthetic plane. Check fluid administration; may be caused by excessive fluid volume.
Systolic Blood Pressure (Adult)	
40 mmHg	Severe hypotension: Needs attention immediately. Decrease anesthetic level and increase fluids.
55–90 mmHg	Normal.
>100 mmHg	Hypertension: Needs attention immediately if surgery is to continue. May be indicative of pain or lightening of anesthetic plane. Check fluid administration; may be receiving excessive fluid.

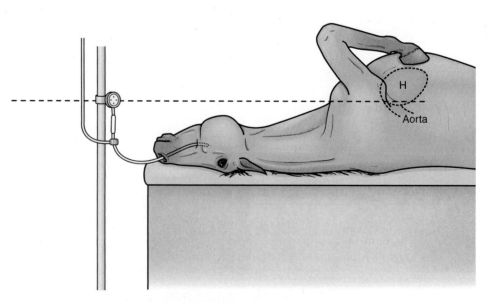

FIG. 10-8 Alignment of manometer in direct blood pressure monitoring.

ELECTROCARDIOGRAPH

An electrocardiograph (ECG) machine produces a waveform tracing that depicts the electrical activity of the heart. Sensors (or leads) may be attached to the skin or incorporated in an esophageal probe. The tracing monitored during anesthesia is lead II. Some monitors may give a choice of leads to observe. All machines come with clips clearly labeled for correct placement on the patient (Fig. 10–9).

Principle

- Detects electrical activity in the heart

Sites

- Factory labeled clips or leads placed on right elbow, left elbow, and left hind leg in small ruminants and pigs
- Lead placement as follows: Right or left front leg in the axilla and the other two leads on the neck large ruminants and horses (Fig. 10–10).

Equipment

- ECG machine and clips or leads
- Isopropyl alcohol 70%
- Conductive gel

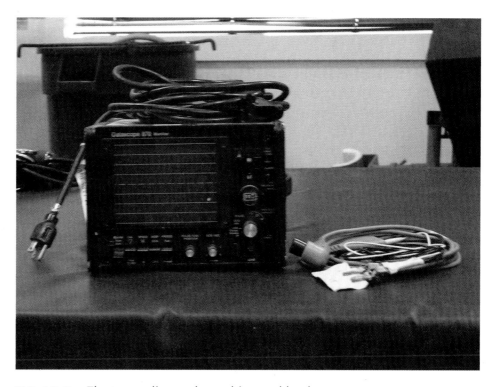

FIG. 10–9   Electrocardiograph machine and leads.

FIG. 10-10 Electrocardiograph leads attached to anesthetized horse.

Procedure for Setting Up the ECG

TECHNICAL ACTION	RATIONALE/AMPLIFICATION
1. Soak skin with 70% isopropyl alcohol.	1a. You may use water instead, but alcohol is a better wetting agent.
2. Apply conduction gel to same spot.	2a. Lubricant jelly or ultrasonography gel work well.
3. Apply metal clip or lead, making sure to have good skin contact (not just hair).	3a. Some practices file down the teeth on the alligator clips so that they don't pinch the skin as badly on small animals. These are difficult to set securely on cattle or pigs.
4. Turn on machine and set to lead I or II.	4a. Lead I often is used in horses and cattle.
5. Difficulties that may be encountered.	5a. Interference from electrical surgery tools will make pattern unreadable.
	5b. Contact of the leads (or even machine) with a metallic object will create a noisy or rough baseline, making it difficult to read the waveforms.

TECHNICAL ACTION	RATIONALE/AMPLIFICATION
6. Findings.	6a. Fig. 10–11.
7. No discernible P, QRS, or T waveforms.	7a. See above.
8. Premature ventricular contraction (PVC).	8a. Usually appears as large, wide waveform not associated with any of the normal waveforms.
	8b. May be drug induced. One or two PVCs per minute are probably not of concern.
9. Heart block (P wave not followed by a QRS complex).	9a. May be drug induced (alpha-2 agonists), but also may be an indicator of anesthetic level being too deep.

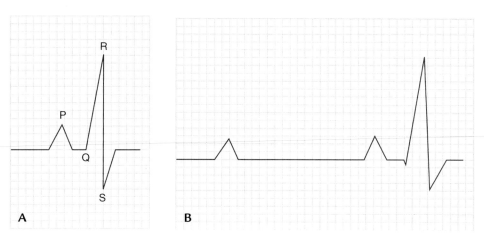

FIG. 10–11 (A) Normal waveform. (B) Example of heart block: P wave with no QRS complex.

PULSE OXIMETRY

Principle

- Measures oxygen saturation of the hemoglobin in red blood cells
- Sensor works by emitting a high-frequency light wave that passes through a capillary bed to a receiver on the other side of the sensor
- Amount of light received is converted by a computer chip in the machine to percentage of oxygen saturation

Sites

- Difficult to find site thin enough for sensor to read through
- Tongue, rectal probe (foal, calf, lamb, cria, kid)

- Ear, lip commissure (adult horse)
- May not find a suitable site—try vulvar lips, tip of tongue, or preputial skin (adult ruminant)
- Ear, tongue, rectal probe (pigs)

Equipment

- Pulse oximeter and probe (Fig. 10-12)

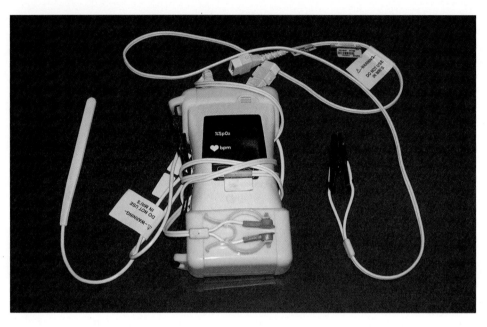

FIG. 10-12 Pulse oximetry machine and probes.

Procedure for Pulse Oximetry

TECHNICAL ACTION	RATIONALE/AMPLIFICATION
1. Turn on the machine and attach a probe.	—
2. Insert probe rectally (if a rectal probe) or attach probe to lip, tongue, ear, or other site.	2a. Clear rectum of feces before inserting probe. 2b. Find any site the probe can read through. Watch the machine; It will tell you if the sensor is in a place that it can get a reading.
3. Check patient's pulse manually and see if it matches that of the machine.	3a. Helps you to determine if the machine is getting an accurate reading of blood oxygenation.
4. May need to secure the probe with tape.	4a. Not usually necessary.

TECHNICAL ACTION	RATIONALE/AMPLIFICATION
5. Readings: Patient receiving supplemental oxygen.	5a. Probe must be well placed to get accurate reading; try several sites and several readings to ensure accuracy.
95–100%	Normal
90–95%	Low
85%	Dangerously low
6. Readings: Patient not receiving supplemental oxygen should be 88–95%.	6a. Normal.
7. Notes.	7a. All readings are affected by blood flow to the site being monitored.
	7b. If peripheral vasoconstriction exists, readings will be artificially lowered.
8. Drugs may affect readings.	8a. Alpha-2 agonists (e.g., xylazine, detomidine).

SEDATIVES AND TRANQUILIZERS USED IN LARGE ANIMAL ANESTHESIA

Classes of sedatives and tranquilizers include phenothiazines, alpha-2 receptor agonists, benzodiazepines, and opioids. The alpha-2 agonists and opioids have the added benefit of providing pain relief (analgesia) and can play an important role in balanced anesthesia. Use of sedatives or tranquilizers identified as controlled substances must be recorded accurately and permanently in a controlled substance log. These drugs must also be kept under lock and key to ensure controlled and limited access because of the potential for human abuse.

Definition

- A sedative calms the animal and may induce a restful sleep state.
- A tranquilizer reduces the physical or psychologic reactivity of an animal to stimuli.
- Some sedatives or tranquilizers may also have pain-relieving (analgesic) effects.

Purpose

- Given to the animal to calm it before administration of any other anesthetic agents
- Reduce the amount of general or local anesthetic agent needed for the surgical procedure

- Ease physical examination or handing of the patient
- Given when individuals are to be added to an already established herd or flock or when intact males are to be held together in confinement
- Ease transportation of an animal either unused to the mode of transportation or with a history of trauma related to transportation
- Sedatives or tranquilizers most often used on horses, ruminants, or wild ungulates

Special Note

Use of drugs in food-producing animals requires concern about drug residues and withdrawal times. Withdrawal time is the period required for the drug to be metabolized and excreted from the body until it falls below detectable levels. Withdrawal times are mandated by Food Animal Residue Abatement Division (FARAD) and listed when available. Use of some of these agents is considered extra-label (not in accordance with FDA approved manufacturer's instructions for use) by the Federal Drug Administration (FDA), so there are no official withdrawal times available. These drugs are identified as Extra-label Usage (ELU). A booklet on extra-label use of drugs is available from the FDA, titled *FDA and the Veterinarian* HHS (FDA) 89-6046.

PHENOTHIAZINE TRANQUILIZERS

- These drugs are neuroleptic agents that are thought to block the release of dopamine in the brain.
- Acepromazine is by far the most commonly used drug in this class.

Acepromazine 10 mg/ml

TABLE 10-2 ACEPROMAZINE

1. Physical effects	1a. Calming, mild muscle relaxant with decreased or slowed response to external stimuli 1b. Improved response to analgesic or anesthetic drugs 1c. Peripheral vasodilation 1d. May induce paraphimosis
2. Uses	2a. Preanesthetic agent 2b. Tranquilizer to ease handling or transportation 2c. Antispasmodic for urinary catheterization 2d. Treatment of azoturia and spasmodic colic in horses 2e. Decreased response to catecholamine-induced dysrhythmias
3. Monitoring	3a. Degree of tranquilization, incoordination, or ataxia 3b. Blood pressure 3c. Body temperature 3d. Heart and respiratory rates
4. Routes	4a. Intravenous (IV) 4b. Intramuscular (IM) 4c. Subcutaneous (SQ) 4d. Per os

TABLE 10-2 ACEPROMAZINE—cont'd

5. Notes	5a. Must be protected from light, heat, and freezing
	5b. Contraindicated when limb fracture suspected but not stabilized
	5c. Contraindicated in neonates, geriatrics, and those patients subject to hypothermia
6. Dosing	6a. Horses: 0.02–0.2 mg/kg IV (high-end dosages IM or SQ), not to exceed 30 mg
	6b. Ruminants: 0.15 mg/kg
	6c. Milk withdrawal time 48 hours
	6d. Meat withdrawal time 7 days. Consult label for requirement changes.

ALPHA-2 RECEPTOR AGONISTS

- Stimulate alpha-2 receptors, which decreases norepinephrine release in the central and peripheral nervous systems

Xylazine 100 mg/ml or 20 mg/ml (Rompun, AnaSed)

TABLE 10-3 XYLAZINE

1. Physical effects	1a. Profound sedation
	1b. Mild to moderate analgesia
	1c. Decreased gastrointestinal motility 3–6 hours
	1d. Muscle relaxation
2. Uses	2a. Component of anesthesia or chemical restraint for short surgical procedures
	2b. Analgesic or antispasmodic in management of colic in horses
	2c. Used alone in restraint or sedation of ruminants for foot trims or brief surgical procedures
	2d. Used in combination with other drugs to induce general anesthesia
3. Monitoring	3a. Degree of ataxia and sedation
	3b. Gastrointestinal motility, rumen motility (bloat or regurgitation)
	3c. Blood pressure
	3d. Heart rate
	3e. Respiratory rate
	3f. Oxygen saturation
4. Routes of administration	4a. Intravenous (IV)
	4b. Intramuscular (IM)
	4c. Epidural
5. Notes	5a. Induces bradycardia, second-degree heart block
	5b. Induces hypoxemia and hypercarbia in most species, with ruminants most severely affected
	5c. Transient hypertension followed by hypotension

Continued

TABLE 10-3 XYLAZINE—cont'd

5. Notes—cont'd	5d. Ruminants assume recumbency and light plane of surgical anesthesia at higher dosages
	5e. Regurgitation a problem in ruminants
	5f. Often mixed with ketamine for induction of general anesthesia in horses, swine, llamas
	5g. Causes transient hyperglycemia, hypoinsulinemia in cattle and sheep
	5h. Oxytocin-like effect on the uterus of pregnant cattle and sheep seen
	5i. High ambient temperatures will cause a prolonged and more profound sedation in ruminants
	5j. Sedation reversible with yohimbine, tolazoline, atipamezole, and idazoxan
	5k. May cause pulmonary edema in sheep
	5l. Causes diuresis sufficient to result in dehydration in an already sick patient
6. Dosing	6a. Horses: 0.3–2.0 mg/kg IV, IM (use high end of dosage IM)
	6b. Ruminants: 0.03–0.1 mg/kg IV or 0.1–0.22 mg/kg IM Milk withdrawal time 72 hours Meat withdrawal time 5 days
	6c. Goats/Sheep: 0.03–0.2 mg/kg IM (repeated dosing for prolonged procedure) Milk withdrawal time 120 hours Meat withdrawal time 10 days
	6d. Ruminants: 0.05 mg/kg epidural Meat withdrawal time 7 days

Detomidine (Dormosedan)

TABLE 10-4 DETOMIDINE

1. Physical effects	1a. Same as xylazine except:
	1b. Sedation and analgesia longer duration (up to 45 minutes)
	1c. Bradycardia more pronounced
2. Uses	2a. Standing analgesia or light anesthesia for minor surgical or dental procedures in combination with other drugs or alone
3. Monitoring	3a. Same as xylazine but for longer time
4. Routes of administration	4a. Intravenous (IV)
	4b. Intramuscular (IM)
	4c. Epidural

TABLE 10-4 DETOMIDINE—cont'd

5. Notes	5a.	High doses may be used to dart free-range cattle and wild ruminants
	5b.	Use more common in horses
	5c.	Possible prolonged sedation and ataxia during recovery from anesthesia
	5d.	No oxytocin-like effect occurs, so thought to be safe in pregnant cattle
	5e.	Reversal agents are same as for xylazine, although atipamezole is drug of choice
6. Dosing	6a.	Horses: 0.005–0.02 mg/kg IV or 0.01–0.02 mg/kg IM
	6b.	Ruminants: 0.02–0.04 mg/kg IM or IV
	6c.	Milk withdrawal time 72 hours Meat withdrawal time 3 days

Medetomidine (Domitor)

TABLE 10-5 MEDETOMIDINE

1. Physical effects	1a.	Same as xylazine
2. Uses	2a.	Used in combination with other drugs for more profound effects
3. Monitoring	3a.	Same as xylazine
4. Routes	4a.	Intravenous
	4b.	Intramuscular (IM)
	4c.	Epidural (cattle)
5. Notes	5a.	May produce recumbency in calves, small ruminants, and adult cattle at higher dosages
	5b.	Approved for use in dogs only in the United States, but used in Japan and Europe on horses
6. Dosing	6a.	Horses: None to date in United States and Canada
	6b.	Ruminants: 25 mcg–35 mcg/kg IM
	6c.	Extra-label use: No withdrawal times available

ALPHA-2 RECEPTOR ANTAGONISTS

- Reverse the effects of alpha-2 agonists
- Generally not used in horses

Reversal Agents (Alpha-2 Antagonists)

TABLE 10-6 REVERSAL AGENTS

Yohimbine (Yobine)	
Dosing in ruminants	0.12–0.25 mg/kg IV Poorly effective in cattle Milk withdrawal time 72 hours Meat withdrawal time 7 days
Atipamezole (Antisedan)	
Dosing in ruminants	20–60 mcg/kg IV, IM, ELU Dose dependent on alpha-2 receptor agonist used and how much used Safer given intramuscularly than intravenously Provide alternative analgesic drug
Tolazoline	
Dosing in ruminants	0.5 to 1.0 mg/kg IV slowly, ELU 2.0 mg/kg IV may cause hyperesthesia in cattle that have not been sedated Not for use in llamas
Doxapram	
Respiratory stimulant	May be used to augment yohimbine or tolazoline (alpha-2 receptor antagonists)
Dosing in ruminants	1.0 mg/kg IV alone or with alpha-2 receptor antagonist, ELU

ELU, extra-label usage; *IM*, intramuscular; *IV*, intravascular.

BENZODIAZEPINES

- Although there are many available on the market today, diazepam remains the most commonly used benzodiazepine in large animal anesthesia.

Diazepam 5 mg/ml (Valium)

TABLE 10-7 DIAZEPAM

1. Physical effects	1a. Muscle relaxation 1b. Anxiolysis 1c. Profound sedation in small ruminants 1d. Excitement possible if used alone in horses
2. Uses	2a. In combination with other drugs to produce muscle relaxation or induce general anesthesia 2b. Reduced amount of induction agent needed 2c. Seizure control
3. Monitor	3a. Blood pressure 3b. Respiratory rate and rhythm 3c. Heart rate and rhythm 3d. Degree of sedation 3e. Degree of muscle relaxation 3f. Degree of ataxia
4. Routes of administration	4a. IV only in horses, llamas 4b. IM in small ruminants, but highly variable effectiveness
5. Notes	5a. Schedule IV controlled substance 5b. Minimal adverse cardiac and respiratory effects 5c. Safe in small ruminants, camelids, and horse 5d. Used with other sedatives and anesthetics 5e. Raises seizure threshold 5f. Suitable for very young animals and those with cardiovascular compromise 5g. Often used in horses in combination with xylazine and ketamine to induce anesthesia
6. Dosing	6a. Give slowly when using IV 6b. Horses: 0.02–0.1 mg/kg IV 6c. Ruminants: 0.25–0.5 mg/kg IV (extra-label use) 6d. 0.55–1.1 mg/kg IM small ruminants, calves.

IM, intramuscular; *IV*, intravenous.

OPIOIDS

- Opioids are known to affect at least three receptors in the central nervous system; mu and kappa being the most relevant to pain control.
- Opioids may be receptor agonists (promoters), antagonists (blockers) or mixed agonists/antagonists.
- Each opioid drug will therefore have different affects within the animal's body in relation to pain control and side-effects.

Butorphanol 10 mg/ml (Torbugesic)

TABLE 10-8 BUTORPHANOL

1. Physical effects	1a. Analgesia (mild to moderate) 1b. Sedation (mild) 1c. Bradycardia and respiratory depression 1d. Reduced gastrointestinal motility or ileus
2. Uses	2a. Provide analgesia and sedation for mild to moderately painful procedures
3. Routes of administration	3a. Intravenous (IV) 3b. Intramuscular (IM) 3c. Subcutaneous 3d. Epidural
4. Monitoring	4a. Heart rate, respiratory rate 4b. Blood pressure 4c. Response to painful stimuli 4d. Gastrointestinal motility 4e. Excitement, muscle twitching (horses)
5. Notes	5a. Schedule IV controlled substance 5b. DO NOT give to horses without prior use of sedative; may elicit excitatory response
6. Dosing	6a. Horses: 0.01–0.02 mg/kg IV 6b. Ruminants: 0.1–0.2 mg/kg IV, IM 6c. Llamas: 0.05–0.1 mg/kg IV, IM

Morphine, Hydromorphone, Oxymorphone, and Fentanyl

TABLE 10-9 MORPHINE, HYDROMORPHONE, OXYMORPHONE, AND FENTANYL

1. Physical effects	1a. Analgesia
	1b. Sedation (not in the horse)
	1c. Bradycardia and respiratory depression
	1d. Reduced gastrointestinal motility or ileus
2. Uses	2a. Provide analgesia and sedation for painful procedures when butorphanol unavailable, inadequate, or inappropriate
3. Routes of administration	3a. Intravenous
	3b. Intramuscular
	3c. Subcutaneous
	3d. Epidural
	3e. Transdermal (fentanyl)
	3f. Intraarticular
4. Monitoring	4a. Heart rate, respiratory rate
	4b. Blood pressure
	4c. Response to painful stimuli
	4d. Gastrointestinal motility, gastric distention
	4e. Excitement (horses)
5. Notes	5a. Schedule II controlled substances
	5b. DO NOT give to horses without prior use of sedative. May elicit excitatory response.
	5c. Duration of effect drug dependent and reversible with naloxone
	5d. Sedation and hypoventilation more pronounced than butorphanol in ruminants
6. Dosing	6a. Horses: 0.1 mg/kg epidural (morphine)
	Fentanyl patch: 100 mcg up to 3 per 50-kg horse
	6b. Ruminants: 0.1 mg/kg epidural (morphine)

ANESTHETIC AGENTS

Injectable anesthetic agents are used commonly in large animal practice for restraint or surgical procedures done outside of a surgical suite. Many of these agents are used off label (ELU); in other words, they are not approved for use in food animals in the United States. Care must be taken to ensure that adequate withdrawal times are observed to prevent contamination of human food sources. A booklet on extra-label usage of drugs is available from the FDA titled *FDA and the Veterinarian* HHS (FDA) 89-6046.

DISSOCIATIVE ANESTHETICS

- These work by disassociating the noxious stimulus from the perception of pain. Animals given a dissociative agent may move their limbs involuntarily or may experience muscle fasciculations; however, they will not remember the procedure.

Ketamine Hydrochloride 100 mg/ml (Ketaset)

TABLE 10-10 KETAMINE HYDROCHLORIDE

1. Physical properties	1a. Drug is a cyclohexylamine causing amnesia and analgesia
	1b. Causes increased heart rate, blood pressure, and muscle rigidity
	1c. Works best when combined with a tranquilizer or sedative
2. Uses	2a. Induction of anesthesia after a sedative or tranquilizer has been given
	2b. Infusion of very low doses, for pain control postoperatively
3. Monitoring	3a. Heart rate and rhythm
	3b. Respiratory rate and rhythm
	3c. Blood pressure
4. Routes of administration	4a. Intravenous (IV)
	4b. Intramuscular (IM) llamas, pigs, calves, and small ruminants
5. Notes	5a. IM injections can be painful
	5b. Give IV only to horses
	5c. May potentiate seizure activity
	5d. Nystagmus common in horses
	5e. Schedule III drug; must be kept in locked site and use recorded in paginated drug log

Tiletamine plus Zolazepam (Telazol)

TABLE 10-11 TILETAMINE PLUS ZOLAZEPAM

1. Physical properties	1a. Similar to ketamine-diazepam combination but longer lasting
	1b. Tiletamine is a cyclohexylamine, and zolazepam is a benzodiazepine
2. Uses	2a. Induction of anesthesia for moderate-length (35–45 minute) procedures
	2b. Anesthesia in swine alone or together with xylazine and ketamine
	2c. Induction of anesthesia in small ruminants
3. Monitoring	3a. Heart rate and rhythm
	3b. Respiratory rate and rhythm
	3c. Blood pressure
4. Routes of administration	4a. Intravenous
	4b. Intramuscular
	4c. Subcutaneous
5. Notes	5a. Rough recoveries noted in horses not pretreated with an alpha-2 receptor agonist
	5b. Anesthesia lasts longer than ketamine and xylazine
	5c. Schedule III controlled substance; must be stored in locked container and use recorded in paginated drug log

OTHER INJECTABLE AGENTS

- Veterinarians may use other injectable compounds to promote muscle relaxation, improve analgesia, or induce anesthesia.

Guaifenesin

TABLE 10-12 GUAIFENESIN

1. Physical effects	1a. Profound muscle relaxation
2. Uses	2a. During induction to smooth the animal's transition to recumbency
	2b. During maintenance of anesthesia to reduce muscular rigidity secondary to ketamine use and to extend anesthesia time
	2c. May smooth recovery following ketamine use
3. Monitoring	3a. Respiratory rate and depth
	3b. Skeletal muscle tone
4. Routes of administration	4a. Intravenous only
5. Notes	5a. Mixed in a 5–10% solution, usually in 1-liter bags
	5b. Solution is thick and requires a large-bore catheter to run into the patient fast enough
	5c. A pressure bag may be placed around the dosing bag to help push the solution more rapidly out of the bag during induction
	5d. Concentrations >10% can cause intravascular hemolysis in horses

Thiopental

TABLE 10-13 THIOPENTAL

1. Physical effects	1a. Ultrashort-acting barbiturate
	1b. Respiratory depressant
	1c. Decreased cardiac output
	1d. Cardiac arrhythmias secondary to catecholamine release
	1e. Protein binding limits barbiturate molecules from entering the tissues
2. Uses	2a. Induction of anesthesia
	2b. Maintenance of anesthesia during short procedures
	2c. Extend anesthesia induced by ketamine, diazepam, and xylazine
	2d. Mixed in guaifenesin infusion for maintenance of anesthesia
3. Monitoring	3a. Respiratory rate and depth
	3b. Heart rate and rhythm
4. Routes of administration	4a. Intravenous only

Continued

TABLE 10-13 THIOPENTAL—cont'd

5. Notes	5a. Difficult to purchase anymore
	5b. May be added to guaifenesin infusion for a deeper anesthesia
	5c. Potency increased in acidemic or hypoproteinemic animals
	5d. Intravenous catheter recommended, because severe tissue irritation will occur if the drug is injected perivascularly
	5e. May experience stormy recoveries due to residual ataxia
	5f. Schedule III drug; careful records must be kept of use and must be stored in a locked container

Propofol 20 mg/ml (PropoFlo)

TABLE 10-14 PROPOFOL

1. Physical effects	1a. A substituted phenol that can be given repeatedly with little or no cumulative effect, so it can be given as a constant infusion
	1b. Rapid onset of anesthesia and rapid smooth recovery
2. Uses	2a. Induction and maintenance of anesthesia
3. Monitoring	3a. Respiratory rate and rhythm
	3b. Heart rate
	3c. Blood pressure
	3d. Pulse oximetry
4. Routes of administration	4a. Intravenous only, given over 60–90 seconds
5. Notes	5a. Cost prohibitive in all but smaller ruminants and ponies
	5b. May be mixed 50:50 with 2.5% thiopental

INHALANT ANESTHETICS

- Inhalant anesthetics used in large animal practice are primarily halogenated compounds such as halothane, isoflurane, enflurane, sevoflurane, and desflurane. They are stored inside a specially designed precision vaporizer in a liquid form. The precision vaporizer allows the liquid to become a gas and allows a set percentage of anesthetic vapor to exit the system. All gas anesthetics are defined by certain properties, including vapor pressure, solubility coefficient, and minimum alveolar concentration.

Properties Common to All Inhalant Anesthetic Agents

TABLE 10-15 PROPERTIES COMMON TO ALL INHALANT ANESTHETIC AGENTS

1. Vapor pressure	1a. Vapor pressure is the measure of the tendency of a liquid to turn into a gas. Vapor pressure is agent and temperature dependent.
	1b. Vapor pressure numbers are reported for temperatures 20°–22°C (68°–72°F).
	1c. Anesthetics currently used in large animal medicine all have high vapor pressures and, if administered to the patient unregulated, would cause a fatal overdose. Therefore they are used only in a precision vaporizer which is controlled for temperature, flow, and back pressure.
2. Solubility coefficient	2a. The physiologic effects of inhalant anesthetics occur because of their solubility characteristics in blood and tissues.
	2b. The blood-to-gas solubility coefficient expresses the speed at which an inhalant anesthetic is taken out of the alveoli and into the blood.
	2c. An inhalant with a low solubility coefficient builds up to higher concentrations in the alveoli, causing rapid induction and recovery.
3. Minimum alveolar concentration (MAC)	3a. The MAC of an inhalant anesthetic is the concentration of anesthetic that produces no clinical response in 50% of patients experiencing a painful stimulus.
	3b. MAC, therefore, is an indicator of potency and provides a guideline for where the vaporizer is set for induction and maintenance of anesthesia.
	3c. Most patients will experience a light plane of anesthesia at 1 × MAC, 1.5 × MAC will produce a surgical plane of anesthesia, and 2 × MAC will produce deep anesthesia.
	3d. This will vary with the individual patient status and other drugs that have been administered, and will serve only as a guideline.

Properties of Halothane

TABLE 10-16 PROPERTIES OF HALOTHANE

1. Physical properties	1a. Minimum alveolar concentration (MAC) 0.88%
2. Attributes	2a. Muscle relaxation and slight analgesia
	2b. Moderately rapid induction and recovery
3. Side effects	3a. Sensitizes heart muscle to catecholamines resulting in dysrhythmias
	3b. Decreased cardiac output
	3c. Increased vagal tone leading to bradycardia and slowed gastrointestinal motility
	3d. Peripheral vasodilation causing low blood pressure and hypothermia
4. Metabolism	4a. Most eliminated via respiratory tract
	4b. 20–45% metabolized by liver, with resultant metabolites excreted via the kidneys
5. Notes	5a. Known to cause hepatotoxicity in humans
	5b. Associated with malignant hyperthermia
	5c. Contains preservative thymol
6. Dosing	6a. To effect
	6b. For induction period set at 3–5%
	6c. Maintenance 1.5–2.5% for most species

Properties of Isoflurane

TABLE 10-17 PROPERTIES OF ISOFLURANE

1. Physical properties	1a. Minimum alveolar concentration (MAC) 1.31%
2. Attributes	2a. Very rapid induction and recovery
	2b. Depth of anesthesia easily, rapidly altered
	2c. Good muscle relaxation in surgical plane of anesthesia
	2d. Safe for use in animals with renal or hepatic disease
	2e. Safe in geriatric and neonatal animals
	2f. No known human toxicity
3. Side effects	3a. Peripheral vasodilation leading to decreased blood pressure and hypothermia
	3b. Respiratory depression at higher dosages (in the horse, 1–2 deep breaths per minute considered normal)
	3c. Dosage-dependent decrease in cardiac output, more profound in the horse
	3d. Dosage-dependent decrease in smooth muscle tone and motility
	3e. Increased intracranial pressure
	3f. Possible trigger of malignant hyperthermia
4. Metabolism	4a. 99% eliminated via respiratory tract
5. Notes	5a. No analgesic properties, so must give injectable analgesics
	5b. Rapid recovery may be stormy and dangerous to patient and personnel if postoperative pain not managed properly
	5c. May cause ileus
	5d. Pungent odor, irritating to mucous membranes
	5e. May cause malignant hyperthermia
6. Dosing	6a. To effect
	6b. Induction, set vaporizer at 4–5%
	6c. Maintenance 1.5–3%

Properties of Sevoflurane

TABLE 10-18 PROPERTIES OF SEVOFLURANE

1. Physical properties	1a. Minimum alveolar concentration (MAC) 2.31–2.84%
	1b. Unstable in CO_2 absorbers, producing compound A which is nephrotoxic in rats
2. Attributes	2a. Ultrarapid acting
	2b. Skeletal muscle relaxation comparable to that achieved with isoflurane
	2c. Reported to provide smooth anesthetic recovery in adult horses
	2d. Does not increase intracranial or intraocular pressure
3. Side effects	3a. Dosage-dependent reduced cardiac output
	3b. Dosage-dependent respiratory depression
	3c. May trigger malignant hyperthermia

TABLE 10-18 PROPERTIES OF SEVOFLURANE—cont'd

4. Metabolism	4a. 95% exhaled	
5. Notes	5a. No analgesic properties, same as isoflurane	
	5b. Not much used in large animal anesthesia	
	5c. No pungent odor	
6. Dosing	6a. To effect	
	6b. Induction, set vaporizer to 5–7%	
	6c. Maintenance 2–4%	

SAMPLE GENERAL ANESTHETIC REGIMENS

Induction and maintenance of anesthesia in large animal practice is complicated by the size and training level of the patients. Food animals most often are induced and maintained with injectable agents. Horses may be relatively safely induced and maintained in the field by injectable means when the procedures are expected to last less than 1 hour. More complicated or invasive surgeries (laparotomy, orthopedic procedures) generally require a surgical suite and gas anesthesia. Neonates of all large animals are easily maintained on small animal gas anesthetic equipment.

The following tables cover the most common means of inducing and maintaining anesthesia in large animal practice in the United States.

Induction and Maintenance of Anesthesia in Horses

	AGENTS	COMMENTS
1. Induction of anesthesia in foals	Xylazine 1.1 mg/kg IV Followed by Butorphanol 0.01 mg/kg IV Ketamine 2 mg/kg IV	**1a.** Suitable for uncomplicated castration or laceration repair. Lasts approximately 20 minutes.
	Or Xylazine 1.1 mg/kg IV Followed by Butorphanol 0.01 mg/kg IV Propofol 2 mg/kg IV	**1b.** Suitable for uncomplicated castration, laceration repair, or cast application. **1c.** Time to standing, approximately 12 minutes.
	Or Xylazine 0.5 mg/kg IV Followed by Mask or nasotracheal intubation with isoflurane administered at 4–5%	**1d.** Anesthesia may be induced in some foals with gas administered by mask without prior sedation. In foals less than 150 kg, induction may be accomplished using a small animal anesthesia machine.

	AGENTS	COMMENTS
2. Maintenance of anesthesia in foals	Halothane 1–2% or Isoflurane 1.5–2.5% or Sevoflurane 3–4%	2a. A small animal anesthesia machine may be used on foals less than 4 months old or less than 150 kg.
3. Induction of anesthesia in adults	Xylazine 1–2 mg/kg IM or IV Followed by Ketamine 2 mg/kg IV Or Xylazine 1.1 mg/kg IV Followed by Butorphanol 0.01 mg/kg Ketamine 2 mg/kg IV Or Xylazine 1.1 mg/kg IV Followed by Diazepam 0.03–0.1 mg/kg IV Ketamine 2 mg/kg IV Or Detomidine 0.02 mg/kg IV Followed by Ketamine 2 mg/kg IV	3a. IM xylazine will help calm a fractious horse before anesthetic induction. 3b. Butorphanol improves sedation and may help decrease some of the muscle rigidity associated with ketamine. 3c. Diazepam improves muscle relaxation but may worsen ataxia during recovery. 3d. Lasts 30–40 minutes; use of detomidine (instead of xylazine) may worsen ataxia.
4. Maintenance of anesthesia in adults	Xylazine 0.55 mg/kg with Ketamine 1 mg/kg every 15–20 minutes as needed or Triple drip: One liter 5% guaifenesin with 1,000 mg ketamine and 500 mg xylazine infused at 1–2 ml/kg/hr Halothane 1.5–2% or Isoflurane 1.75–2.75% or Sevoflurane 3.5–4.5%	4a. Not suitable for very painful procedures or those expected to last longer than 1 hour. 4b. Not recommended for longer than 1 hour of anesthesia. Nasal oxygen at 10–15 L/min will decrease hypoxemia. 4c. All the gas anesthetics require endotracheal intubation. Injectable agents such as xylazine and ketamine may be used to supplement gas anesthesia.

Induction and Maintenance of Anesthesia in Ruminants

	AGENTS	COMMENTS
1. Induction of anesthesia in calves	Diazepam 0.2 mg/kg IV Followed by Butorphanol 0.1 mg/kg IV Ketamine 2–4 mg/kg IV	**1a.** Xylazine 0.1–0.2 mg/kg IM may be used in calves >3 months old to facilitate restraint and placement of IV catheter. **1b.** Mask induction with gas anesthetic may be done for calves <5 months old instead of using ketamine.
2. Maintenance of anesthesia in calves	One liter of 5% guaifenesin with 1,000 mg of ketamine. Infuse at 1.5 ml/kg/hr IV Halothane 1–2% or Isoflurane 1.5–2.5% or Sevoflurane 2.5–4%	**2a.** Infuse via IV catheter. Monitor drip carefully along with clinical signs, because it is easy to cause an overdose. **2b.** A small animal anesthesia machine may be used for calves <50 kg. A mask may be used, but endotracheal intubation is safest.
3. Induction of anesthesia in adult cattle	Xylazine 0.1 mg/kg IV Followed by Ketamine 2 mg/kg IV	**3a.** Xylazine may be given as premedication to facilitate IV catheter placement. **3b.** Detomidine 0.01 mg/kg IV may be given instead of xylazine. **3c.** Ketamine is given after the cow becomes recumbent as a single-bolus dose.
4. Maintenance of anesthesia in adult cattle	Ketamine 1 mg/kg IV repeated every 10–15 minutes as needed Or One liter of 5% guaifenesin with 1,000 mg of ketamine; infuse at 2–2.5 ml/kg/hr IV Xylazine 50 mg may be added for more profound sedation	**4a.** Supplemental oxygen may be given intranasally (10–15L/min) or via endotracheal tube (8–10 L/min) if procedure is expected to last more than 30 minutes. **4b.** Provides more profound muscle relaxation allowing a deeper anesthetic plane. **4c.** Intubation is recommended, as is supplemental oxygen administration.

	AGENTS	COMMENTS
	Halothane 2–3% or Isoflurane 2.5–3.5% or Sevoflurane 2.5–4.5%	**4d.** Small tidal volume and rapid respiratory rate requires a higher setting on the vaporizer than with other same-sized species. **4e.** Requires intubation and large animal anesthesia machine.
5. Induction of anesthesia in sheep	Xylazine 0.2 mg/kg IV Followed by Ketamine 2–4 mg/kg IV once patient is recumbent Or Diazepam 0.2 mg/kg IV Butorphanol 0.1 mg/kg IV Followed by Ketamine 2–4 mg/kg IV	**5a.** Not for use in sheep with cardiovascular compromise. **5b.** Safer in debilitated sheep.
6. Maintenance of anesthesia in sheep	Ketamine 1–2 mg/kg IV bolus every 10–15 minutes as needed Or Halothane 1–2% or Isoflurane 1.5–2.5% or Sevoflurane 2.5–4%	**6a.** Administer supplemental oxygen intranasally or by mask. **6b.** Intubate or maintain with mask using small animal anesthesia equipment.
7. Induction of anesthesia in goats	Xylazine 0.05 mg/kg IV Followed by Ketamine 2–4 mg/kg IV once patient is recumbent Diazepam 0.2 mg/kg mixed with ketamine 3 mg/kg in same syringe given as IV bolus	**7a.** May give up to 0.1 mg/kg xylazine if goat is highly agitated and difficult to restrain. **7b.** Author's experience is that this provides a very smooth, safe induction.
8. Maintenance of anesthesia in goats	Ketamine 1–2 mg/kg IV bolus every 10–15 minutes as needed Halothane 1–2% or Isoflurane 1.5–2.5% or Sevoflurane 2.5–4%	**8a.** Maintains light anesthesia for castration or simple surgical procedure. **8b.** Intubate or maintain with mask using small animal anesthesia equipment.

Induction and Maintenance of Anesthesia in Llamas

	AGENTS		COMMENTS
1. Sedation of adults	Xylazine 0.2–0.5 mg/kg IM May be combined with Butorphanol 0.05 mg/kg IM	1a.	For fractious (especially intact male) llamas and alpacas will provide restraint for up to 30 minutes
2. Induction of anesthesia in adults	Xylazine 0.04 mg/kg IM Butorphanol 0.3–0.4 mg/kg IM Followed by Ketamine 3–4 mg/kg IM Or	2a.	Appropriate for induction of fractious llamas or alpacas. Time to recumbency is 4–7 minutes, with time to standing 22–63 minutes
	Xylazine 0.25 mg/kg IV Followed by Ketamine 2–4 mg/kg IV Or	2b.	Best given via IV catheter
	Diazepam 0.2 mg/kg IV or Midazolam 0.1 mg/kg IV Followed by Ketamine 2–4 mg/kg IV Or	2c.	Safer for debilitated or compromised camelids than using xylazine
	Xylazine 0.25 mg/kg IV or Diazepam 0.2 mg/kg IV Propofol 2 mg/kg IV	2d.	Requires IV catheter
	Xylazine 0.4 mg/kg IV Followed by Thiopental 8–10 mg/kg IV	2e.	Requires IV catheter
	Mask with gas anesthetic	2f.	Requires smaller or very cooperative llama
3. Maintenance of anesthesia in adults	Propofol 0.2–0.4 mg/kg/min or Ketamine 1–2 mg/kg IV bolus every 10–15 minutes as needed or Or	3a.	Apnea may occur, necessitating intubation
	Halothane 1–2% or Isoflurane 1.5–2.5% or Sevoflurane 2.5–4%	3b.	Most alpacas and many llamas may be maintained on small animal anesthesia equipment
4. Induction of anesthesia in cria	Diazepam 0.1 mg/kg Butorphanol 0.05 mg/kg Followed by Ketamine 1 ml/kg Or Mask induction with halothane, isoflurane or sevoflurane		—
5. Maintenance of anesthesia in cria	Halothane 1–2% or Isoflurane 1.5–2.5% or Sevoflurane 2–3.5%	5a.	Mask may be used instead of intubation

Induction and Maintenance of Anesthesia in Pigs

	AGENTS	COMMENTS
1. Sedation or restraint	Acepromazine 0.25–0.5 mg/kg combined with morphine 0.5–1.0 mg/kg IM Medetomidine 0.008 mg/kg IM Butorphanol 0.2–0.4 mg/kg IM	1a. IM injections are best given using an 18g × 1.5 inch needle just caudal to the ear and about 2 inches off midline.
2. Induction	TKX (tiletamine-ketamine-xylazine) IM mixed by adding 250 mg (2.5 ml) ketamine and 250 mg (2.5 ml) xylazine to a bottle of tiletamine plus zolazepam powder instead of sterile water Results in mixture containing tiletamine plus zolazepam 100 mg/ml, ketamine 50 mg/ml, and xylazine 50 mg/ml. Pigs <100 kg: 1 ml TKX IM per 25 kg Pigs >100 kg: 3 ml TKX given IM, followed by 2 ml IV once sedation has been achieved	2a. Pigs are extremely difficult to anesthetize and care must be taken to limit stress placed upon them prior to induction.
3. Maintenance	Xylazine-guaifenesin-ketamine mixed together as follows: Add to 500 ml of 5% guaifenesin, 500 mg ketamine (5 ml) and 500 mg of xylazine (5 ml of 100 mg/ml xylazine). Infuse at a rate of 2 ml/kg/hr Or Halothane 1.5–2.5% or Isoflurane 2–3% or Sevoflurane 2–4%	3a. Requires IV catheter. 3b. Monitor anesthesia carefully and adjust rate as needed. 3c. Nasal oxygen will help maintain adequate oxygenation. 3d. Pig may vomit during recovery. 3e. Recovery occurs approximately 30–45 minutes after the infusion is stopped. 3f. May be maintained by mask, nasal tube, or tracheal intubation. 3g. Porcine stress syndrome or malignant hyperthermia may be triggered by any inhalant anesthetic, often resulting in death of the animal.

FIELD ANESTHESIA

Performed outside a surgical suite, field anesthesia generally relies on injectable anesthetics to keep the animal recumbent while the surgeon operates. Because a source of electricity may not be close at hand, clippers are often battery operated and some electronic monitoring equipment may not be available. The technician's duty is to the patient's and surgeon's safety. Care must be taken to monitor the patient and not be distracted by the surgery.

Purpose

- Most common reasons for field anesthesia in horses are castration, laceration repair, and bandage or cast application.
- Adult cattle undergo field anesthesia less commonly than horses, most commonly for foot surgery, preputial or penile lacerations, and correction of a left displaced abomasum.
- Small ruminants (including young calves) most commonly are anesthetized for applying a limb cast or surgical repair of inguinal hernia.
- Llamas may be castrated or have a limb cast applied under general anesthesia in the field, but more extensive surgeries or procedures would more commonly be done in a surgery suite.
- Swine may undergo general anesthesia for castration of boars over 200 lb or cesarean section on a large sow, but it is rare because of the difficulty in achieving adequate anesthesia without endangering the life of the animal.

INDUCTION OF FIELD ANESTHESIA IN THE HORSE

Equipment

- Heavy duty flat nylon halter
- Cotton or nylon lead rope 8–10 feet long
- 18g × 1.5 inch needles
- 12 cc syringes, 6 cc syringes, 3 cc syringes
- Items for jugular catheter

    14-18g intravenous catheter

    Clippers with #40 blade

    Surgical scrub, 70% isopropyl alcohol, and surgical solution

- Anesthetic induction drugs (see Induction and Maintenance of Anesthesia in Horses)

Complications

- Injury to anesthetist, surgeon, or handler from the horse falling on someone or waking up during surgery
- Injury to horse from rough induction or failure to control collapse to the ground
- Anesthesia-related death from overdose or inattention to vital signs

Procedure for Method One: Horse Field Induction and Recovery

TECHNICAL ACTION	RATIONALE/AMPLIFICATION
1. Restraint	1a. Halter horse with heavy duty, flat nylon halter and a long, soft cotton or nylon lead rope.
	1b. Secure horse in cross-ties or a stock.
2. Anesthetic preparation	2a. Insert jugular catheter (Chapter 6).
	2b. Wash mouth out with water.
	2c. Lead the horse to the induction area.
3. Induction	3a. Secure the horse to a heavy, stout fence post (if available) by wrapping lead rope 1-½ times around the post and having an assistant hold the rope from the other side of the fence (Fig. 10–13).
	3b. Do not tie the rope, because you will need to allow it to slowly slip free as the horse lies down.
	3c. The person holding the rope should be wearing heavy leather gloves to prevent rope burns.
	3d. Fractious horses may need to be snubbed up close to the fence to prevent injury to the person injecting the drugs.
	3e. Inject xylazine 1.1 mg/kg IV in catheter or jugular vein.
	3f. Wait approximately 3–5 minutes for nose to drop to ground and horse to be unsteady on his feet.
	3g. Inject ketamine 2.2 mg/kg IV rapidly into catheter or jugular vein.

TECHNICAL ACTION	RATIONALE/AMPLIFICATION
	3h. Step away from the horse as it lies down, staying level with shoulder. Horse will begin to tremble.
	3i. Person holding rope should hold firmly as horse pulls back against the rope, and then slowly let off the pressure as the horse lies or falls down. Many horses rear back on their haunches as the ketamine takes effect; it is critical that the rope holder not allow the horse to flip over backward.
	3j. Position the horse as required for procedure to be done. Never walk behind or between the hind legs, because muscle spasms or light plane of anesthesia may cause deadly kicks.

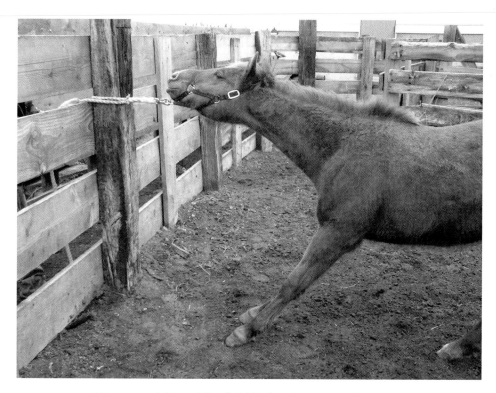

FIG. 10-13　Horse positioned for field induction using the post technique. Note that the rope is wrapped 1-½ times around a very solid post. Horse is sitting back as the drugs take effect.

TECHNICAL ACTION	RATIONALE/AMPLIFICATION
	3k. Cover horse's eyes with large soft towel to protect them from injury, drying, and light. Towel also eliminates visual stimulation, which could cause premature arousal.
	3l. Monitor eye signs every few minutes.
	3m. Begin monitoring vital signs.
4. Recovery (20–40 minutes)	**4a.** Lay horse in lateral recumbency.
	4b. Leave halter on or put back on.
	4c. Attach and leave lead rope stretched out in front of horse.
	4d. Leave towel loosely draped over eyes.
	4e. Monitor heart rate every 5 minutes until ear twitches in response to noise or movement, and monitor respirations.
	4f. Step away from the horse and monitor respirations from a distance.
	4g. Allow horse to attain sternal recumbency on its own; do not encourage the horse to stand.
	4h. When horse begins to rise, you may steady its head by letting it lean against the lead rope; otherwise allow it to rise on its own.
	4i. Approach the horse's left shoulder and keep it calm until it has fully regained its balance.
	4j. Remove IV catheter.
	4k. Return horse to stall or paddock and remove the halter.

Procedure for Method Two: Horse Field Induction and Recovery

TECHNICAL ACTION	RATIONALE/AMPLIFICATION
1. Restraint	**1a.** Halter horse with heavy duty, flat nylon halter and a long, soft cotton or nylon lead rope.
	1b. Secure horse in cross-ties or a stock.
2. Anesthetic preparation	**2a.** Insert jugular catheter (Chapter 6).
	2b. Wash mouth out with water.
	2c. Lead the horse to the induction area. A grassy surface is best, because it is soft and provides good traction during recovery.
3. Induction	**3a.** Inject xylazine intravenously. Flush catheter (if using one) with 3 cc saline. If no catheter is being used, insert needle in vein first to ensure placement in vein not artery. Blood will be dark and drip slowly from hub only if vein is compressed. An arterial stick will yield spurting bright red blood.
	3b. Wait approximately 3–5 minutes for nose to drop to ground and horse to become unsteady on his feet.
	3c. Inject diazepam as an IV bolus injection.
	3d. Inject ketamine IV as a rapid bolus injection.
	3e. Circle the horse in a tight circle in the direction opposite to the side you want down (Fig. 10–14A).
	3f. Take care to not pull the horse over on top of you when circling. Take care to fold lead rope in holding hand; do not leave it dragging on the ground or wrap it around your hand.
	3g. Ease the horse's head to the ground as he lies down.
	3h. Position the horse as required for procedure to be done.

TECHNICAL ACTION	RATIONALE/AMPLIFICATION
	3i. Cover the eyes with a large, soft towel. Towel protects eyes from injury and light, and eliminates any visual stimulation which could cause premature arousal.
	3j. Monitor eye signs every few minutes.
	3k. Begin monitoring vital signs.
4. Recovery (30–40 minutes)	**4a.** Lay horse in lateral recumbency.
	4b. Leave halter on or put back on.
	4c. Attach and leave lead rope stretched out in front of horse.
	4d. Leave towel loosely draped over eyes.
	4e. Monitor heart rate every 5 minutes until ear twitches in response to noise or movement, and monitor respirations.
	4f. Step away from the horse and monitor respirations from a distance.
	4g. Allow horse to attain sternal recumbency on its own; do not encourage the horse to stand (Fig. 10–14B).
	4h. When horse begins to rise, you may steady its head by letting it lean against the lead rope, but otherwise allow it to rise on its own.
	4i. Approach the horse's left shoulder and keep it calm until it has fully regained its balance.
	4j. Remove IV catheter.
	4k. Return horse to stall or paddock and remove the halter.

FIG. 10–14 (A) Circling the horse for induction of field anesthesia when a post is not available. (B) Recovering horse in sternal recumbency. Towel for head has fallen to the side.

INDUCTION OF FIELD ANESTHESIA IN THE RUMINANT

Equipment

- Halter and lead rope
- Nose tongs
- 18g × 1.5 inch needles
- 12 cc syringes, 6 cc syringes, 3 cc syringes
- Jugular catheter items (optional)
  14–18g intravenous catheter
  Clippers with #40 blade
  Surgical scrub, 70% isopropyl alcohol, and surgical solution
- Ropes
- Cushions (Chapter 8)
- Orogastric tube

Complications

- Injury to anesthetist, surgeon, or handler during induction process or from patient waking up during surgery
- Aspiration pneumonia or death from regurgitation of ruminal contents
- Rumen tympany causing asphyxiation or ischemia
- Overdose of anesthetic agents causing delay in recovery or death

Procedure for Induction of Ruminants

TECHNICAL ACTION	RATIONALE/AMPLIFICATION
1. Restraint	1a. Restrain patient in chute with a side gate, stock, or head lock.
	1b. Small calves may be restrained manually.
	1c. Apply halter and tie head off to the side.
2. Anesthetic preparation	2a. Insert a jugular catheter (Chapter 6).
	2b. If jugular catheter not being used, xylazine may be injected IM or into the tail vein.
	2c. Administer tranquilizer or sedative such as acepromazine or xylazine (Tables 10–2 and 10–3).

TECHNICAL ACTION	RATIONALE/AMPLIFICATION
3. Induction	3a. Inject ketamine by IV bolus.
	3b. Place patient on cushioning (Chapter 8).
	3c. Position the patient in sternal recumbency if possible.
	3d. Patients in lateral or dorsal recumbency must be positioned with the head lower than the body to allow ruminal juices to drain and to decrease risk of aspiration.
	3e. Place orogastric tube if bloating is a concern.
	3f. Begin monitoring vital signs.
4. Recovery	4a. All ruminants should be recovered in sternal recumbency. Straw bales may be used to prop the animal in this position.
	4b. Monitor vital signs until the patient is able to rise under its own power.
	4c. Return patient to stall or pen separate from herd mates.

INHALANT ANESTHESIA IN LARGE ANIMAL SURGERY

Inhalant anesthesia is most commonly reserved for horses, neonates, and very valuable ruminants, and performed in surgical referral hospitals. Small patients (under 150 kg) may be maintained on a typical small animal anesthesia machine, but larger patients require a large animal machine because of their greater lung capacity.

Purpose

- Inhalant anesthesia requires a highly controlled drug delivery system.
- Inhalant anesthesia has the added benefit of administering 100% oxygen while the patient is under anesthesia.

Equipment

- Medical grade compressed oxygen
- Large animal anesthesia machine

- Endotracheal tube or nasotracheal tube or mask
- Inhalant anesthetic agent

Complications

- Insufficient oxygen in the tank
- Insufficient anesthetic in the vaporizer
- Improper maintenance of the anesthesia machine
- Overdose of anesthetic to patient
- Injury to patient's lung tissue

Parts of the Anesthetic Delivery System

TABLE 10-19 PARTS OF THE ANESTHETIC DELIVERY SYSTEM

1. Compressed gas supply	1a. Valve handle on top opens and closes tank flow.
	1b. Pressure gauge tells how much oxygen is left in the tank.
	1c. Reducing valve and gauge decrease pressure from the tank to 50 psi in the gas line, so the oxygen flow doesn't overwhelm the controls on the anesthesia machine.
	1d. Large (G or H) cylinders are often green or white. Smaller (E) tanks do not hold enough oxygen for most large animal surgery (Fig. 10-15A).
	1e. Tank must be secured by chain to wall or carrying cart to prevent accidental falling.
2. Gas lines	2a. Specialized line that is protected from static charge carries oxygen to the anesthesia machine.
	2b. May be hooked directly to the machine or run through the walls and ceiling for hook-up in room separate from oxygen tank.
3. Anesthesia machine	3a. Mixes oxygen with anesthetic vapor to deliver anesthetic to the patient (Fig. 10-15B).
	3b. Also removes carbon dioxide from the inhaled recirculated gases and eliminates waste gas.
	3c. Provides a means to ventilate patient and deliver oxygen.
4. Flow meter	4a. Indicates the actual number of liters of oxygen flowing to the patient.
	4b. Consists of a dial and a glass cylinder marked off in milliliters or liters per minute.
	4c. For flow meter with a ball indicator, read at the middle of the ball to determine flow rate.
	4d. Further reduces oxygen flow pressure to atmospheric pressure or approximately 15 psi.

Continued

 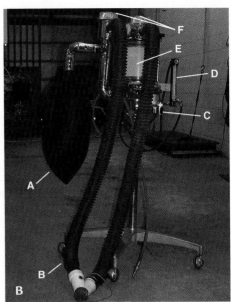

FIG. 10-15 (A) H-size oxygen tank secured to carrier. (B) Anesthesia machine. Rebreathing bag (**A**), Y-piece hoses (**B**), vaporizer (**C**), flow meter (**D**), CO_2-absorbent canister (**E**), unidirectional flutter valves (**F**).

TABLE 10-19 PARTS OF THE ANESTHETIC DELIVERY SYSTEM—cont'd

5. Precision vaporizer	5a. Converts the liquid anesthetic into a vapor allowing a carefully controlled amount of anesthetic to be delivered to the patient.
	5b. Dial is in percentage (%) of vapor delivered and has a locking feature in the off position to prevent accidentally turning on vaporizer.
	5c. Flow through the vaporizer is unidirectional.
6. Fresh gas inlet	6a. Delivers the mixture of oxygen and vaporized anesthetic to the breathing circuit.
7. Circle breathing circuit	7a. Delivers oxygen and vaporized anesthetic to the patient while removing carbon dioxide from exhaled gases.
8. Fresh gas hose	8a. Contains oxygen and vaporized anesthetic from vaporizer.
9. Unidirectional flutter valves	9a. These direct the flow of gases to and from patient. When the patient inhales, the inhalation valve opens, allowing oxygen and anesthetic to flow into the patient. When the patient exhales, the exhalation valve opens, allowing exhaled gasses to flow into the carbon dioxide absorber canister where carbon dioxide is removed.
10. Breathing hoses	10a. Attach to unidirectional flutter valves and to patient.
	10b. Large diameter hoses allow flow and decrease back pressure for large animals.
	10c. Hoses attach to Y-piece, which then connects to the endotracheal tube.

TABLE 10-19 PARTS OF THE ANESTHETIC DELIVERY SYSTEM—cont'd

11. Reservoir bag	11a. Large black rubber bag holds 15–30 liters of gas on a large animal anesthesia machine.
	11b. Often contains a mix of exhaled gases (from which carbon dioxide has been removed) and fresh gas from the vaporizer. Bag empties as patient inhales and fills as patient exhales or as fresh gas flows into it.
	11c. Functions:
	1. Store gas
	2. Monitor depth and rate of respirations by movement of bag
	3. Allow controlled delivery of oxygen (with or without anesthetic) to the patient by manually squeezing the bag and forcing gas into the patient's lungs. This is known as bagging the patient.
	11d. Bag should be sized to hold 30–60 ml/kg of patient weight, so an adult horse would require a bag size of 60 ml $\times$ 500 kg = 3,000 ml or a 30-liter bag.
	11e. Overfilling bag will cause back pressure to the patient, making it difficult or impossible for the patient to exhale and for blood to leave the lung tissue. Ultimately then, pronounced back pressure (from failure to leave pop-off valve open) would kill the patient.
12. Pop-off valve	12a. Also known as a pressure-relief valve, it usually is located above the exhalation valve.
	12b. Allows excess gas to exit from the circuit and to enter the scavenging system and to prevent buildup of back pressure.
	12c. Normally kept open unless you wish to ventilate the patient.
13. Carbon dioxide absorbent	13a. A canister containing an absorbing chemical such as soda lime (calcium hydroxide, sodium hydroxide, and potassium hydroxide) or barium hydroxide lime.
	13b. The absorbing chemicals interact with carbon dioxide to produce heat, water, and calcium or barium carbonate.
	13c. Granules saturated with carbon dioxide initially turn blue and hard, and these should be removed as soon as the surgery is complete.
	13d. In large animal anesthesia the granules may need to be changed after every surgery. Do not rely on color change, because the blue color will fade even though the granules are no longer absorbent.
14. Oxygen flush valve	14a. Delivers straight oxygen at a high flow rate to the breathing circuit bypassing the vaporizer.
	14b. Used when patient is hypoxic or to cleanse system before recovering the patient at the end of anesthesia.
15. Pressure manometer	15a. Measures the pressure of gases within the breathing circuit.
	15b. Pressure in excess of 20 cm H_2O indicates buildup of back pressure, usually because the pop-off valve has been left in the closed position.
	15c. Watch this gauge when ventilating the patient. A normal healthy adult horse should receive about 25 cm H_2O pressure when ventilating.
	15d. Bloated animals may require pressures up to 50 cm H_2O pressure to deliver an adequate tidal volume. Relief of bloat is paramount, because pressures this high may damage lung tissues.

Operation of Anesthesia Machine

TECHNICAL ACTION	RATIONALE/AMPLIFICATION
1. Turn on oxygen tank.	1a. Turn valve counterclockwise.
2. Make sure vaporizer is full.	2a. Indicator window is on the front of the vaporizer.
3. Attach hoses and reservoir bag to machine.	3a. Select reservoir bag that is appropriate for the size of the patient. Remember that 60 ml × body weight in kg = minimum size needed, so a 150-kg calf would need 60 × 150 = 9.5 L bag.
4. Check CO_2 absorber granules.	4a. Change them if hard or discolored.
5. Check for leaks from the machine.	5a. Hold your hand over the Y-piece to seal it, turn pop-off valve to closed, and fill the reservoir bag until the manometer reads 20 cm H_2O. Turn off oxygen flow, and pressure should remain at 20 cm H_2O for 2 minutes.
	5b. Most leaks are located at the canister or the reservoir bag.
6. Calculate oxygen flow rate (6–10 ml/kg/min).	6a. For example, a 500-kg horse would need a flow rate of 10 ml/min × 500 = 5,000 ml/min or 5 L/min.
7. Induce anesthesia.	7a. Injectable anesthetic most commonly is used (Table 10–14, 10–15, or 10–16).
8. Intubate, verify placement and inflate cuff.	8a. See Procedure for Orotracheal Intubation Or
	8b. See Procedure for Nasotracheal Intubation
9. Position patient on table.	9a. Chapter 8.
10. Attach Y-piece and breathing hoses to endotracheal tube.	10a. Make sure the pop-off valve is in the open position.
11. Set desired oxygen flow rate.	11a. Approximately 10 ml/kg/min.
12. Turn on vaporizer by depressing lock mechanism while turning dial. Set at desired percentage.	12a. Estimate initial level, which varies with patient need: Halothane 3.5–4% Isoflurane 4–5% Sevoflurane 4–6%
	12b. As the anesthetic equilibrates, the vaporizer flow may be adjusted downward.

TECHNICAL ACTION	RATIONALE/AMPLIFICATION
13. Begin monitoring patient and hook up any monitoring equipment.	13a. Procedure for Monitoring Horses Procedure for Monitoring Ruminants Procedure for Monitoring Pigs Procedure for Doppler Blood Pressure Monitoring Procedure for Oscillometric Blood Pressure Monitoring Procedure for Direct Blood Pressure Monitoring Procedure for Setting Up the ECG Procedure for Pulse Oximetry
14. Facilitate recovery.	14a. Turn off vaporizer when surgery is complete and restraints are released. 14b. Allow straight oxygen to run to the patient until it is ready to be moved to recovery area or until anesthesia is light enough that the endotracheal tube can be removed. 14c. Turn off flowmeter, and unhook Y-piece. 14d. Deflate cuff and extubate patient when it is able to swallow.

INTUBATION: ACQUIRING ACCESS TO THE AIRWAYS

Purpose

- Endotracheal tubes are heavy silicone or polyvinyl chloride tubes placed in the trachea to allow direct access to the pulmonary system (Fig. 10–16A).

- Most often they are used when anesthesia is to be maintained by inhalant anesthetic.

- Other reasons for intubation include protection of the airway from aspiration of gastric or ruminal juices and to support ventilation of patient.

- Nasotracheal intubation may be used to access or maintain airway in foals and small horses (<100 kg).

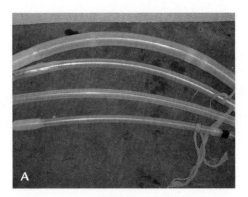 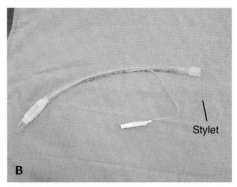

FIG. 10-16 (A) Variety of endotracheal tubes for large animals. (B) Endotracheal tube with stylet.

Equipment

- Nasotracheal tube (for foals): Internal diameter (ID) 7–12 mm, length 50–60 cm
- Endotracheal tube

    Adult horses: 24–30 mm ID

    Adult cattle: 24–30 mm ID

    Calves: 10–14 mm ID

    Small ruminants: 8–12 mm ID

    Llama: 10–14 mm ID

    Swine: 6–10 mm ID

- Mouth speculum (for horse): Can be made with 2-inch heavy polyvinyl chloride tube cut to approximately 3 inches long and wrapped with elastic bandage for cushioning
- Laryngoscope with 15–25 cm blade (for ruminants)
- 2% lidocaine in a 3 cc syringe
- Wire stylet
- 2-inch nonelastic tape or 3-inch rolled gauze

Complications

- Trauma to larynx and associated structures
- Placement of tube in esophagus instead of trachea
- Ischemic necrosis of the tracheal mucosa secondary to overinflated cuff
- Hypoventilation secondary to inadequate tube size

Procedure for Nasotracheal Intubation

TECHNICAL ACTION	RATIONALE/AMPLIFICATION
1. Restraint	**1a.** Manually restrain foal at mare's side. Keeping the foal near the mare until intubation is complete eases handling.
	1b. Some foals or small horses may require chemical restraint (xylazine 0.3–0.5 mg/kg intravenously) to pass nasotracheal tube.
2. Preparation for tube insertion	**2a.** Numb nostrils with 4% topical lidocaine gel.
	2b. Measure tube from nostril to midneck and mark length.
3. Tube insertion	**3a.** Insert tube into ventromedial nasal meatus.
	3b. Gently pass tube to level of nasopharynx.
	3c. Extend head and neck and pass tube into trachea.
	3d. Confirm placement by: 1. Cough reflex 2. Feeling puffs of air in synchrony with respirations 3. Palpating neck to verify there is only one tube-shaped structure (the trachea) palpable
	3e. Secure tube by taping it to muzzle.
	3f. Inflate cuff using 20 cc syringe to the point at which resistance is first felt.
	3g. Once general anesthesia is induced, the nasotracheal tube may be replaced with a larger orotracheal tube if desired.

Procedure for Orotracheal Intubation

TECHNICAL ACTION	RATIONALE/AMPLIFICATION
1. Restraint	1a. Induce anesthesia by chemical means.
	1b. Remove any halter or cranial restraint device.
2. Preparation for tube insertion	2a. Palpate trachea to estimate correct diameter. This is best done before induction of anesthesia.
	2b. In small ruminants, neonates, and swine, measure from tip of snout to midneck to determine how far to insert the endotracheal tube.
	2c. Mark tube with tape at rostral end to show correct length.
3. Blind intubation used for most ruminants and all adult horses	3a. Place patient in sternal recumbency. Horses may be lateral or dorsal, as well.
	3b. Insert mouth speculum or use gauze ties to help open the mouth and allow visualization.
	3c. Place head and neck in extension.
	3d. Insert endotracheal tube through speculum and into mouth.
	3e. Advance the tube to the oropharynx and then gently into larynx, rotating the tube in a clockwise manner as you advance.
	3f. You may palpate the epiglottis and larynx digitally in smaller ruminants and horses to ease placement.
	3g. Verify placement.
4. Laryngoscope-guided intubation used in small ruminants	4a. Procedure is best done with patient in sternal recumbency, with head and neck extended.
	4b. Gently open mouth and pull tongue out.
	4c. Place long laryngoscope blade in mouth.
	4d. Visualize the soft palate and epiglottis (Fig. 10–17A).

TECHNICAL ACTION	RATIONALE/AMPLIFICATION
	4e. Drip 1 ml of 2% lidocaine on epiglottis and arytenoid cartilages.
	4f. Using tip of blade, gently hook the tip of the epiglottis and pull it down to the floor of the pharynx. You should see the opening to the larynx and arytenoid cartilages (Fig. 10–17B).
	4g. Gently insert the tracheal tube into the opening by following the blade of the laryngoscope.
	4h. Verify placement.
5. Laryngoscope-guided intubation used in pigs	**5a.** Requires excellent muscle relaxation induced by chemical means.
	5b. Prepare endotracheal tube by:
	1. Placing a wire stylet into the tube to make it more rigid
	2. Making a 90-degree bend where the wire exits the rostral end of the endotracheal tube to keep the wire from sticking out the distal end of the tube and damaging the larynx (Fig. 10–16B)
	3. Alternatively, a polyurethane urinary catheter may be used as a stylet and guide to passing the tube

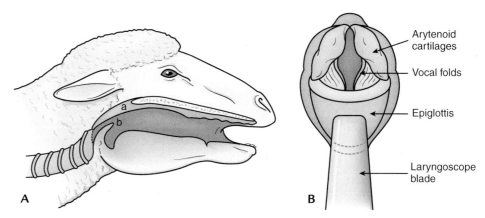

FIG. 10–17 (A) Lateral view of anatomy of larynx and associated structures: Soft palate (a) and epiglottis (b). (B) Appearance of larynx, epiglottis, and arytenoid cartilages after application of laryngoscope blade.

TECHNICAL ACTION	RATIONALE/AMPLIFICATION
	5c. Position pig with head and neck extended in lateral or sternal recumbency.
	5d. Place a small animal mouth speculum on canines to hold mouth open.
	5e. Pass laryngoscope blade into the mouth to visualize the oropharynx and epiglottis.
	5f. Drip 1–2% lidocaine onto larynx to decrease laryngospasm.
	5g. Wait 1–3 minutes.
	5h. Pass tube into mouth with scope blade depressing the tongue ventrally.
	5i. Using end of tube or laryngoscope blade, hook the tip of the epiglottis and pull it down to floor of pharynx.
	5j. Use scope to visualize passage of tube into the dorsal opening to the larynx then into the trachea.
	5k. Remove stylet and verify placement.
	5l. Secure the tube with tape around the snout.
6. Placement verification within the trachea	**6a.** Feel air exhaled from tube in time with animal's respirations.
	6b. Palpate neck. If you can feel the tube, it is in the esophagus.
7. Securing tube and inflating cuff	**7a.** Use 2-inch nonelastic tape (1-inch for most neonates) or 3-inch gauze tied tightly around the tube and then around the patient's muzzle.

LOCAL AND REGIONAL ANESTHETIC BLOCKS

The term, local anesthesia, refers to the use of a chemical agent to disrupt nerve impulse transmission causing a temporary lack of sensation and, at times, voluntary muscle control. Regional anesthesia is the use of that same chemical at a nerve root or in the epidural space to block transmission at the level of the spinal cord. On occasion an agent of this same chemical class may be injected into the vascular system of a leg that has had a tourniquet applied, to anesthetize everything below the level of the tourniquet.

Purpose

- Local and regional anesthetics are used often in food animal medicine in order to perform surgical procedures without using more risky and logistically complicated general anesthesia.
- In equine medicine and surgery, local blocks often are used to help diagnose the cause of a limb lameness and perform minor surgical procedures.
- Local or regional anesthesia may also be used as an adjunct to general anesthesia, blocking pain impulses so that less general anesthetic is required to keep the animal recumbent.

Mechanism of Action

- Local anesthetics block nerve conduction by inhibiting the influx of sodium ions through sodium channels in the nerve membrane.
- When sodium cannot pass through the membrane, an action potential cannot be generated, so any electrical impulse to that nerve is blocked and prevented from reaching the central nervous system.

Systemic and Toxic Effects of Local Anesthetic Agents

TABLE 10-20 SYSTEMIC AND TOXIC EFFECTS OF LOCAL ANESTHETIC AGENTS

SIGNS	CAUSES
1. Bradycardia, increased PR interval, and widened QRS complexes	1a. Blockage of sodium channels in heart muscle
2. Cardiac arrest	2a. Accidental intravenous injection of large amounts of local anesthetic
3. Hypotension	3a. May be seen with epidural use of local anesthetics due to sympathetic blockade
4. Sedation	4a. Early sign of toxicity from overdose
5. Twitching, seizures, coma, and death	5a. Late signs of overdose, but may also be seen with accidental administration into cerebrospinal fluid
6. Respiratory arrest	6a. Secondary to paralysis of respiratory muscles either from an epidural extending too high up the spinal cord or excessive injection into pleural space
7. Methemoglobinemia	7a. Dose-dependent response to prilocaine.
	7b. Some individual animals may have this adverse reaction to lidocaine or procaine given well within the normal dose range.

Local Anesthetic Agents Commonly Used in Large Animal Medicine and Surgery

TABLE 10-21 LOCAL ANESTHETIC AGENTS COMMONLY USED IN LARGE ANIMAL MEDICINE AND SURGERY

1. Lidocaine (Xylocaine)	1a. Available as 1% or 2% injectable solution, 4% topical solution or gel. May have 0.01 mg/ml epinephrine (1:100,000) in it. 1b. Epinephrine constricts capillary and venous circulation, thereby extending duration of action. 1c. Onset of action is up to 5 minutes. 1d. Duration of action is 1 to 2 hours, depending on site injected and whether epinephrine was used. 1e. May cause local swelling and irritation when injected into tissues. This is particularly problematic in horses when used with laceration repair or lower limb blocks. 1f. May be mixed with sodium bicarbonate (8.4% or 1 mEq/ml) at the rate of 0.8 cc bicarbonate to 10 cc 2% lidocaine to reduce burning sensation when injected.
2. Bupivacaine (Marcaine)	2a. Available as 0.25%, 0.5%, and 0.75% injectable solution with and without epinephrine. 2b. Onset of action is 20 minutes. 2c. Prolonged duration of action (up to 8 hours when used with epinephrine). 2d. Expensive, but excellent analgesia postoperatively.
3. Mepivacaine (Carbocaine)	3a. Available as 1% and 2% injectable solution. 3b. Causes minimal postinjection edema. 3c. Very rapid onset of action (less than 5 minutes). 3d. Duration of action is 90–180 minutes. 3e. Popular with equine veterinary specialists.
4. Prilocaine (Citanest)	4a. Available. 4b. Slower onset of action, and it does not spread as well as lidocaine. 4c. May be used by an equine specialist for a very specific nerve block when accuracy is important.
5. Proparacaine (Ophthaine)	5a. Available as 0.5% ophthalmic solution. 5b. Used for anesthetizing the cornea. 5c. Onset of action 1 minute. 5d. Duration of action 15–30 minutes. 5e. Not irritating to ocular tissues and does not inhibit pupillary musculature.
6. Procaine (Novocaine)	6a. No longer used in veterinary medicine except combined with the antibiotic, penicillin, to decrease the pain associated with intramuscular injection of penicillin.

Methods of Producing Local Anesthesia

TABLE 10-22 METHODS OF PRODUCING LOCAL ANESTHESIA

1. Topical blocks	1a. Most commonly used to desensitize the larynx for intubation.
	1b. Desensitize the cornea for ocular examination or minor surgery.
	1c. Topical preparations of lidocaine applied to the skin to ease placement of intravenous catheter.
2. Intrasynovial blocks	2a. Placement of a local anesthetic into a joint, tendon sheath or bursa.
	2b. Diagnostic use.
	2c. Pain relief.
	2d. Extreme care must be taken to use aseptic technique to ensure sterility. Injection of bacteria into those spaces could result in devastating infection.
3. Infiltrative blocks	3a. Injection of local anesthetic in and around the actual site of operation numbs nerve endings. May result in inflammation at surgical site, excessive edema, and delay of wound healing.
	3b. Line block: Injection of local anesthetic into tissues between spinal cord and surgical site creates a wall of anesthesia proximal to the surgery site. This technique prevents damage to the surgery site and subsequent delay in wound healing, and induces muscle relaxation or paralysis.
	3c. Ring block: Injection of local anesthetic completely encircles an anatomic part such as a digit or teat. Allows surgery to be done distal to the block.
4. Intravenous regional	4a. Application of a tourniquet to a distal portion of a limb completely shuts off arterial supply, followed by injection of a local anesthetic into a vein distal to the tourniquet.
	4b. Local anesthetic diffuses throughout the distal limb in 3–5 minutes, completely numbing it.
	4c. Allows also for relatively blood-free surgery.
	4d. Care must be taken to release the tourniquet slowly when surgery is finished, to prevent an excessive dose of anesthetic from entering the systemic circulation.
5. Regional nerve blocks	5a. Injection of a local anesthetic into a major nerve plexus or near a nerve root.
	5b. Results in blockage to a relatively large area.
	5c. Both sensation and muscle function are blocked.
6. Spinal and epidural blocks	6a. Technically a form of regional anesthesia.
	6b. Local anesthetic injected into the epidural space to block transmission of neural impulses up the spinal cord.
	6c. Results in loss of sensation and muscular function to limbs and tissues supplied by that segment of spinal cord.
	6d. In large animals, most epidural blocks are performed caudal to the sacrum to avoid ataxia or paralysis.
	6e. Ataxia or paralysis of caudal limbs is undesirable in large animals (horses, cattle) because of potential for injury to the patient or personnel.
	6f. Analgesics such as opioids or alpha-2 receptor agonists may be mixed in with the local anesthetic to improve analgesia while minimizing ataxia or paralysis.

INTRASYNOVIAL BLOCKS

Purpose

- Diagnose intrasynovial pathology
- Provide analgesia for surgery or therapy

Sites

Joints of Horses (Fig. 10-18)

- Distal interphalangeal (coffin) joint
- Proximal interphalangeal (pastern) joint
- Metacarpophalangeal/metatarsophalangeal (fetlock) joint
- Carpal joints
- Cubital (elbow) joint
- Scapulohumeral joint
- Tarsocrural (hock) joints
- Stifle joint
- Coxofemoral joint: Difficult to do, so rarely done

Joints of Other Species

- Same as horse except that cloven footed animals have medial and lateral digits

Tendon Sheaths

- Superficial and deep digital flexor tendons
- Carpal synovial sheath

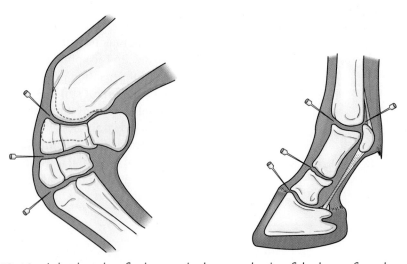

FIG. 10-18 Injection sites for intraarticular anesthesia of the horse front lower leg.

Bursas

- Navicular bursa (horses)
- Olecranon bursa (horses): Rarely done
- Bicipital bursa (near shoulder joint)
- Cunean bursa (near hock joint)
- Trochanteric bursa

Equipment

- Sterile 12 cc or 6 cc syringes
- 18 g × 1 inch needle for local anesthetic bottle
- Needles (Table 10–23)
- Clippers with #40 blade
- Surgical scrub, 70% alcohol, and surgical preparation solution
- Local anesthetic agent
- Sterile surgical gloves

TABLE 10-23 NEEDLE SIZES AND LENGTHS FOR INTRASYNOVIAL BLOCKS

SITE	GAUGE	LENGTH
Navicular bursa	20 gauge	2 inches
Coffin joint	18 or 20 gauge	1 to 1.5 inches
Pastern joint	20 gauge	1 inch
Fetlock joint	20 gauge	1 inch
	18 gauge	1.5 inches
Carpal joints	20 gauge	1 inch
Elbow joint	18 gauge	2 inches
Shoulder joint	18-gauge spinal needle	3.5 inches
Tarsocrural portion of hock joint	20 gauge	1 inch
Stifle joint	18 gauge	2 inches
Hip joint	16 gauge	6 inches

Complications

- Swelling
- Needle breakage
- Sepsis
- Trauma to articular cartilage

Procedure for Intrasynovial Blocks

TECHNICAL ACTION	RATIONALE/AMPLIFICATION
1. Shave site.	1a. Using sharp blades and shaving against the lay of the hair, clip a 2-inch square centered over the site to be injected. A safety razor may be used afterward if desired (Fig. 10–19).
2. Perform surgical preparation.	2a. At least 10 minutes should be spent carefully scrubbing the site.
	2b. This is described in Chapter 8 in Procedure for Three-step Surgical Scrub.
3. Using aseptic technique, place sterile syringes on sterile field.	3a. Carefully open and invert the syringe case over the sterile towel or tray so that the syringe falls out and onto the sterile field. Do not touch the syringe unless wearing sterile gloves.
4. Wash and dry hands.	4a. Wash hands 30–60 seconds with soap and water. Dry thoroughly with paper towel.
5. Apply surgical gloves.	5a. See Chapter 8 Procedure for Personal Preparation for Surgery.

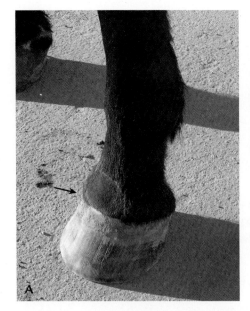 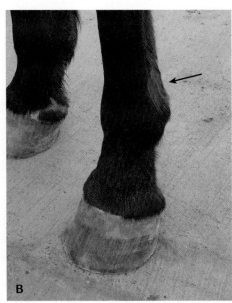

FIG. 10–19 (A) Preparation for coffin joint block. (B) Preparation for fetlock joint block. Arrows indicate shave sites.

TECHNICAL ACTION	RATIONALE/AMPLIFICATION
6. Draw up local anesthetic.	6a. Amount of anesthetic depends on species and joint space.
	6b. Do not exceed 2 mg/kg lidocaine in the goat.
	6c. The following amounts apply to an average size horse: Phalangeal joints: 5 ml Carpal joint: 10 ml Elbow joint: 10 ml Shoulder joint: 10–20 ml Hock joint: 20 ml Stifle joint: 10–20 ml Hip joint: 10–15 ml Bursas: Generally 5–10 ml
7. Ensure excellent restraint is being applied to the patient.	7a. It is vital that the patient not move when the needle is placed in a joint.
	7b. Damage to the joint surface could occur, or the needle may be broken and part of it left in the intraarticular space if the patient moves while the needle is inside the joint.
8. Insert sterile needle into synovial space.	8a. Watch for synovial fluid to drip or spurt from hub. It is normally clear yellow and slightly sticky or stringy.
9. Attach a sterile syringe to needle and withdraw a volume of fluid equal to that being injected.	9a. Withdrawing fluid gives room for the anesthetic to be injected with a minimum of discomfort to the patient or damage to the synovial space.
10. Remove syringe and attach syringe containing local anesthetic agent.	10a. Steady the needle hub when changing syringes so that the needle does not come out or scrape against the joint surface.
11. Inject anesthetic and withdraw needle and syringe.	11a. Inject carefully. Too much pressure on the syringe may cause the syringe to separate from the needle, spraying anesthetic and potentially contaminating the synovial space.

HORSE LOWER LIMB BLOCKS

The technician's role in these blocks is usually to prepare and restrain the animal for the veterinarian. Therefore, a general description of perineural local anesthesia will be given in order to help the technician meet the veterinarian's needs most effectively.

Purpose

- To locate specific site of pain in a limb
- Generally starts with the most distal site and progresses up the limb until the lameness has been eliminated

Equipment

- Needles 25–20g × 1 inch

  Size of needle depends on veterinarian preference. The veterinarian will use a fresh needle for each block to be performed, so it is wise to have at least a half dozen.

- Syringes 3 cc to 12 cc

  Most nerve blocks require less than 6 cc of anesthetic agent per site.

  A fresh syringe will be used for each nerve block.

- Clippers with #40 blade (optional)
- Surgical scrub, 70% isopropyl alcohol, and surgical solution

  The above are needed to cleanse site whether or not is shaved first.

  A scrub brush comes in handy to thoroughly clean the site.

- Bottle of local anesthetic (usually 2% lidocaine or mepivacaine)
- One 18g × 1 inch needle for withdrawing the anesthetic from the bottle

  Always use a fresh bottle and sterile needle.

  Once the needle is placed in the bottle, you can leave it there until the veterinarian has completed giving all the blocks, provided you protect the bottle from contamination.

Complications

- Swelling: Lidocaine can be irritating to tissue; swelling can be minimized by wrapping the limb afterward.
- Sepsis: Always use a fresh needle and syringe for each nerve block, and scrub the injection site with a new scrub brush when preparing for the nerve block.
- Needle breakage: This is more likely to happen if syringe is attached to the needle when it is placed in the skin.

- Accidental intravascular injection: Always aspirate before injecting.
- Injury to operator or handler: This can be minimized by the person injecting the horse and the person restraining the horse standing on the same side of the horse.

Procedure for Administering Diagnostic Lower Limb Blocks to Horses

TECHNICAL ACTION	RATIONALE/AMPLIFICATION
1. Restrain horse with halter and lead rope.	1a. Nose twitch or skin twitch may be necessary.
	1b. Lifting opposite limb may help keep leg being blocked on the ground.
	1c. Using very fine gauge needles (22–25g) decreases reactivity.
2. Clip area to be blocked in a 1-inch square.	2a. Again, clipping is optional.
3. Scrub area to be blocked vigorously with scrub brush and disinfectant soap.	3a. Use a scrub brush if area to be injected has not been clipped or shaved.
4. Wipe soap away with 70% alcohol.	4a. Use plenty of alcohol to remove completely any soapy residue.
5. Spray with disinfectant solution.	5a. Optional.
6. Wash and dry hands.	6a. Some veterinarians wear examination or surgery gloves.
7. Draw up 3–5 ml local anesthetic.	7a. Amount to draw up depends on site to be blocked. Posterior digital nerve block: 1–2 ml; Abaxial sesamoid block: 2–4 ml; Low palmar or plantar nerve block: 2–4 ml; High palmar or plantar nerve block: 3–5 ml
8. Locate nerve to be blocked by palpating digitally.	8a. Nerves run with an artery and vein, so it is often easier to palpate for the arterial pulse and then locate the nerve (Fig. 10–20).
	8b. The nerve will feel like a firm, rubbery piece of spaghetti when you roll it back and forth under your finger (Fig. 10–21).

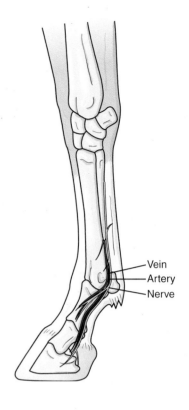

FIG. 10-20 Lower leg of horse. Note relation of digital vein, arteries, and nerves.

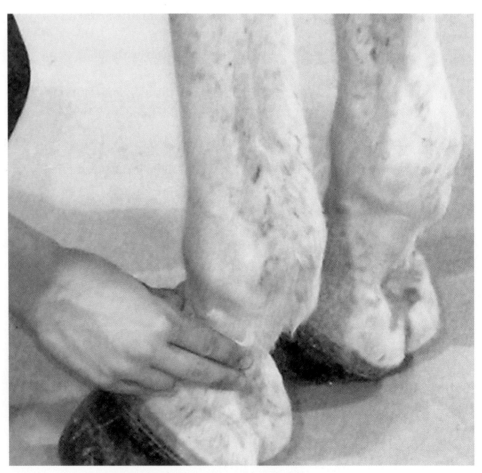

FIG. 10-21 Palpating lateral palmar digital nerve.

TECHNICAL ACTION	RATIONALE/AMPLIFICATION
	8c. Some nerves are located by bony, tendinous, or ligamentous landmarks instead (Fig. 10–21).
9. Ensure good restraint.	**9a.** See above. Wrap the tail or knot it up out of the way if working on the hind legs.
10. Insert 20–25g needle into the skin overlying the nerve to be blocked.	**10a.** If the horse jumps, the needle usually will remain in place if it is not attached to a syringe.
	10b. Observe needle hub for presence of blood. If blood is in the hub, acquire a new needle and place in a slightly different spot.
11. Attach syringe containing blocking agent.	**11a.** Steady the hub as you attach the syringe to avoid irritating the horse.
	11b. Make sure syringe is firmly attached.
12. Inject anesthetic agent.	**12a.** Inject at a medium rate.
	12b. The needle may be redirected while still under the skin if resistance is met and to ensure good coverage of the nerve at that site.
13. Wait 5–10 minutes for agent to take effect.	**13a.** Time to wait depends on anesthetic used.
14. Using blunt probe, test below block for effectiveness.	**14a.** A ballpoint pen or the pointed jaws of an old pair of mosquito forceps make good sensation testers.
15. Perform postprocedure care.	**15a.** Horse should be confined to stall or small pen until the block has worn off (1 to 2 hours).
	15b. Some veterinarians prefer a light stable-type wrap be applied to the limb once the examination is over to prevent swelling.

RING BLOCK

A ring block is the injection of local anesthetic completely encircling an anatomic part such as a digit or teat.

Purpose

- Allows the veterinarian to perform surgery or some other painful procedure on that digit or teat without causing distress to the animal
- Most commonly done on ruminants and may be done by a competent technician under supervision

Equipment for Teat Block

- Clippers with #40 blade
- Several 22–20g × 1 inch needles
- 6–12 ml syringe
- Surgical scrub and solution (alcohol generally not used, to avoid contamination of milk and chapping of teat skin)
- Bottle of 2% lidocaine

Complications from Teat Block

- Mastitis may occur secondary to poor aseptic technique or secondary to irritation from the anesthetic. Milk should be withheld from the tank for 72 hours.
- Operator may be injured because poor restraint techniques are used.

Procedure for Teat Block

TECHNICAL ACTION	RATIONALE/AMPLIFICATION
1. Restrain cow in stock or head lock.	1a. Beef cows may be placed in squeeze chute.
	1b. An anti-kick device may be put on a dairy cow to prevent her from kicking.
	1c. Alternatively, an assistant may tail-up the cow (Chapter 2) to prevent kicking.
2. Wash teat and ventral udder with surgical scrub.	2a. Dishwashing detergent may be used instead.
3. Clip away any long hairs.	3a. Most dairy cows already have hair-free udders for sanitation during milking.
4. Scrub again, rinse with water, and towel dry.	—

TECHNICAL ACTION	RATIONALE/AMPLIFICATION
5. Apply surgical solution.	5a. Alcohol generally is not used, to avoid contamination of milk and chapping of teat skin.
6. Wash hands.	6a. Wash for 30–60 seconds in disinfectant soap and dry thoroughly.
7. Draw up 3–5 ml of lidocaine.	7a. May need more in an aged animal. Never use more than 2 mg/kg lidocaine in a goat.
8. Ensure appropriate restraint.	8a. Have someone tail-up the cow standing in stock, chute, or stanchion. Goats may be held with one hand under the chin and one hand holding up the tail.
9. A right-handed person should grab hold of teat with left hand and gently pull straight down.	9a. This will make the skin at the base of the teat taut, making it easier to slide the needle in painlessly.
10. Insert needle under the skin at base of teat, all the way to the hub.	10a. Syringe may or may not be attached to the needle at this point. Most people leave the syringe attached to reduce the chance of contamination (Fig. 10–22).
11. Inject lidocaine as you slowly withdraw needle.	11a. Aspirate before injecting to ensure that the needle is not in a vein.
12. Repeat until entire base of teat is encircled.	12a. Wait 5 minutes to check for adequate anesthesia.

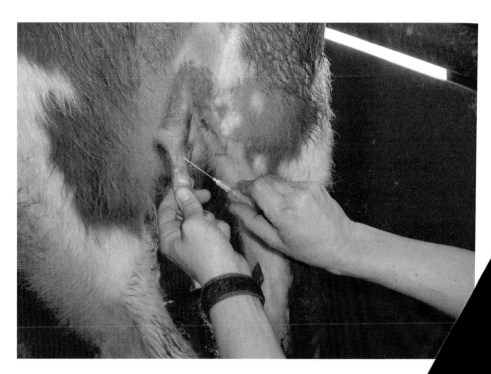

FIG. 10-22 Injection of cow's teat for ring block.

Equipment for Digit Block in Ruminants

- Needles: 20–22g × 1 inch, 20g × 1.5 inch, and one 18g × 1 inch needle for bottle.
- Syringes: 12–20 ml, two syringes
- Surgical scrub, 70% isopropyl alcohol, and disinfectant solution
- Fresh scrub brush for scrubbing manure and debris from foot and pastern
- Clippers with #40 blade

Complications

- Broken needle
- Sepsis
- Operator injury

Procedure for Digit Block in Ruminants

TECHNICAL ACTION	RATIONALE/AMPLIFICATION
Preparation	
1. Secure foot to be blocked and or administer chemical restraint.	1a. A standing cow may have her foot tied securely to the side of a stock or in foot-holding device.
	1b. Goats and many cattle are sedated with xylazine to ease restraint.
2. Scrub heels, pasterns, and interdigital areas with soap.	2a. Washing away debris before clipping helps preserve the sharpness of the blades. Soap acts as a lubricant.
3. Clip away hair.	3a. Clip against the lay.
4. Using scrub brush, scrub foot from coronary band to fetlock with surgical scrub.	4a. The same preparation is used for surgery, so it is important to do a thorough surgical preparation.
5. Rinse with 70% isopropyl alcohol.	5a. Water may be used instead of alcohol.
6. Repeat twice more.	—
7. Soak whole area with surgical preparation solution.	7a. Doctor may prefer to not have solution applied at this stage.
Method	
8. Ensure animal is restrained appropriately.	8a. See above. Safety is paramount for operator as well as patient.
9. Fill two 12-ml syringes with 10–12 ml of anesthetic.	9a. Use half that amount for an adult sheep and a quarter that amount for a goat.

TECHNICAL ACTION	RATIONALE/AMPLIFICATION
	9b. Keep the syringes clean by having someone hold them, or lay them on a clean towel within reach.
10. Insert a 22g needle under the skin to the hub at midpastern level, parallel to the coronary band in the direction of dorsal to palmar or plantar (Fig. 10–23A).	**10a.** Sliding a needle through the thick hide of a cow requires a fresh, sharp needle. **10b.** Some prefer to use a 20g or even an 18g needle to reduce chance of bending or breaking it.
11. Check hub for blood.	**11a.** If blood is in the hub, pull out the needle, discard, and start over with a fresh one.
12. Attach first syringe.	**12a.** Steady hub as you firmly attach syringe.
13. Inject anesthetic as you slowly withdraw needle.	**13a.** You should see the skin rise a bit where the anesthetic was injected.
14. Repeat procedure with slight overlap until you have encircled the entire digit, using up to 15 ml.	**14a.** Slightly overlapping injections will decrease discomfort for the patient without using excessive amounts of anesthetic. **14b.** Use no more than 2 mg/kg of lidocaine in the goat.
15. Insert 20g × 1.5-inch needle into the interdigital skin and ligament (Fig. 10–23B).	**15a.** May require an 18g needle in larger bulls and cows.
16. Inject the remaining 5 ml of anesthetic into the interdigital ligament.	**16a.** Be sure to aspirate and/or check hub for blood before injecting anesthetic.

FIG. 10–23 (A) Injection of pastern area for digital ring block. (B) Injection of interdigital space.

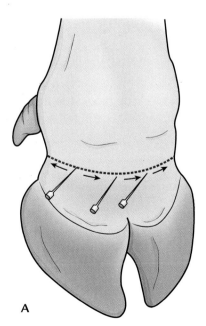
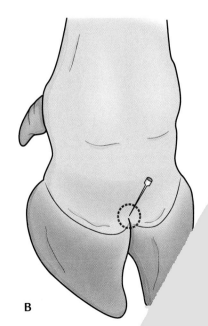

A

B

INTRAVENOUS REGIONAL BLOCK (BIER BLOCK)

Purpose

- May be performed by a veterinarian as an adjunct to general anesthesia or profound sedation
- Generally done on adult cattle undergoing surgery on the distal limb, but could be applied to other species
- Should always be done under the close supervision of a veterinarian due to the risk of toxicity to the patient if done improperly
- Desensitize the entire lower limb for a surgical procedure, with the added benefit of providing a bloodless field on which to operate

Equipment

- Esmarch bandage (heavy rubber bandage 3–4 inches wide and 6–8 feet long)
- Tourniquet
- Surgical scrub, 70% alcohol, and surgical solution
- Clippers with #40 blade
- 18–20g over-the-needle intravenous catheter
- Tape or glue to affix catheter
- Handful of 18–20g × 1 inch needles
- 6–20 ml syringes, two or three
- Lidocaine without epinephrine (epinephrine will constrict vasculature and prevent even distribution of anesthetic throughout the limb)

Complications

- Cardiovascular collapse secondary to flood of local anesthetic entering circulation
- Inadequate block due to insufficient volume of lidocaine, dislodged intravenous catheter, or use of lidocaine with epinephrine
- Collapse of patient secondary to anoxic waste when tourniquet is released
- Tourniquet left on too long causing ischemic tissue necrosis

Procedure for Intravenous Limb Block

TECHNICAL ACTION	RATIONALE/AMPLIFICATION
Preparation	
1. Animal is chemically or physically restrained, or both.	1a. It is safest for patient and operator to have the patient in sternal or lateral recumbency.
2. Limb is shaved and prepped for placement of intravenous catheter.	2a. Any distal, accessible vein may be used, but the catheter is placed most often in either the radial vein (front leg) or the lateral branch of the lateral saphenous vein (hind leg).
3. Place and affix intravenous catheter.	3a. Chapter 6.
Method	
4. Apply the Esmarch bandage.	4a. Starting from the foot, apply the Esmarch bandage, pulling as tightly as you can. Wrap all the way to the carpus or tarsus.
5. Apply a tourniquet just proximal to the carpus or tarsus.	5a. Pull tightly enough to shut off arterial blood supply (Fig. 10–24).
6. Unwrap lower limb carefully and flush intravenous catheter with 2 ml of saline.	6a. Watch to make sure the catheter is still in the vein and that the saline isn't making a skin bleb.
7. Inject lidocaine into catheter.	7a. Use 10–15 ml for an adult cow. Use 2–6 ml for small calves, sheep, and goats. Do not exceed 2 mg/kg lidocaine in the goat. Takes 15 minutes for full effect.

FIG. 10–24 Bier block, tourniquet applied. Front leg (A) hind leg (B).

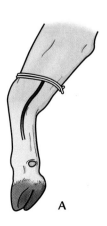

TECHNICAL ACTION	RATIONALE/AMPLIFICATION
8. Verify anesthesia after waiting 15 minutes.	8a. Poke proposed surgery site with 18g needle and observe for reaction.
9. When the surgical procedure is complete, slowly release the tourniquet.	9a. Release tourniquet over 30 seconds. Normal limb sensation returns within 5 minutes after lidocaine has had a chance to be distributed to the body.

INFILTRATIVE BLOCKS

Purpose

- Also known as line blocks, they can be used to block sensation directly at the site of incision or laceration, or they can be used in a more regional manner to block an area where an incision is to be made.
- Examples of line blocks include ring blocks, inverted L block, and simple line block.
- They may be used on any species, so long as care is taken to ensure total maximum dosage (per kilogram) is not exceeded.
- The same technique is used for all these blocks, so in the table that follows an inverted L block will be covered.

Equipment

- Clippers with #40 blade
- 16–18g × 3 inch needles (adult cattle, horses); 18–20g × 1.5 inch needles (small ruminants)
- 35–60 ml syringes (adult cattle, horses); 12–20 ml syringes (small ruminants)
- Surgical scrub and solution, 70% isopropyl alcohol, and scrub brush
- Bottle of lidocaine
- Use 2% lidocaine for adult cattle and horses
- Dilute lidocaine with sterile saline to a 1% solution when administering blocks to small ruminants or calves

Complications

- Lidocaine overdose: Not to exceed 2 mg/kg lidocaine in goats or 13 mg/kg in cattle and horses
- Inadequate desensitization from either not blocking deeply enough or not using enough volume of lidocaine

Procedure for Performing an Inverted L Regional Block

TECHNICAL ACTION	RATIONALE/AMPLIFICATION
Preparation	
1. Clip and scrub surgical area and area to be injected.	1a. Standard three-step surgical scrub is described in Chapter 8.
2. Draw up at least two syringes of lidocaine, leaving a needle in the bottle for rapid refill.	2a. An adult beef cow may take up to 100 ml of 2% lidocaine to attain complete desensitization.
Method	
3. Start at point just below transverse process of the second lumbar vertebra and next to the last rib.	3a. Fig. 10–25A.
4. Slide needle (without syringe) under the skin and direct it parallel to the transverse processes in a caudal direction.	—
5. Attach syringe and inject lidocaine as you gradually withdraw the needle.	5a. In adult beef cow, inject approximately 5 ml per site.
6. Repeat superficial injections until all the skin below the transverse lumbar processes two through four is blocked.	6a. This will block superficial tissues for several inches ventral to the lumbar processes but will not affect the muscular layers.
7. Go back to the site of the first needle stick, slide needle under skin, and direct it ventrally, parallel to the last rib.	7a. Try to penetrate the skin where it is already desensitized so as not to cause pain to the cow.
	7b. Fig. 10–25B.

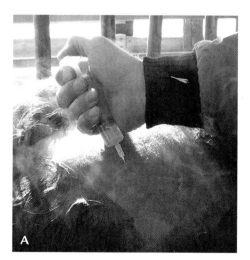

FIG. 10–25 (A) Injecting lidocaine into the flank of a cow. (B) *Lines* the bl[ock] should follow for complete anesthesia of the paralumbar fossa.

TECHNICAL ACTION	RATIONALE/AMPLIFICATION
8. Inject lidocaine as directed above until a vertical line is blocked to a level below that of the proposed incision.	8a. The vertical limb of this block desensitizes the ventral half of the flank area.
9. Using a 2–3 inch needle (adult beef cow), inject lidocaine deep into the musculature along the same lines you have injected subcutaneously.	9a. An adult beef cow in good flesh will have a body wall thickness 2 inches or more.
	9b. Most goats and ewes will need only a 1.5-inch needle to reach all the way through the body wall.
	9c. Very thin dairy goats may require only a 1-inch needle.

PARAVERTEBRAL BLOCK

Purpose

- Nerves to the flank region may be blocked as they exit the vertebral column
- Excellent for standing or recumbent celiotomy in ruminants, because they provide maximal muscle relaxation and deep anesthesia to the incisional area

Equipment

- Clippers with #40 blade
- Surgical scrub and solution, 70% isopropyl alcohol, scrub brush
- Needles: One 18g × 1.5 inch, one 16g × 4–6 inches, and one 16g × 1 inch needle for bottle
- Syringes: Two 20–35 ml syringes
- Bottle of 2% lidocaine (1% for goats and sheep)

Complications

- Lidocaine toxicity: Not to exceed 2 mg/kg in goats or 13 mg/kg in adult cattle
- **Ataxia** and scoliosis possible if the anesthetic reached the nerve roots
- Duration generally 90 minutes

Procedure for Paravertebral Block

TECHNICAL ACTION	RATIONALE/AMPLIFICATION
Preparation 1. Clip and scrub surgical area and area to be injected.	**1a.** When clipping the flank, go all the way to the dorsal spinous lumbar processes. **1b.** The sites to be injected are the lateral ends of the transverse processes of the first, second, and fourth lumbar vertebrae (L1, L2, and L4) (Fig. 10–26A, 10–26B).
2. Draw up at least two syringes of lidocaine, leaving a needle in the bottle for rapid refill.	**2a.** Each syringe should contain 10–20 ml lidocaine 2% for an adult cow.
Method 3. Palpate the transverse processes of the first, second, and fourth lumbar vertebrae.	**3a.** The transverse process of the first lumbar vertebra is tucked behind the last rib as it leaves the thirteenth thoracic vertebra and sometimes is difficult to feel.
4. Inject a small bleb (1 ml) of lidocaine at the tip of each process.	**4a.** Use an 18g needle on a 3 cc syringe to create the skin blebs.

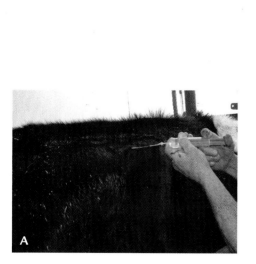
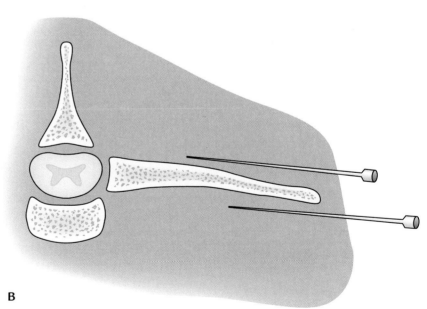

FIG. 10–26 (A) Insertion of a 6-inch needle dorsal to the caudal aspect of the first lumbar lateral process. (B) Diagram of needle placement for paralumbar block.

TECHNICAL ACTION	RATIONALE/AMPLIFICATION
	4b. Creating the blebs makes it less painful for the cow when the longer needle is inserted, and you can use the puncture site as the entry site for the long needle.
5. Starting with the first lumbar vertebra, insert a 4–6 inch needle into the skin bleb and pass it parallel to and dorsal to the transverse process.	**5a.** Fig. 10–26A. **5b.** Push the needle in all the way to the hub to ensure adequate exposure of the nerves to the anesthetic agent. **5c.** A goat or small ruminant requires a much shorter needle.
6. Attach a syringe and inject 5–10 cc of lidocaine as the needle is withdrawn slowly.	**6a.** It may be necessary to hold the hub of the needle onto the syringe to prevent them from pulling apart while you are injecting. This is particularly necessary in dehydrated cattle.
7. Remove syringe, insert needle through previous skin entry site, and this time direct the needle parallel to and ventral to the lumbar process.	**7a.** Fig. 10–26.
8. Attach syringe and inject 10–15 ml lidocaine as needle is withdrawn slowly.	**8a.** Try to fan out the anesthetic as you inject to ensure adequate exposure of the nerves to the substance.
9. Repeat process on second and fourth lumbar processes.	**9a.** Because anatomic variations occur, the third lumbar transverse process may need to be blocked for complete anesthesia.
10. Wait 10 minutes to test for desensitization.	**10a.** Failure to block the dorsal branches will result in skin sensitivity dorsal to the flank fossa, while inadequate blockage of ventral nerve branches will result in skin sensitivity from flank fossa ventrally.

CAUDAL EPIDURAL ANESTHESIA

Purpose

- Desensitize the perineum and caudal thighs of a ruminant to ease rectal palpation, calving, reduction of a rectal, vaginal, or uterine prolapse, or to perform surgery
- May use larger volumes to paralyze and desensitize the pelvic limbs, udder, and abdominal wall for surgery
- Analgesics (xylazine, butorphanol) may be used for analgesia without causing paresis or paralysis

Equipment

- Clippers with #40 blade
- Surgical scrub and solution, 70% isopropyl alcohol, and scrub brush
- Needles: 18g x1.5 inch (adult cattle and horses); 20g × 1 inch (small ruminants)
- Syringes: 10–12 cc (adult cattle and horses); 3–6 cc (small ruminants)
- Bottle of 2% lidocaine (adult cattle and horses)
- Bottle of 1% lidocaine (small ruminants, llamas)

Complications

- Uneven block: Seen as the tail held to one side, indicating that the needle was not placed in the epidural space or that scar tissue is preventing even distribution within the vertebral canal
- Meningitis: May be chemical or septic, prevented by always using aseptic technique
- Ataxia or even falling down if the block advanced further than expected or if xylazine is used with or without lidocaine
- Injury to patient or handler from ataxia, paresis, or paralysis

Procedure for Caudal Epidural in the Ruminant

TECHNICAL ACTION	RATIONALE/AMPLIFICATION
Preparation	
1. Clip tail head in rectangular shape from third coccygeal vertebra to midsacrum.	1a. Fig. 10–27A.
2. Scrub area well with surgical soap, rinse with water or isopropyl alcohol, and apply surgical preparation solution.	—

TECHNICAL ACTION	RATIONALE/AMPLIFICATION
Method (Hanging Drop Technique)	
3. Wash hands.	3a. Or put on fresh examination gloves.
4. Palpate the first obvious articulation caudal to the sacrum with your first finger.	4a. Lift the base of the tail up while palpating with your finger just caudal to the sacrum. You will feel the joint space open and close (Fig. 10–27B).
5. Fill the hub of a 18g × 1.5 inch needle with lidocaine.	—
6. Insert the needle directly on the midline into the articulation palpated, keeping the needle perpendicular to the skin.	6a. Keep the needle in a vertical position to keep the drop from spilling out of the hub.
7. Observe the drop of lidocaine as you advance the needle.	7a. When the needle penetrates the ligamentum flavum and enters the epidural space, the drop will be sucked out of the hub and into the needle (Fig. 10–27C).
8. Attach syringe and inject 3–6 ml of anesthetic.	8a. Standard dose given to an adult cow for perineal desensitization. Llama: 1–3 ml; goat or sheep: 1–2 ml (use 1% lidocaine)
9. Check tail tone for effectiveness of block.	9a. Tail should be flaccid within 2 minutes.

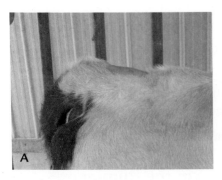  

FIG. 10–27 (A) Preparation of tail head for caudal epidural block. (B) Palpation of joint space for caudal epidural placement. (C) Needle in position for caudal epidural injection.

REVIEW QUESTIONS

1. What is a minimum data base?
2. What needs to be done to prepare a horse for general anesthesia or heavy sedation?
3. Why is it important to rinse the animal's mouth before administering general anesthesia?
4. List some of the benefits of using an alpha-2 receptor agonist.
5. Name a dissociative anesthetic commonly used in veterinary practice.
6. Is heart rate or blood pressure a more sensitive indicator of anesthetic depth in the horse?
7. How can we decrease bloat and the chance of aspiration in anesthetized ruminants?
8. Why are neonates allowed to suckle right before general anesthesia?
9. What parameters can be monitored in the anesthetized patient without the use of machinery?
10. List the purposes of using an endotracheal tube during general anesthesia.
11. What is a ring block and where is it used?
12. What is the maximum amount of lidocaine a goat can take before exhibiting signs of toxicosis?
13. What are the signs of lidocaine toxicity?
14. When is a caudal epidural performed in large animal practice?
15. What is a paravertebral block?

BIBLIOGRAPHY

Aubin, M. L., & Mama, K. R. (2002). Field anesthetic techniques for use in horses. *Compend Contin Educ Pract Vet, 24,* 411–417.

Bennett, R. C., & Steffey, E. P. (2002). Use of opioids for pain and anesthetic management in horses. *Vet Clin North Am Equine Pract, 18,* 47–60.

Benson, G. J., Hartsfield, M. H., Reidesel, D. H., Dodman, N. H., Haskins, S. C., Sawyer, D. C. (1988). *Analgesic and anesthetic applications of butorphanol in veterinary practice: Proceedings of a roundtable discussion.* Lawrenceville, NJ: VLS Co., Inc.

Brunson, D. B. (1988). *A Compendium on aerane (isoflurane, USP) for equine anesthesia,* Anaquest Form No. 07-0012-A01 1-88-5 Rev.

Caulkett, N. (2003). Anesthesia of ruminants. *Large Animal Veterinary Rounds, 3,* Retrieved February 7, 2006, from http://www.canadianveterinarians.net/larounds Search terms: anesthesia, ruminants.

Gavier, D., Kittelson, M. D., Fowler, M. E., Johnson, L. E., Hall, G., Nearenberg, D. (1988). Evaluation of a combination of xylazine, ketamine and halothane for anesthesia in llamas. *Am J Vet Res, 49,* 2047–2055.

George, L. W. (2003). Pain control in food animals. In E. P. Steffey (Ed.). *Recent advances in anesthetic management of large domestic animals,* Ithaca NY: IVIS. Retrieved February 7, 2006, from http://www.ivis.org Search term: A0615.1103.

Haskell, S. R. R. (2001). *Small ruminant clinical diagnosis and therapy,* S' University Minnesota, College of Veterinary Medicine.

Kronen, P. W. (2003). Anesthetic management of the horse: Inhalation anesthesia. In E. P. Steffey (Ed.). *Recent advances in anesthetic management of large domestic animals,* Ithaca NY: IVIS. Retrieved February 7, 2006, from http://www.ivis.org Search term: A0605.0103.

Mama, K. Personal correspondence, 2004–2005.

Mama, K. R. (2000). Anesthetic management of camelids. In E. P. Steffey (Ed.). *Recent advances in anesthetic management of large domestic animals,* Ithaca NY: IVIS. Retrieved February 7, 2006, from http://www.ivis.org Search term: A0608.0900.

Mama, K. R. (2000). Anesthetic management of the horse: Intravenous anesthesia. In E. P. Steffey (Ed.). *Recent advances in anesthetic management of large domestic animals,* Ithaca NY: IVIS. Retrieved February 7, 2006, from http://www.ivis.org Search term: A0604.1000.

Mama, K. R. (2002). Traditional and nontraditional uses of anesthetic drugs—An update. *Vet Clin North Am Equine Pract, 18,* 169–180.

McKelvey, D., & Hollingshead, K. W. (2003). *Veterinary anesthesia and analgesia,* (3rd ed.). St. Louis: Mosby.

Merck & Company. (2003). *Merck veterinary manual,* (8th ed.). New York: John Wiley & Son.

Natalini, C. C., & Robinson, E. P. (2003). Effects of epidural opioid analgesics on heart rate, arterial blood pressure, respiratory rate, body temperature and behavior in horses. *Vet Ther, 4,* 364–375.

Pypendop, B. H., & Steffey, E. P. (2001). Focused supportive care: Ventilation during ruminant anesthesia. In E. P. Steffey (Ed.). *Recent advances in anesthetic management of large domestic animals,* Ithaca NY: IVIS. Retrieved February 7, 2006, from http://www.ivis.org Search term: A0611.1001.

Reibold, T. (2003). *Supportive therapy in the anesthetized horse.* Eugene, Oregon: Oregon State University, Veterinary Teaching Hospital, College of Veterinary Medicine. Retrieved February 7, 2006, from http://www.surgivet.com. Click on Support, then SMART, and then Archived Articles.

Riebold, T. W. (2001). Anesthetic management of cattle. In E. P. Steffey (Ed.). *Recent advances in anesthetic management of large domestic animals,* Ithaca NY: IVIS. Retrieved February 7, 2006, from http://www.ivis.org Search term: A0603.0201.

Reibold, T. W., Geiser, D. R., & Goble, D. O. (1995). *Large animal anesthesia,* (2nd ed.). Ames, IA: Iowa State Press.

Reibold, T. W., Kaneps, A. J., & Schmotzer, W. B. (1989). Anesthesia in the llama. *Vet Surg, 18,* 400–440.

Robinson, E. P., & Claudio, C. N. (2002). Epidural anesthesia and analgesia in horses. *Vet Clin North Am Equine Pract, 18,* 61–82.

Stashak, T. S. (1987). *Adam's lameness in horses,* (4th ed.). Philadelphia: Lea & Febiger.

Steffey, E. P. (2002). Recent advances in inhalation anesthesia. *Vet Clin North Am Equine Pract, 18,* 159–168.

Thurmon, J. C., Sarr, R., & Denhart, J. W. (1999). Xylazine sedation antagonized with tolazoline. *Compend Contin Educ Pract Vet, 21,* S11–S20.

Wagner, A. (2000). Focused supportive care: Blood pressure and blood flow during equine anesthesia. In E. P. Steffey (Ed.). *Recent advances in anesthetic management of large domestic animals,* Ithaca NY: IVIS. Retrieved February 7, 2006, from http://www.ivis.org Search term: A0612.0900.

Appendix Tables

TABLE A-1 CONVERSION CHART

LIQUIDS (approximates)	U.S. LIQUIDS	WEIGHT	LENGTH
1 tsp = 5 ml	1 Tb = 3 tsp	1 oz = 28.4 gm	1 cm = 10 mm
1 Tb = 15 ml	1 oz = 2 Tb	1 lb = 454 gm	1 inch = 2.5 cm
1 oz = 30 ml	1 cup = 8 oz	1 kg = 2.2 lb	1 foot = 30.5 cm
1 pt = 473 ml	1 qt = 32 oz	1 lb = 16 oz	1 mile = 1.61 km
1 gal = 3.8 L	1 gal = 128 oz	1 ton = 2,000 lb	—

TABLE A-2 NORMAL TEMPERATURE, PULSE, AND RESPIRATION

	HORSE	CATTLE	SHEEP	GOAT	LLAMA	PIG
Temperature	99°–101°F 36°–38.5°C	100°–102.5°F 37.8°–39.2°C	101°–103.5°F 38.2°–39.8°C	101°–103°F 38.2°–39.5°C	99°–101.5°F 37.2°–38.5°C	101.5°–103°F 38.5°–39.5°C
Pulse rate beats per minute (bpm)	28–42 adult 70–80 foal	60–80 adult 100–120 calf	70–90	70–90	60–90	70–90 100–130 piglet
Respirations per minute	10–14	10–30	12–20	12–20	10–30	8–18

TABLE A-3 AGING BY ERUPTION OF PERMANENT INCISORS AND CANINES

	HORSE	CATTLE	SHEEP	GOAT	LLAMA
Central incisors	2.5–3 years	2 years	1–1.5 years	1 year	2 years
Second incisors	3.5–4 years	2.5 years	1.5–2 years	1–2 years	3 years
Lateral incisors	4.5–5 years	3.5 years	2.5–3 years	2–3 years	3–5 years
Corner pair	—	4.5 years	3.5–4 years	3–4 years	N/A
Canines	4 years	N/A	N/A	N/A	3 years

TABLE A-4 BASIC HUSBANDRY PERIODS

	HORSE	CATTLE	SHEEP	GOAT	LLAMA	PIG
Age to first breed the female	2.5–3.5 years	12–15 months	6–8 months	7–10 months	2.5–3.5 years	7–9 months
Gestation	337–365 days	281 days (9 months)	148 days (5 months)	150 days (5 months)	335–365 days	115 days (3 months, 3 weeks, and 3 days)
Weaning	6 months	3 months for dairy 4–6 months for beef	3 months	3 months	6 months	3–6 weeks
Castration (ideal time)	1–2 years	0–6 months	0–3 months	Less than 4 weeks	2 years	4–6 weeks
Dehorn	N/A	4 weeks to 8 months	N/A	Less than 4 weeks	N/A	N/A

TABLE A-5 PERITONEAL FLUID ANALYSIS OF CATTLE

	APPEARANCE	TOTAL PROTEIN	SPECIFIC GRAVITY	WHITE BLOOD CELLS	RED BLOOD CELLS	OTHER
Normal	Clear, amber colored	0.1 g/dl to 3.1 g/dl Does not clot	1.005–1.015	PMN-to-monocyte ratio is 1:1 Total count 300–5,000/mcl	Rare	No bacteria No plant material
Moderate inflammation	Amber to pink May be cloudy or turbid	2.8 g/dl to 7.3 g/dl May clot	1.016–1.025	Non-toxic PMNs 50–90% unless chronic 2,700–40,000/mcl	100,000–200,000/mcl	No bacteria No plant material
Severe inflammation	Bloody, cloudy, viscous	3.1–5.8 g/dl Clots readily	1.026–1.040	Presence of toxic or bacteria-laden neutrophils 2,000–32,000/mcl	300,000 to 500,000/mcl	Bacteria usually present Plant material (feces) may be present

PMN, polymorphonuclear.

TABLE A-6 PERITONEAL FLUID ANALYSIS OF HORSES

	APPEARANCE	TOTAL PROTEIN	SPECIFIC GRAVITY	WHITE BLOOD CELLS	RED BLOOD CELLS	OTHER
Normal	Pale yellow, clear	0.5-1.5 g/dl Does not clot	1.000-1.015	PMNs to monocytes 1:1 ratio 500-5,000/mcl	None	No bacteria No plant material
Suspect	Slightly cloudy Still yellow	1.6-2.5 g/dl	1.016-1.020	Segmented neutrophil 50-60% May see mesothelial cells 5,000-15,000 mcl	50,000-100,00/mcl	Known as a modified transudate No plant material
Moderate inflammation	Orange to pink, cloudy, and viscous	2.6-4.0 g/dl May clot	1.020-1.025	Segmented neutrophils 70-80% Occasional toxic neutrophil seen 15,00-60,000 mcl	1,000,000-2,000,000/mcl	Bacteria may be present No plant material
Severe inflammation	Pink to serosanguineous, cloudy, and thick May contain flocculent material	4.0-7.0 g/dl Often clots readily	Greater than 1.025	Segmented neutrophils 80-90% Degenerated neutrophils with bacteria More than 60,000/mcl	600,000-1,000,000/mcl	Bacteria usually present Plant material (feces) may be present

PMN, polymorphonuclear.

TABLE A-7 COMPLETE BLOOD COUNT NORMAL VALUES

	HORSE	CATTLE	SHEEP	GOAT	LLAMA	PIG
RBC (millions/mcl)	6.5–13.5	5–10	8–16	8–18	9–19.5	5–8
Hemoglobin	11–18 g/dl	8–15 g/dl	8–16 g/dl	8–14 g/dl	12.5–18 g/dl	10–16 g/dl
PCV	32–55%	25–46%	25–50%	25–48%	28–45%	32–50%
MCV	37–58 fl	40–60 fl	23–48 fl	19.5–37 fl	17–28 fl	50–68 fl
MCH	12.3–19.7 pg	11–17 pg	9–12 pg	N/A	7–19 pg	17–21 pg
MCHC	31–38 g/dl	30–36 g/dl	29–38 g/dl	30–35 g/dl	36–50 g/dl	30–34 g/dl
WBC/dl	5,400–14,000	4,000–10,000	4,000–12,000	4,000–13,000	7,000–14,000	11,000–22,000
Segmented neutrophils	30–75%	15–35%	10–50%	30–48%	50–70%	30–50%
Bands	0–2%	0–1%	Rare	Rare	0–5%	0–2%
Lymphocytes	20–70%	60–70%	40–75%	50–70%	25–45%	45–65%
Monocytes	0–10%	0.5–7%	0–6%	0–4%	0–2%	2–10%
Eosinocytes	0–10%	0–20%	0–10%	1–8%	0–15%	0–12%
Basophils	0–3%	0–2%	0–3%	0–1%	0–2%	0–2%
Total protein	6.5–7.8 g/dl	6–7.5 g/dl	6–7.5 g/dl	6.5–7 g/dl	5.8–7 g/dl	6–8 g/dl
Fibrinogen	100–400 g/dl	300–700 g/dl	100–500 g/dl	100–400 g/dl	100–400 g/dl	100–500 g/dl

fl, femtoliters; *MCH*, mean corpuscular hemoglobin; *MCHC*, mean corpuscular hemoglobin concentration; *MCV*, mean corpuscular volume; *PCV*, packed cell volume; *RBC*, red blood cells; *WBC*, white blood cells.

TABLE A-8 BLOOD CHEMISTRY NORMAL VALUES

	HORSE	CATTLE	SHEEP	GOAT	LLAMA	PIG
Sodium mEq/L	134–143	132–152	139–152	135–154	140–155	140–150
Potassium mEq/L	3.2–5.2	3.9–5.8	3.9–5.4	4.6–6.4	4–6.5	4.7–7.1
Chloride mEq/L	95–107	95–110	95–103	105–120	100–118	100–105
Calcium mg/dl	11.2–13.6	8–10.5	11.5–12.8	8.6–10.6	8.8–10.4	11–11.3
Phosphorous mg/dl	3.1–5.6	4–7	5–7.3	4.2–9.8	4.5–8.5	4–11
Magnesium mg/dl	2.2–2.8	1.2–3.5	2.2–2.8	2.8–3.6	1.5–3.0	1.9–3.9
BUN mg/dl	12–24	6–27	8–20	13–28	10–30	8–24
Creatinine mg/dl	0.9–2.0	1.0–2.7	1–2.7	0.9–1.8	2.0–8.0	1.0–2.7
Glucose mg/dl	75–115	35–55	42–76	60–100	80–145	65–95
Albumin g/dl	2.8–4.8	2.1–3.6	2.4–3	3–3.4	3.0–5.0	1.9–2.4
Total bilirubin mg/dl	0.5–1.8	0–1.9	0.14–0.32	0–0.9	0.1–0.3	0–0.2
AST IU/L	101–290	60–150	87–256	41–62	10–280	—
GGT IU/L	4–13	0–31	25–59	24–39	3–30	0–25
CPK IU/L	50–150	65	42–62	<38	10–200	65
Alkaline phosphatase IU/L	143–395	35–350	238–440	123–392	10–100	9–20
LDH IU/L	162–412	697–1,445	238–440	123–392	50–300	—

AST, aspartate aminotransferase; *BUN*, blood urea nitrogen; *CPK*, creatine phosphokinase; *GGT*, gamma-glutamyltransferase; *LDH*, lactic dehydrogenase.

TABLE A-9 MILK COMPOSITION CHART

	HORSE	COW	SHEEP	GOAT	LLAMA	PIG
% Fat	1.6	3.6	10.4	3.5	3.2	7.9
% Sugar	2.4	4.5	3.7	4.6	3.9	4.9
% Protein	6.1	3.3	6.8	3.1	5.6	5.9

TABLE A-10 COMMON NAMES OF LARGE ANIMALS

	HORSE	CATTLE	SHEEP	GOAT	LLAMA	PIG
Intact male	Stallion or stud	Bull	Ram	Buck or billy	N/A	Boar
Castrated male	Gelding	Steer	Wether	Wether	Gelding	Barrow
Juvenile male	Colt	Bull or steer calf	Ram or wether lamb	Buck kid	Cria	Piglet
Adult female	Mare	Cow	Ewe	Doe	Female or harem (if a group of females)	Sow
Juvenile female	Filly	Heifer	Ewe lamb	Doe kid	Cria	Gilt

TABLE A-11 COMMON NAMES FOR LEG ANATOMY

ANATOMIC TERM	LAY TERM
Third or distal phalanx	Coffin bone
Second or middle phalanx (P2)	Short pastern
First or proximal phalanx (P1)	Long pastern
Distal sesamoid	Navicular
Proximal sesamoid	Sesamoid
P1–P2	Pastern
Metacarpal 3 (MC3) or metatarsal 3 (MT3)	Cannon
Metacarpals 2,4 or metatarsals 2,4	Splint bones
MC3-P1 or MT3-P1 joint	Fetlock
Carpus	Knee
Tarsus	Hock

TABLE A–12 COMMON NAMES FOR FOOD ANIMAL ANATOMY

ANATOMIC TERMS	CATTLE	LAY TERMS SHEEP/GOATS	PIG
Sacral-caudal junction	Tailhead	Dock (sheep)	Rump
Tuber sacrale	Hooks	Hooks	—
Tuber ischii	Pins	Pins	—
Metatarsals/metacarpals	Shanks	Shanks	Shanks
Rostral pectoralis muscles	Brisket	Brisket	Chest
Paralumbar fossa	Flank	Thurl (goat)	Flank
Occiput	Poll	Poll	—
Lumbar epaxial muscles	Tenderloin	Tenderloin	Tenderloin
Lateral lumbar processes	Short ribs	Short ribs	Short ribs
Gluteal muscles, quadriceps, and biceps femoris	Hind end	Rump	Ham
Hooves	Hooves	Hooves	Trotters
Vestigial digits 1 and 4	Dewclaws	Dewclaws	Dewclaws
Rostrum	Nose	Nose	Snout
Masseter muscles	Cheeks	—	Jowls
Widest part of thorax	Barrel	Barrel	Barrel
Dorsal thoracic vertebral processes (T1–T5)	Withers	Withers	Withers

TABLE A-13 HORSE VACCINES

COMMON NAME	ORGANISM PROTECTED AGAINST	ANIMALS VACCINATED
Tetanus / Lockjaw	*Clostridium tetani*	All horses
Flu	Influenza virus types A1, A2	All horses
Rhino / Rhinopneumonitis	Equine herpes virus types 1, 4	All horses
Sleeping sickness / EEE, WEE	Eastern and Western equine Encephalomyelitis virus	All horses
Strangles / Distemper	*Streptococcus equi*	All horses
West Nile	West Nile virus	All horses
Rabies	Rabies (rhabdovirus)	Animals with regional risk
Potomac horse fever	*Ehrlichia risticii*	Animals with regional risk
Botulism (Shaker foal syndrome)	*Clostridium botulinum*	Pregnant mares and foals at risk
Rotavirus	Rotavirus	Pregnant mares with anticipated at-risk foal
Salmonella	*Salmonella typhimurium*	Known exposure
Viral arteritis	Equine arteritis virus	Limited use for broodmares under regulation
Equine protozoal myelitis (EPM)	*Sarcocystis faculata*	Limited use
Anthrax	*Bacillus anthracis*	Limited use under regulation

EEE, Eastern equine encephalitis; *WEE*, Western equine encephalitis.

TABLE A-14 CATTLE VACCINES

COMMON NAME	ORGANISM PROTECTED AGAINST	DISEASES PREVENTED	AGE CATTLE VACCINATED
BVD types 1 and 2	Bovine virus diarrhea type 1 Bovine virus diarrhea type 2	Abortion Mucosal disease Chronic bloat Immune suppression	*Calves >2 weeks old Weaned calves Replacement heifers and bulls *Adult cows Adult bulls
(IBR) Infectious Bovine Rhinotracheitis	*Mycoplasma bovis*	Upper respiratory infection Pneumonia Pinkeye Mastitis	Calves >2 weeks old Weaned calves Replacement heifers/bulls *Adult cows
PI3	Parainfluenza-3 virus	Pneumonia Shipping Fever	Calves >2 weeks old Weaned calves Replacement heifers/bulls Feedlot cattle on arrival Adult cows
BRSV	Bovine Respiratory Syncytial Virus	Severe pneumonia	Weaned calves Replacement heifers/bulls Feedlot animals
Somnus	*Haemophilus somnus*	Pneumonia Abortion Septic Arthritis	Replacement heifers/bulls Feedlot animals
Pasteurella	*Pasteurella multocida* *Mannheimia haemolytica*	Fibrinous pneumonia Pericarditis	Weaned calves Feedlot animals
Scour vaccine	*Escherichia coli* Rotavirus Coronavirus	Neonatal diarrhea	Cows 30 days prior to calving
Salmonella	*Salmonella newport*	Calfhood diarrhea Cholangitis Human zoonosis	Calves >2 weeks old Feedlot animals Dairy herd (all ages in outbreak)
Vibrio	*Campylobacter fetus*	Abortion Infertility	Replacement heifers/bulls Adult cows prior to breeding Adult bulls prior to breeding

...ea; *BRSV*, bovine respiratory syncytial virus; *PI3*, parainfluenza 3 virus.
...st be used in pregnant cattle and in calves nursing on pregnant cattle.

TABLE A-14 CATTLE VACCINES—cont'd

COMMON NAME	ORGANISM PROTECTED AGAINST	DISEASES PREVENTED	AGE CATTLE VACCINATED
Lepto Leptospirosis	Leptospira	Abortion Pyelonephritis	Replacement heifers and bulls Adult cows Adult bulls
Trichomonas	*Tritrichomonas foetus*	Infertility Metritis	Adult cows prior to breeding Adult bulls prior to breeding
Clostridial 7 or 8 way	Clostridial bacteria	Blackleg (gangrene) Hemorrhagic enteritis Hemorrhagic hepatitis	Calves >10 days old Weaned calves Feedlot animals Replacement heifers/bulls Adult cattle
Pinkeye	*Moraxella bovis*	Pink eye	Calves >30 days old Feedlot animals
Bang's vaccine	*Brucella abortus*	Brucellosis Infertility Zoonosis	Calves 3-12 months *Depends on local and state laws*
Anthrax	*Bacillus anthracis*	Hemorrhagic diathesis Septicemia Pneumonia Zoonosis	In face of outbreak, by state or federal permission

BVD, bovine virus diarrhea; *BRSV*, bovine respiratory syncytial virus; *PI3*, parainfluenza 3 virus.
*Killed BVD virus vaccine must be used in pregnant cattle and in calves nursing on pregnant cattle.

TABLE A-15 GOAT VACCINES

COMMON NAME	ORGANISM PROTECTED AGAINST	ANIMALS VACCINATED
Tetanus Lockjaw	*Clostridium tetani*	All goats
Over-eaters disease Enterotoxemia	*Clostridium perfringens* types C and D	All goats
E. coli	*Escherichia coli*	Limited specific indication use
Pneumonia	Pasteurella	Limited specific indication use
Rabies	Rabies (rhabdovirus)	Limited specific indication use

Special note:

Vaccines labeled specifically for caprine use are not available. Vaccinations administered to goats are those labeled for cattle, sheep or horses.

TABLE A-16 LLAMA VACCINES

VACCINE	ORGANISM	DISEASES PREVENTED	ANIMALS VACCINATED
Clostridium vaccines	*Clostridium* spp	Neonatal enterotoxemia Tetanus Blackleg Septicemia	Crias Jill 4 weeks prepartum Any adult
Leptospirosis	*Leptospira interrogans* spp	Abortion Anterior uveitis (Usage of the vaccine is disputed. There is no consensus on whether the vaccine prevents or lessens the severity of anterior uveitits.)	All adults
Rhinopneumonitis	Equine herpes-1	Abortion	All adults
Infectious bovine rhinotracheitis	*Mycoplasma bovis*	Mild respiratory disease Abortion	All adults exposed to cattle
Bovine virus diarrhea	Bovine virus diarrhea types 1 and 2	Abortion Weak cria Neonatal loss	All adults exposed to cattle
Ovine enzootic abortion	Chlamydia	Abortion	All adults exposed to sheep
Rabies	Rhabdovirus	Death	Cria 3–6 months All adults in endemic areas

TABLE A–17 PIG VACCINES

VACCINE	ORGANISM	DISEASES PREVENTED	AGE OF ANIMAL
Atrophic rhinitis	*Bordetella bronchiseptica*	Wry nose sinusitis Poor growth	Piglets 7–10 days Replacement gilts and boars Breeding sows
Mycoplasma	*Mycoplasma pneumoniae*	Acute pneumonia Chronic pneumonia Poor growth	Piglets 7–10 days Breeding sows
Erysipelas	*Erysipelothrix rhusiopathiae*	Diamond skin disease	Piglets preweaning Piglets postweaning Replacement animals Breeding sows
Contagious pleuropneumonia	*Actinobacillus pleuropneumoniae*	Acute pneumonia Death	Weaner pigs
Lepto Leptospirosis	Leptospira	Abortion Neonatal loss	Replacement gilts and boars Breeding adults
Parvo	Porcine parvovirus	Abortion Infertility	Replacement gilts
Transmissible gastroenteritis (TGE)	Corona virus	Vomiting Diarrhea Neonatal pig loss	Replacement gilts Breeding sows
E. coli	*Escherichia coli*	Diarrhea Poor growth Neonatal death	Weaner pigs Sows
Glasser's disease	*Haemophilus suis*	Encephalitis Cardiomyopathy Septic arthritis	Weaner pigs

Bibliography

Blood, D. C., & Radostits, O. M. (1989). *Veterinary medicine: A textbook of the diseases of cattle, sheep, pigs, goats and horses* (7th ed.). London: Balliere & Tindall.

Robinson, N. E. (Ed.). (1992). *Current therapy in equine medicine 3*. Philadelphia: W. B. Saunders.

Sherman, D. M., & Robinson, R. A. (1983). Clinical examination of sheep and goats. *Vet Clin North Am, 5,* 403–426.

Smith, B. P. (1990). *Large animal internal medicine.* St. Louis: Mosby.

Glossary

abdominocentesis – Puncture of the abdominal cavity to aspirate (withdraw) fluids for diagnosis.

analgesia – Reduction in the sensation of pain without loss of other sensations.

anesthesia – Complete loss of sensory input (can be local or general).

antiseptic – A substance that tends to inhibit growth and reproduction of microorganisms.

apnea – Complete absence of respiratory effort.

artifact – An inaccurate structure or feature not normally present but visible as a result of an external agent or action, such as one seen on a radiograph.

aseptic technique – Using all the steps necessary to prevent contamination of a surgical site, wound, or puncture.

aspiration – Removal of fluid or cells with the aid of suction. Inhalation of fluid or feed into the lungs.

ataxia – Failure to move in a coordinated manner.

auscultate – Examine an organ system or part of the body by listening, usually with a stethoscope.

autoclave – A device that sterilizes items by creating steam under pressure.

balling gun – An instrument used to administer medication or magnets orally.

ballottement – Maneuver for palpating solid objects suspended in a liquid by pushing in a rhythmic manner. May be used to feel the solid portion of rumen contents or to diagnose pregnancy.

bight – Fold of rope.

biopsy – Removal, for pathologic examination, of specimens in the form of small tissue samples.

blind spot – An area of impaired vision directly caudal to the animal.

borborygmi – The rumbling noises created by the progressive movement of air through the intestines.

BQA – Beef Quality Assurance; procedures ensuring that beef has no residues or loss of marketability.

bradycardia – Slower than normal heart rate.

California mastitis test – A test used to determine somatic cell counts in a milk sample.

cannon – The lower portion of the leg extending from the hock or carpus to the fetlock.

carotid artery – The artery located in the neck though which blood from the heart goes to the brain.

carpus – The group carpal bones forming the joint between the radius and metacarpals. Also called the knee.

cassette tunnel – A device made of radiolucent material that encases a cassette, used to protect the cassette when it must be under the hoof.

casting – The use of ropes to lay down cattle.

castration – The removal of both testicles.

catheterization – Insertion of a tubular device into a duct, blood vessel, hollow organ, or body cavity for injecting or withdrawing fluids for diagnostic or therapeutic purposes.

caudal – A directional term or reference point describing toward the hind end.

chute – An adjustable restraint device used for cows. Similar in appearance to a stock.

coagulation studies – Tests that measure the blood's ability to clot.

coccygeal vein – Vessel located on the ventral aspect of tail which is commonly used for blood collection.

coffin bone – The third phalanx, which is enclosed within hoof. Also called the pedal bone.

colostrum – The yellowish fluid secreted by the mammary glands at the time of parturition that is rich in antibodies and minerals, and precedes the production of true milk. Also called foremilk or first milk.

cow kick – To kick cranially using a hind limb.

cranial – A directional term or reference point describing toward the head.

cria – A baby llama.

curry – To rub down and clean the coat of a horse. A circular stiff rubber comb.

cyanosis – Blue color of the mucous membranes secondary to lack of oxygen in the red blood cells.

dermatophyte – A fungus that can cause parasitic skin infections.

diagnosis – The identification of a disease process and the ability to distinguish one disease from another.

DLPM – Dorsal, lateral, palmar, medial, a radiographic position describing a beam that enters the dorsal lateral area of the limb and exits the palmar medial area.

DMPL – Dorsal, medial, palmar, lateral, a radiographic position describing a beam that enters the dorsomedial area of the limb and exits the palmar lateral area.

dorsal – A directional term relating to the back or dorsum of an animal. The opposite of ventral.

dorsopalmar – A radiographic view in which the beam enters the dorsocranial area and exits the ventral-palmar aspect.

dosimetry badge – A device used to monitor the amount of radiation a person is exposed to.

dyspnea – Difficulty breathing.

dysrhythmia – Abnormal or irregular rhythm, as in an abnormal heart beat.

electrolyte – An element or compound that, when dissolved in water or another solvent, dissociates into ions. Common electrolytes include calcium, potassium, and sodium.

emaciated – Excessively thin, to the point of muscle wasting and protein loss.

enema – The introduction of liquid into the rectum for the purpose of evacuation of feces (meconium in the foal).

enucleation – Removal of the globe of the eye.

eructate – To belch or burp. A normal release of gas from the rumen through the esophagus and mouth.

exploratory – A diagnostic surgery done to determine the cause of a disease or set of clinical signs.

fasciculations – Waves of small involuntary muscular contractions in major muscle groups.

fetlock – The joint between the cannon bone and the pastern.

fetus – The unborn offspring of any mammal, generally after it has developed the body parts which identify it as belonging to that species.

fly mask – A mesh mask that protects eyes from insects, dust, dirt, and other irritants.

halter – A restraint device that fits around the head and is used to lead or secure an animal.

hay flake – A section (part) of a bale of hay.

hay net – A mesh net that suspends hay above the ground, enabling a horse easy access to food.

hondo – Loop through which the rope passes in a lariat.

hot shot – A probe with an electrical shock used for encouraging animals to move.

hyperthermia – An abnormal elevation of body temperature, usually as a result of a pathologic or stressful process.

hypothermia – Abnormally low body temperature.

icteric – Yellow-tinged tissues or serum, secondary to exc[ess biliru]bin in the blood.

impression smear – A tissue examination techni[que] making an imprint on a microscope slide with th[e tissue]

intramuscular – Within a muscle; an intra[muscular]

intravenous – Within or administered into a vein.

intubation – Introduction of a tube into a hollow organ to restore or maintain patency.

ischemia – Lack of blood flow to and from tissues, body part, or organ.

isotonic – A solution that has the same concentration of solute particles as another solution.

jugular – The blood vessel (vein) in the neck which drains blood from the head and conveys it toward the heart.

kid – A baby goat.

laparotomy – Surgical entry into the abdomen of a mammal.

lariat – Coated, braided rope designed for catching livestock out in the field, pasture, or pen. Used by cowboys on the ranch or in the rodeo.

lateral – A directional term describing toward the left or right side of the body, away from the midline.

legume – Plant in the family *Leguminoseae* such as alfalfa or clover that stores nitrogen.

loop – An ovoid or circular shape formed by crossing the ends of a rope.

malignant – Tending to spread or become progressively worse, often resulting in death of the patient.

meconium – The first fecal material produced by the neonate.

medial – A directional term describing toward the middle or midline of the body.

nasogastric intubation – The insertion of a tube from the nasal cavity to the stomach.

nasolacrimal duct – The passage leading downward from the eye to the nose, through which tears are conducted.

navicular bone – The distal sesamoid bone located within the hoof.

neonate – A radiographic positioning aid.

nose ... of restraint ... examination.

... nasal septum for purposes ...

nosocomial infection – An infection acquired during hospitalization.

nystagmus – An involuntary rapid movement of the eyeball. Can be horizontal, vertical, or mixed.

obese – Excessively fat to the point of interfering with the health or productivity of the animal.

oblique view – A radiographic view, such as the DLPM or DMPL, which is not transverse or longitudinal.

obstetrical – Anything related to the management of pregnancy, labor, and the immediate postparturient period.

ocular – Of or relating to the eye.

orogastric intubation – The passage of a tube from the oral cavity to the stomach.

ovariectomy – Removal of an ovary.

palmar – A directional term used instead of caudal when describing the limb from the knee distally. Also the ventral aspect of the front hoof.

palpate – The art of examination by touch or feel.

palpebral – Pertaining to the eyelid. Palpebral reflex is the same as a blink reflex in response to the touch of an eyelid.

parenteral – Administration of substances through injection.

parenteral nutrition – The intravenous infusion of nutrient solutions.

parturition – The process of giving birth.

pastern – The part of the limb between the fetlock and the hoof.

pig board – A board used to restrain a pig.

placenta – A membranous vascular organ that develops in female mammals during pregnancy, lining the uterine wall and partially enveloping the fetus, to which it is attached by the umbilical cord. Following birth, the placenta is expelled.

plantar – A directional term used instead of caudal when describing the hind limb from the hock distally. Also the ventral aspect of the hind hoof.

prognosis – An educated opinion on the probable outcome of a disease.

prolapse – Protrusion of a viscous organ through an opening, such as the rectum through the anus or the vagina through the vulva.

radiograph – Examination of any part of the body for diagnostic purposes by means of roentgen rays, recording the image on a sensitized surface (such as photographic film).

recumbency – In a lying down position. Lateral recumbency would be lying on one's side, and dorsal recumbency would be lying on one's back.

regurgitation – A backward flowing of gastric or rumen contents without contraction of abdominal muscles.

rug – A heavy blanket used to protect horses from the (winter) elements.

rumenocentesis – Puncture of the rumen to withdraw fluids.

sedative – A chemical agent given to allay excitement or irritability.

sepsis – Condition or syndrome caused by the presence of microorganisms or their toxins in the tissue or bloodstream.

sheath – The skin, or prepuce, that covers the penis.

sheet – A lightweight fine blanket that offers protection to horses from insects and ultraviolet light during the summer.

sisal – Fiber obtained from an agave plant used to make rope.

snout snare – A restraint device that goes around a pig's snout.

stable blanket – A blanket used to protect a horse, placed over the back and secured with straps, made of fleece or other material.

stall stripping – Removal of all bedding (straw, sawdust, shavings) from stall and replacing with new, clean bedding.

sterile – Free of microorganisms.

stock – A nonadjustable restraint device consisting of vertical pillars arranged in a rectangular shape connected by horizontal bars. Used to keep livestock restrained in a standing position.

stridor – A harsh high-pitched respiratory sound associated with upper airway obstruction.

stylet – A fine wire that is run through a catheter, cannula, or hollow needle to keep it stiff or clear of debris.

subcutaneous – Located, found, or placed just beneath the skin.

sweep tub – A round pen in which the gate pushes forward, allowing for cows to be crowded into an alleyway.

sweet feed – A commercially balanced horse feed that has molasses added to oats and other ingredients in the grain mix.

syringe gun – A syringe that holds multiple doses and releases a set amount each time the trigger is pulled.

tachycardia – An abnormally fast heart rate.

tachypnea – An abnormally fast rate of breathing.

tailing-up – Grabbing the base of a cow's tail and elevating it vertically to restrain the animal.

tarsus – The joint between the tibia and metatarsus. Also called the hock.

tensile strength – Amount of load or stress a given substance can bear without breaking, when applied lengthwise.

thoracocentesis – Procedure to remove fluid from the space between the lining of the outside of the lungs (pleura) and the wall of the chest cavity.

torsion – Twisting of a viscous organ. Most commonly occurs with the uterus or intestines.

transtracheal wash – A technique used to collect bronchial exudate samples.

twitch – A restraint device used on the horse's nose.

tympany – Distention of the abdomen due to either free abdominal gas or, more commonly, gas in the rumen or other viscous organ. Has a drumlike, resonant sound when percussed with a finger-flick while listening with a stethoscope.

venipuncture – Technique used to draw blood from a vein for diagnostic purposes or treatment.

ventral – Directed toward or situated on the abdominal surface. The opposite of dorsal.

Index

A

Abdomen
  auscultation of
    in cattle, 88–89
    in horses, 85–86, 86f
  palpation of, in pigs, 101
Abdominal wraps, 252
Abdominocentesis, 150–155
  complications of, 155
  in cows, 153
  in horses, 151–153
  in llamas, 154–155
Acepromazine, 464t–465t
Acetylcysteine, for meconium impaction, 295
Adhesive drape, 320
Alcohol–dry ice mixture, for freeze branding, 263–264
Alfalfa, 65–67, 66f, 66t
Alley ways, 31, 31f
Alpha-2 receptor agonists, 465t–466t
Alpha-2 receptor antagonists, 468t
AnaSed (xylazine), 465t–466t
Anatomic vs. lay terms, 533t–534t
Anemia, equine infectious, Coggins test for, 121, 122f, 123
Anesthesia, 426–524
  agents for, 471–482
    dissociative, 471t–472t
    for horses, 477–479
    inhalant, 474t–477t
    injectable, 471t–474t
    for pigs, 482
    reversal, 468t
    for ruminants, 479–481
    sedatives and tranquilizers, 463–471
  consent form for, 427, 428f
  documentation of, 438f
  field, 483–490
    in horses, 483–488, 485f, 488f
    in ruminants, 489–490
  hypertension in, 454, 458
  hyperthermia in
    in horses, 444
    in pigs, 451
    in ruminants, 448
  induction and maintenance of
    in horses, 477–479
    in pigs, 482
    in ruminants, 479–481
  inhalant
    agents for, 474t–477t
    anesthetic delivery system for, 491–495, 491t–493t
    complications of, 491
    intubation in, 495–500
    technique of, 490–500
  local and regional, 500–524
    Bier block in, 516–518, 517f
    digit blocks in, 514–516, 515f
    horse lower limb blocks in, 508–512
    infiltrative (line) blocks in, 503t, 518–520
    intrasynovial blocks in, 503t, 504–508, 504f, 505t
    intravenous blocks in, 503t, 516–517
    inverted L regional block in, 519–520, 519f
    mechanism of action in, 501
    methods of, 503t
    paravertebral block in, 520–522, 521f
    purpose of, 501
    regional nerve blocks in, 503t, 516–522
    ring block in, 512–516
    spinal and epidural nerve blocks in, 503t, 523–524, 524f
    systemic effects of, 501t
    teat block in, 512–513, 513f
    topical blocks in, 503t
  minimum data base for, 427–429
  monitoring in
    blood pressure, 452–458
    electrocardiography in, 459–461, 460f
    equipment for, 451–463
    in horses, 438–444
    in pigs, 449–451
    pulse oximetry in, 461–463, 462f
    in ruminants, 445–448
  preanesthesia period in, 427–436
  preparation for
    for horses, 429–431
    for llamas, 433–434
    for pigs, 435–436
    for ruminants, 431–433
  reversal agents in, 468t
  sedatives and tranquilizers for, 463–471
  stages of, 436–437
  veterinary technician's role in, 427
Anesthesia machine, 491t, 492f
  operation of, 494–495
Anesthetist, veterinary, roles of, 427
Apnea, in anesthesia
  in horses, 441
  in ruminants, 446
Arrhythmias, in anesthesia. *See* Heart rate monitoring, in anesthesia
Arterial puncture, 140–142
  carotid artery, in horses, 140–141
  coccygeal artery, in cattle, 142
  complications of, 140
  equipment for, 140
  facial artery, in horses, 141, 141f
  greater metatarsal artery, in horses, 141, 141f
  inadvertent, 211
  median auricular artery, in cattle, 142, 142f
  purpose of, 140
  transverse facial artery, in horses, 141, 141f

Note: Page numbers followed by f indicate figures; those followed by t indicate tables.

547

Artificial vagina, 112
Aseptic technique, 302. *See also* Sterilization
Aspiration biopsy, 171–174
Atipamezole (Antisedan), 468t
Auscultation
　of abdomen
　　in cattle, 88–89, 93
　　in horses, 85–86, 86f
　　in llamas, 97
　of heart
　　in cattle, 88, 92
　　in horses, 85, 86f
　　in llamas, 96
　　in pigs, 99–100
　of lungs
　　in cattle, 88, 92
　　in horses, 85, 86f
　　in llamas, 96–97
　　in pigs, 100
　of rumen
　　in beef cattle, 93
　　in dairy cattle, 88–89
Autoclave, 311t, 312f, 313–315

B

Bacterial skin cultures, 179, 181
Balling guns, 196–197, 197f
Ballottement, of rumen, 88
Bandaging, 247–252. *See also* Wraps
Barley, 67, 67t
Barnes dehorner, 361f, 362–363, 362f
Bedding, 61–62
Beef quality assurance, injections and, 202–203, 205
Benzodiazepines, 469t
Bier block, 516–518, 517f
Biopsy, 163–174
　dermal, 170–171
　fine-needle aspiration, 171–174
　liver, 165–170
　　in horses, 167–169, 168f
　　in llamas, 169–170
　uterine, 163–165
Blankets, 57–59, 59t
Bleeding, nasal, in nasogastric intubation, 223
Bloat, 93, 93f
　nasogastric intubation for, 221–224, 223f
　orogastric intubation for, 221–222, 224–228, 225f, 226f
Blood chemistry, normal values for, 532t
Blood clotting studies
　blood collection for, 143–144
　normal values for, 531t
Blood collection
　arterial puncture for, 140–142. *See also* Arterial puncture
　for coagulation studies, 143–144
　venipuncture for, 126–140. *See also* Venipuncture
Blood collection tubes, 128, 128f
Blood count, normal values for, 531t
Blood pressure monitoring
　direct, 456–458, 458f
　Doppler, 452–454, 453f
　oscillometric, 454–456, 455f

Blood studies
　normal values for, 531t–532t
　preanesthesia
　　in horses, 430
　　in ruminants, 433
Boars. *See* Pig(s)
Body condition score
　for beef cattle, 91, 92f
　for dairy cattle, 87
　for pigs, 99
Borborygmi
　in cattle, 89
　in horses, 85
Bottle feeding, 272–277
　of calves, 273–274, 274f
　of foals, 272–273
　of kids, 275, 275f
　of pigs, 276
Bouin's solution, 165
Bounding pulse, 440
Bovine somatostatin (BST), 78
Bowline, 6–7, 7f
Bradycardia, in anesthesia
　in horses, 439
　in ruminants, 445
Braiding, of eye splice, 10–11, 11f
Branding
　freeze, 262–264
　hot, 261–262
Breath sounds
　in cattle, 88, 92
　in horses, 85, 86f
　in llamas, 96–97
　in pigs, 100
Breeding soundness examination. *See also* Reproductive examination
　for bulls, 113–117
　for goats, 118
　for llamas, 118–119, 119f
　for sheep, 118
　for stallions, 110–113, 111f
Brushes, 53f
BST (bovine somatostatin), 78
Buhner needle, 372f
Bulls. *See also* Cattle
　breeding soundness examination for, 113–117
　nose rings for, 41
　safety guidelines for, 30
Bupivacaine (Marcaine), 502t
Burley casting method, 35
Butorphanol (Torbugesic), 470t
Butterfly needle, 228t

C

California mastitis test, 183–184
Calves. *See* Cattle
Calving interval (cycle), 78
Canine teeth, eruption of, 528t
Cannulation. *See also* Catheters/catheterization
　nasolacrimal, 259–261, 260f
Cap, surgical, 305
Capillary refill time
　in anesthesia monitoring
　　in horses, 442
　　in pigs, 450
　　in ruminants, 447
　in pigs, 100, 450

Carbocaine (mepivacaine), 502t
Cardiac arrhythmias, in anesthesia. *See* Heart rate monitoring, in anesthesia
Carotid artery puncture, in horses, 140–141
Carpus, radiography of, 405t, 410–416
　dorsopalmar view in, 405t, 410–411, 411f
　extended/flexed lateral views in, 405t, 411–412, 412f, 413–414, 413f–415f
　skyline views in, 405t, 415, 416f
Casting, 35, 36f
Castration, 329–339
　of calves, 330–332, 331f
　of horses, 330, 332–335, 334f
　of llamas, 330, 332, 335–336
　of pigs, 330, 337–339, 338f
　　restraint for, 46, 46f, 337–338, 338f
　timing of, 329, 528t
Catheters/catheterization, 228–247. *See also* Cannulation
　butterfly, 228t, 229f
　care of, 230
　intraarterial, 244–247
　intravenous, 228t, 229f, 230–244, 231f
　over-the-needle, 228t, 229f
　　for horses, 232–236, 233f, 235f
　size of, 229t
　through-the-needle, 228t, 229f
　　in horses, 236–237
　types of, 228t, 229f, 229t
　urinary
　　in horses, 146–147
　　in llamas, 149–150, 150f
Cattle
　abdominocentesis in, 153, 155
　anatomic and lay terms for, 534t
　anesthesia in. *See* Anesthesia
　arterial puncture in
　　coccygeal artery, 142
　　median auricular artery, 142
　basic husbandry periods for, 528t
　beef
　　body condition score for, 91, 92f
　　history taking for, 76–77
　　physical examination of, 91–95, 93f, 94f
　bloat in, 93, 93f
　blood chemistry values for, 532t
　branding of, 261–264
　calves
　　bottle feeding of, 273–274, 274f
　　castration of, 330–332, 331f
　　dehorning of, 360–364, 361f, 362f, 528t
　　flanking of, 36–37, 37f
　　orogastric intubation in, 281–282, 282f
　　postpartum care for, 285, 288, 288f
　chutes for, 31, 31f, 32
　common names of, 533t
　complete blood count in, 531t
　dairy
　　body condition score for, 87
　　history taking for, 77–79
　　milk composition and, 532t
　　physical examination of, 87–90, 89f
　　reproductive cycle of, 78
　ear tags for, 266–267, 267f
　ear tattoos for, 264–266
　enucleation in, 367–370, 369f
　eye care in, 256–261
　eye disorders in, 94, 256–261, 367–370, 369f

fecal sample collection from, 145
foot care in, 37–38, 38f, 74–75
grooming of, 56–57
handling of, 30
health certification for, 120–121
insurance examinations for, 121
intramammary infusion for, 255
intravenous catheterization in, 237–240, 239f
labor and delivery in, 288, 288f
    assisted vaginal delivery in, 350f, 351–352
    cesarean section in, 353, 354t
laparotomy for
    flank, 339, 343–345, 344f
    ventral, 339, 345–348, 346f, 347f
liver biopsy in, 166–167, 166f
milk composition in, 532t
necroscopy of, 185–191, 186f, 190f
peritoneal fluid analysis in, 529t
processing facilities for, 31, 31f
reproductive examination in, 105–106, 107f
restraint of, 30–43. *See also* Restraint, of cattle
rope halter for, 12–14, 13f, 14f
rumenocentesis in, 158–159, 159f
safety guidelines for, 30
tailing-up for, 34, 34f
thoracocentesis in, 156–157
tooth eruption in, 528t
transport of, 120–121
urine collection from, 148
uterine culture for, 177–178
uterine prolapse repair in, 357–360, 358f
vaccines for, 536t–537t
vaginal prolapse repair in, 353–357, 355f
venipuncture in
    alternative sites for, 140
    coccygeal (tail vein), 132–133, 132f
    jugular, 131–132, 131f
vital signs in, normal values for, 527t
Caudal epidural anesthesia, 503t, 523–524, 524f
Cecal tympany, in horses, 85
Centesis, 150–159
    abdominocentesis, 150–155
    rumenocentesis, 158–159, 159f
    thoracocentesis, 155–158
Cesarean section, 353, 354t
Chains, for horses, 22–23, 23f
Chlorhexidine, for umbilical stump, in cria, 290
Chutes
    for cattle, 31, 31f, 32
    for llamas, 50, 50f
Circulator, during surgery, 302–304
Citanest (prilocaine), 502t
Cleaning. *See* Disinfection; Sterilization
Clinches, 382
Coagulation studies
    blood collection for, 143–144
    normal values for, 528t
Coccygeal artery puncture, in cattle, 142
Coccygeal venipuncture, in cows, 132–133, 132f
Coffin, radiography of, 386–389
    45-degree dorsopalmar view in, 385t, 387, 387f
    horizontal 0-degree dorsopalmar view in, 385t, 388, 388f
    lateral view in, 385t, 386–387, 386f
    oblique views in, 389, 389f
Coggins test, 121, 122f, 123
Cold sterilization, 311t, 312f
Colic
    nasogastric intubation for, 221–224, 223f
    orogastric intubation for, 221–222, 224–228, 225f, 226f
Collaring, of goats, 43–44
Colostrum, 75–76, 79, 89, 278, 285. *See also* Milk
    immunoglobulin content of, 285
Combs, 53f, 54
Complete blood count, normal values for, 531t
Consent form, for anesthesia, 427, 428f
Constipation, meconium, 76, 287, 293–295
Continuous sutures, 328, 328f
Conversion chart, for units of measure, 527t
Corn, 67, 67t
Corneal lesions, in cattle, 94
Corneal reflex, in anesthesia monitoring
    in horses, 443
    in pigs, 450
    in ruminants, 448
Corner pair teeth, eruption of, 528t
Cotton rope, 3
Cows. *See* Cattle
Cradling, of foals, 29, 29f
Cria. *See* Llamas
Cross-ties, for grooming, 54
Cruciate mattress sutures, 328–329, 328f
Cultures
    milk, 182–183, 182f
    skin, 179–181
        bacterial, 179, 181
        dermatophyte, 180–181
    uterine, 176–178
Curry comb, 53f, 54
Cushioning, for surgery, 320–321
Cutting needles, 323, 323f
Cyanosis, in anesthesia
    in horses, 442
    in pigs, 450
    in ruminants, 442

D

Dairy animals. *See* Cattle, dairy; Goats, dairy
Dandy brush, 53f
Dehorning
    of calves, 360–364, 361f, 362f
    of goats, 297–298, 297f, 364–367, 366f
    timing of, 528t
Delivery. *See* Labor and delivery
Dermal biopsy, 170–171
Dermatophytosis
    in cattle, 94
    culture in, 180–181
Detomidine (Dormosedan), 466t–467t
Deworming, of horses, 75
Diazepam (valium), 469t
Digital pulse, in horses, 86
Digit block, in ruminants, 514–516, 515f
Dip Quick stain, 176
Directional terms, in radiography, 384, 385f, 385t
Disbudding, in kids, 297–298, 297f, 528t
Disinfection. *See also* Sterilization
    of stalls, 62–63
Distal phalanx, radiography of, 386–387, 386f
Documentation
    for anesthesia, 427, 428f, 438f
    of consent, 427, 428f
Domitor (medetomidine), 467t
Doppler blood pressure monitoring, 452–454, 453f
Dormosedan (detomidine), 466t–467t
Double half hitch, 8–9, 9f
Doxapram, 468t
Draping, 318–320
Dressings, bandaging for, 247–251, 249f. *See also* Wraps
Drug(s). *See also specific drugs*
    extra-label use of, 464
    sedative/tranquilizing, in anesthesia, 463–471
    withdrawal times for, 464
Drug administration
    intramuscular, 199–210. *See also* Intramuscular injections
    intraocular, 256
    intravenous, 211–217. *See also* Intravenous injections
    oral, 194–197. *See also* Oral administration
    parenteral, 197–221
    subcutaneous, 217–220. *See also* Subcutaneous injections
Dysrhythmias, in anesthesia. *See* Heart rate monitoring, in anesthesia

E

Ear notching, 267–268, 269f
Ear tags, 266–267, 267f
Ear tattooing, 264–266
Ear twitch, in anesthesia monitoring
    in horses, 443
    in ruminants, 448
Elastrator method, for tail docking, 299
Electrocardiography, 459–461, 459f
Electroejaculation
    in bulls, 115, 116f
    in llamas, 119
Emasculator
    for castration, 330f, 331–332, 331f
    for tail docking, 300
Endoscopy, in horses, 373–375, 375f
Endotracheal intubation, 495–500
    complications of, 496
    equipment for, 496, 496f
    nasotracheal, 495–497
    orotracheal, 497–500, 499f
    purpose of, 495
Enemas, for meconium passage, in foals, 287, 293–295
Enucleation, 367–370, 369f
Epidural nerve blocks, 503t, 523–524, 524f
Epistaxis, in nasogastric intubation, 223
Equine infectious anemia, Coggins test for, 121, 122f
Ethylene oxide sterilization, 311t
Ewes. *See* Sheep
Exercise, postcastration, in horses, 335
Extra-label use, 464
Eye care, 256–261
Eye disorders, in cattle, 94
    enucleation for, 367–370, 369f

Eye position, in anesthesia monitoring
   in horses, 441–442
   in ruminants, 447
Eye reflexes, in anesthesia monitoring
   in horses, 443
   in pigs, 450
   in ruminants, 448
Eye splice, 10–11, 11f

F

Facial sinus venipuncture, in horses, 130, 130f
Feathers, 55
Fecal sample collection, 144–145
Feeding
   bottle, 272–277
      of calves, 273–274, 274f
      of foals, 272–273
      of kids, 275, 275f
      of pigs, 276
   grains in, 67, 67t
   hay in, 65–67, 66f
   history taking for
      for cattle, 76–77, 78
      for horses, 74–75
   preanesthesia
      in horses, 430
      in llamas, 434
      in pigs, 435
      in ruminants, 433
Feed tubs, care of, 63–64
Feet. *See* Foot; Hoof
Femoral pulse, in pigs, 100
Fenestrated drape, 319–320
Fentanyl, 470t–471t
Fertility examination. *See* Reproductive examination
Fetlock, radiography of, 399–405, 399t
   dorsopalmer view in, 399–400, 399t, 400f
   extended lateral view in, 399t, 401, 401f
   flexed lateral view in, 399t, 402, 402f
   flexor surface (skyline) view in, 399t, 404–405, 404f
   oblique views in, 399t, 403–404, 403f
Fetus
   palpation of, in cows, 106
   ultrasonography of, in mares, 103
Film identification marks, 384
Fine-needle aspiration biopsy, 171–174
First milk. *See* Colostrum
Flagging, 113
Flanking, 36–37, 37f
Flank laparotomy
   for horses, 339–343, 341f
   for ruminants, 339, 343–345, 344f
Fluorescence staining, ocular, 257–258
Fly masks, 57, 60
Fly repellent, 57
   for cattle, 57
   for horses, 55
Foals. *See* Horses
Foot. *See also* Hoof
   examination of
      in cattle, 37–38, 38f, 89–90, 94
      in horses, 86
      in llamas, 97
      in pigs, 100–101

Foot baths, 74
Foot care, 54–55, 55f
   in cattle, 37–38, 38f, 74–75
   history taking for, 73–74
   in horses, 53f, 55
Forceps, thumb, 324f
Ford-interlocking sutures, 328f, 329
Forms
   for breeding soundness in bulls, 116, 117f
   consent, for anesthesia, 427, 428f
Four-towel technique, 318–319, 319f
Freeze branding, 262–264

G

Gas sterilization, 311t
Gastric endoscopy, in horses, 373–375
Gastric intubation, 221–228
Genital examination. *See* Reproductive examination
Gestation, length of, 528t
Gestational age estimation, in mares, 103
Glossary of terms, 542–546
Gloves, surgical, 305, 306
Goats
   anatomic and lay terms for, 534t
   anesthesia in. *See* Anesthesia
   basic husbandry periods for, 528t
   blood chemistry values in, 532t
   collaring and leading of, 43–44
   common names of, 533t
   complete blood count in, 531t
   dairy
      history taking for, 73
      milk composition and, 532t
   dehorning of, 297–298, 297f, 364–367, 366f, 528t
   drug administration in. *See* Drug administration
   handling of, 43–44, 45f
   intramammary infusion for, 255
   intravenous catheterization in, 240–242, 241f
   jugular venipuncture in, 135–136, 136f
   kids
      bottle feeding of, 275, 275f
      disbudding of, 297–298, 297f, 364–367, 366f
      orogastric intubation in, 282–283, 283f
      postpartum care for, 285, 289
   labor and delivery in, 289
      assisted vaginal delivery in, 351–352
      cesarean section in, 353, 354t
   laparotomy for
      flank, 339, 343–345
      ventral, 339, 345–348
   milk composition in, 532t
   placing in stanchions, 44, 45f
   reproductive examination in
      in females, 107–108
      in males, 118
   restraint of, 43–44, 45f
   tooth eruption in, 528t
   vaccines for, 538t
   ventral laparotomy for, 345–348
   vital signs in, normal values for, 527t
Gowns, surgical, 305, 306–307
Grains, 67, 67t

Grass hay, 65–67, 66f, 66t. *See also* Hay
Grooming, 53–57
   of cattle, 56–57
   equipment for, 53, 53f
   of horses, 54–56, 56f
   of llamas, 57
Growth hormone implants, 220–221
Guaifenesin, 473t

H

Hair removal, preoperative, 316–317
Halothane, 475t
Halters
   for cattle
      application of, 33, 33f
      fashioning of, 12–14, 13f, 14f
   for horses
      application of, 20–21, 20f
      for foals, 28, 28f
      temporary rope, 14–15, 15f
   for llamas, 48–49f
   for sheep, 12–14, 13f, 14f
Hand washing, for surgery, 305–307, 306f
Hay
   flakes of, 67
   grass, 65, 66f, 66t
   legume, 65, 66f, 66t
   quality of, 67
   quantity of, 67
Hay nets, 64–65, 65f
Health certificate, 120–121, 120f
Heart, auscultation of
   in cattle, 88, 92
   in horses, 85, 86f
   in llamas, 96
   in pigs, 99–100
Heart rate monitoring, in anesthesia
   electrocardiography in, 459–461, 459f
   in horses, 439–440, 440f
   in pigs, 449, 449f
   in ruminants, 445
Heaves, feeding in, 64
History taking
   for cattle, 76–79
   for dairy animals, 73, 77–79
   general guidelines for, 71
   for hoof care and health, 73–74
   for horses, 74–76
   for llamas, 80
   for pigs, 80–81
   for vaccinations, 72–73
Hogs. *See* Pig(s)
Hog snare, 47–48, 47f
Hondos, 3, 10–11, 11f
Hoof. *See also* Foot
   care of
      in cattle, 37–38, 38f
      equipment for, 53f, 55
      history taking for, 73–74
      in horses, 54–55, 55f
   clinches in, 382
   examination of
      in cattle, 37–38, 38f, 86, 89–90, 94
      in horses, 54–55, 55f, 86
      in pigs, 100–101
   horseshoe removal from, 382
   preparation for radiography, 381–382

Hoof picks, 53f, 55
Horn removal. *See* Dehorning
Horses
  abdominocentesis in, 151–153, 155
  anatomic and lay terms for, 533t
  anesthesia in. *See* Anesthesia
  arterial puncture in
    carotid artery, 140–141
    facial artery, 141, 141f
    greater metatarsal artery, 141, 141f
    transverse facial artery, 141, 141f
  auscultation in
    of abdomen, 85–86, 86f
    of heart, 85, 86f
    of lungs, 85, 86f
  basic husbandry periods for, 528t
  biopsy in
    of liver, 167–169, 168f
    of uterus, 163–165
  blankets for, 57–59, 59t
  blood chemistry values in, 532t
  castration of, 330, 332–335, 334f
  catheterization in
    intraarterial, 244–247, 246f
    intravenous, 232–237
  Coggins test for, 121, 122f
  common names of, 533t
  complete blood count in, 531t
  deworming of, 75
  drug administration for. *See* Drug administration
  endoscopy in, 373–375, 375f
  fecal sample collection from, 144–145
  flank laparotomy for, 339–343, 341f
  fly masks for, 57, 60
  foals
    bottle feeding of, 272–273
    cradling of, 29, 29f
    enemas for, 287, 293–295
    handling of, 28–29, 28f
    history taking for, 75–76
    meconium in, 287, 293–295
    nasal oxygen for, 292–293
    nasogastric intubation in, 279–280
    parenteral nutrition for, 291–292
    postpartum care for, 285–287, 286f, 291–295
    tailing-up for, 29, 29f
  grooming of, 54–56, 56f
  halter for
    putting on, 20–21, 20f
    temporary rope, 14–15, 15f
  health certification for, 121, 122f
  history taking for, 74–76
  insurance examination for, 121
  labor and delivery in, 286–287, 286f
    assisted vaginal delivery in, 350
    cesarean section in, 353, 354t
  leading of, 21
  loading in trailer, 26–27, 27f
  milk composition in, 532t
  nasolacrimal cannulation in, 259–261, 260f
  necroscopy of, 185–191, 186f
  ocular procedures for, 256–261
  peritoneal fluid analysis in, 530t
  physical examination for, 83–86
  prepurchase examination for, 121–123
  reproductive examination for
    in mare, 102–104, 102f
    in stallion, 110–113, 111f
  respiratory sampling in, 159–163
  restraint of, 17–30. *See also* Restraint, of horses
  safety guidelines for, 17–18
  stallions
    breeding soundness examination in, 110–113, 111f
    handling of, 29–30
  temperature taking in, 84, 86f
  thoracocentesis in, 156–157, 156f
  tooth eruption in, 528t
  transtracheal wash in, 161–163, 162f
  urine collection from
    by catheterization, 146–147
    free-catch, 147–148
  uterine culture for, 177
  vaccines for, 535t
  venipuncture in
    alternative sites for, 140
    facial sinus, 130, 130f
    jugular, 128–129, 129f
  vital signs in, normal values for, 527t
Horseshoes, removal of, 382
Horse trailers, loading of, 26–27, 27f
Hot branding, 261–262
Hot docking iron, 299
Hot shots, 39–40, 39f
Humane twitches, 25–26, 25f
Husbandry periods, 528t
Hydromorphone, 470t–471t
Hypertension, intraoperative, 454, 458
Hyperthermia, in anesthesia
  in horses, 444
  in pigs, 451
  in ruminants, 448
Hypotension, intraoperative, 454, 458
Hypothermia, in anesthesia
  in horses, 444
  in pigs, 451
  in ruminants, 448

I

Icteric skin, in pigs, 99
Identification techniques, 261–269
  ear notching, 267–268, 269f
  ear tags, 266–267, 267f
  ear tattooing, 264–266
  freeze branding, 262–264
  hot branding, 261–262
Immunization. *See* Vaccination
Impression smears, 175–176
Incisors, eruption of, 528t
Infiltrative blocks, 503t, 518–520
Inhalant anesthesia. *See* Anesthesia, inhalant
Injections. *See* Intramuscular injections; Intravenous injections; Subcutaneous injections
Instrument pack, preparation of, 307–309, 309f, 310f
Instruments, sterilization of, 310–315
Insurance examinations, 121
  postmortem, 191
International transport, health certification for, 120, 120f
Interrupted sutures, 327–328, 328f
Interstate transport, health certification for, 120, 120f
Intraarterial catheterization, 244–247, 246f
Intramammary infusion, 255
Intramuscular injections, 199–210
  in cows, 202–205, 204f
    beef quality assurance and, 202–203, 205
  in goats, 206–207, 207f
  in horses, 199–202, 200f, 201t, 202f
  hubbing needle in, 200
  in llamas, 209–210, 210f
  in pigs, 208, 209f
  syringe gun for, 204, 205–206, 206f
Intrasynovial blocks, 503t, 504–508, 504f, 505t
Intravenous catheterization, 230–244
  catheters for, 228f, 229t, 230
    removal of, 244
  in cows, 237–240, 239f
  equipment for, 228t, 229f, 230, 231–232, 231f
  in goats, 240–242, 241f
  guidelines for, 230
  in llamas, 237–240, 239f
  in pigs, 242–243
Intravenous injections, 211–217. *See also* Venipuncture
  in goats, 214, 215f
  in horses, 211–213, 212f
  in llamas, 216–217, 217f
  in pigs, 215–216, 216f
Intubation
  nasogastric, 221–224, 223f
    in foals, 279–280
  nasotracheal, 495–497, 496t
    complications of, 496
    equipment for, 496, 496f
    purpose of, 495
  orogastric, 221–222, 224–228, 225f, 226f
    in calves, 281–282, 282f
    in cria, 283–284
    in kids, 282–283, 283f
  orotracheal, 497–500, 499f
    complications of, 496
    equipment for, 496, 496f
    purpose of, 495
Inverted L regional block, 519f, 519–520
Iodine, for umbilical stump
  in calves, 288
  in foals, 287
  in kids, 289
Isoflurane, 476t

J

Jugular venipuncture
  in cows, 131–132, 131f
    for anesthesia, 433
  in goats, 135–136, 136f
    for anesthesia, 433
  in horses, 128–129, 129f, 211–213, 212f
    for anesthesia, 430
  in llamas
    high-neck, 133–134, 133f
    low-neck, 134–135, 134f
  in pigs, 136–138, 137f, 138f
  in sheep, for anesthesia, 433

K

Kendal humane twitch, 25–26
Ketamine hydrochloride (Ketaset), 471t–472t
Kids. *See* Goats
Knots, 5–9
    bowline, 6–7, 7f
    double half hitch, 8–9, 9f
    quick-release, 5, 6f
    surgeon's, 325
    suture, 323–326, 325f
    tail tie, 9–10, 10f
    tomfool, 6–7, 7f

L

Labels, film, 384
Labor and delivery. *See also* Postpartum care
    in cows, 288, 288f, 289
        assisted vaginal, 351–352
        cesarean section, 353, 354t
    in ewes
        assisted vaginal, 351–352
        cesarean section, 353, 354t
    in goats, 289
        assisted vaginal, 351–352
        cesarean section, 353, 354t
    in mares, 286–287, 286f
        assisted vaginal, 350
        cesarean section, 353, 354t
    in sows, 289
        assisted vaginal, 352
        cesarean section, 353t
Lactation cycle, in cattle, 77
Lambs. *See* Sheep
Laparotomy, 339–348
    flank
        for horses, 339–343, 341f
        for ruminants, 339, 343–345, 344f
    ventral, 345–348
Lariats, 3
    hondos in, 3, 10–11, 11f
Lay vs. anatomic terms, 533t–534t
Leading
    of goats, 43–44
    of horses, 21, 21f
    of llamas, 48–49, 49f
Leg. *See* Lower limb
Legume hay, 65, 66f, 66t. *See also* Hay
Lidocaine (Xylocaine), 502t
Limb. *See* Lower limb
Lime, for stall disinfection, 62
Line blocks, 503t, 518–520
Liquid nitrogen, for freeze branding, 263–264
Liver biopsy, 165–170
    in cattle, 166–167, 166f
    in llamas, 169–170
Llamas
    abdominocentesis in, 154–155
    anatomic and lay terms for, 534t
    anesthesia in. *See* Anesthesia
    basic husbandry periods for, 528t
    blood chemistry values in, 532t
    castration of, 330, 332, 335–336
    common names of, 533t
    complete blood count in, 531t
    cria
        bottle feeding of, 276
        orogastric intubation in, 283–284
        postpartum care for, 285, 290
    defecation habits of, 63
    drug administration in. *See* Drug administration
    fecal sample collection from, 145
    grooming of, 57
    haltering and leading of, 48–49, 49f
    handling of, 48
    history taking for, 80
    intravenous catheterization of, 237–240, 239f
    jugular venipuncture in
        high-neck, 133–134, 133f
        low-neck, 134–135, 134f
    liver biopsy in, 169–170
    milk composition in, 532t
    ocular procedures for, 256–261
    physical examination of, 95–97, 95f
    reproductive examination in
        in females, 107–108
        in males, 118–119, 119f
    restraint of, 48–50
    stall care for, 63
    stocks for, 49–50, 50f
    thoracocentesis in, 157–158
    tooth eruption in, 528t
    urine collection from, 149–150, 150f
    vaccines for, 539t
    vital signs in, normal values for, 527t
Local anesthesia, 500–524. *See also* Anesthesia, local and regional
Lower limb
    anatomic and lay terms for, 533t–534t
    nerve blocks in, 508–512, 510f
    radiography of, 378–424. *See also* Radiography
    ultrasonography of, 423–424
    wraps for, 247–249, 249f
Lungs, auscultation of
    in cattle, 88, 92
    in horses, 85, 86f
    in llamas, 96–97
Lymph nodes, palpation of, 83

M

Magnets, balling guns for, 197
Malignant hyperthermia
    in horses, 444
    in pigs, 451
    in ruminants, 448
Mane combs, 53f
Manometric blood pressure monitoring, 456–458, 458f
Manure fork, 61, 61f
Marcaine (bupivacaine), 502t
Marginal ear vein venipuncture, in pigs, 139
Mask, surgical, 305
Mastitis
    in beef cattle, 95
    California test for, 183–184
    in dairy cattle, 89
    intramammary infusion for, 255
    milk culture for, 182–183, 182f
Mats, cleaning of, 62–63
Mattress sutures, 328–329, 328f
Meconium, 76, 287
    enema for, 287, 293–295
Medetomidine (Domitor), 467t
Median auricular artery puncture, in cattle, 142, 142f
Medications. *See* Drug(s)
Mepivacaine (carbocaine), 502t
Metacarpals/metatarsals, radiography of
    dorsopalmar view in, 405t, 406, 406f
    lateral view in, 405t, 407, 407f
    oblique views in, 405t, 408–409, 408f
Milk
    composition of, 532t
    culture of, 182–183, 182f
    first. *See* Colostrum
Milking technique, 182, 182f
Milk production, history of, 77
Minimum alveolar concentration (MAC), 475t
Morphine, 470t–471t
Mucous membrane color, in anesthesia monitoring
    in horses, 441–442
    in pigs, 449–450
    in ruminants, 446–447

N

Names, common, of large animals, 533t
Nasal oxygen, for foals, 292–293
Nasal swabs, 160, 160f
Nasogastric intubation, 221–224, 223f
    in foals, 279–280
Nasolacrimal duct patency, assessment of, 257–258
Nasolacrimal flushing, 258
Nasotracheal intubation, 495–497, 496t
    complications of, 496
    equipment for, 496, 496f
    purpose of, 495
Navicular, radiography of, 390–394
    lateral view in, 390t, 392, 392f
    oblique views in, 390t, 393–394, 394f
    palmarproximal (skyline) view in, 390–391, 390t, 391f
    65-degree dorsopalmar (upright pedal) view in, 390t, 392–393, 393f
Necroscopy sampling, 184–191
    complications of, 184
    equipment for, 185
    procedure for, 185–191
    purpose of, 184
Needle(s), 198t
    Buhner, 372f
    butterfly, 228t, 229f
    cutting, 323, 323f
    hubbing of, 200
    for intrasynovial blocks, 505t
    for suturing, 323, 323f
    taper, 323, 323f
Needle holders, 324f
Needle method, for abdominocentesis, 151–153
Needle teeth clipping, in piglets, 295–296, 296f
Neonatal clinical procedures, 271–300
    bottle feeding, 272–277
    colostrum, 278

milk replacers, 277–278, 278t
nasogastric intubation, 279–284
postpartum care, 284–290
routine processing, 295–300
Nerve blocks. *See* Anesthesia, local and regional
Nitrogen, liquid, for freeze branding, 263–264
Nosebleed, in nasogastric intubation, 223
Nose rings, for bulls, 41
Nose tongs, 40–41, 41f
Nosocomial infections, 60
Novocaine (procaine), 502t
Nylon rope, 3, 5. *See also* Rope(s)

O

Oats, 67, 67t
Observation, patient, 81
Obstetrical procedures, 349–360. *See also* Labor and delivery; Postpartum care
    assisted vaginal delivery
        of dead fetus, 349, 350f
        equipment for, 350f
        of horses, 350
        of pigs, 352
        of ruminants, 351–352
    cesarean section, 353, 354t
    uterine prolapse repair, 357–360, 358f
    vaginal prolapse repair, 353–357, 355f
Ocular procedures, 256–261
    drug administration, 256, 259–261, 260f
    fluorescence staining, 257–258
Ointments, ocular, 256
Open gloving technique, 306–307
Ophthaine (proparacaine), 502t
Opioids, 470t–471t
Oral drug administration
    balling guns for, 196–197, 197f
    of pastes, 194–196, 195f
Orogastric intubation, 281–284
    in calves, 281–282, 282f
    in cria, 283–284
    decompressive, 221–222, 224–228, 225f, 226f
    in kids, 282–283
Orotracheal intubation, 497–500, 499f
    complications of, 496
    equipment for, 496, 496f
    purpose of, 495
Oscillometric blood pressure monitoring, 454–456, 455f
Over-the-needle catheter, 228t, 229f
    for horses, 232–236, 233f, 235f
Oxygen, nasal, for foals, 292–293
Oxymorphone, 470t–471t

P

Padding, for surgery, 320–321
Paddles, 39–40, 39f
Palpation
    of abdomen, in pigs, 101
    of digital pulse, in horses, 86
    of femoral pulse, in horses, 86
    of lymph nodes, 83
    of reproductive organs
        in cows, 105–106
        in mares, 103, 104
        in stallions, 111–112

Palpebral reflex, in anesthesia monitoring
    in horses, 443
    in pigs, 450
    in ruminants, 447
Parasitic infections
    deworming for, 75
    fecal sample collection in, 144–145
Paravertebral block, 520–522, 521f
Parenteral drug administration, 197–221
    intramuscular, 199–210
    needles for, 198t
Parenteral nutrition, for foals, 291–292
Pastern, radiography of
    dorsopalmar view in, 395–396, 395t, 396f
    oblique views in, 395t, 397–398, 398f
Pastes, oral, 194–196, 195f
Pelvic measurement, in cows, 106, 107f
Pelvimeter, 106, 107f
Penis
    examination of
        in small ruminants, 118–119
        in stallion, 111, 111f
    washing of, 110–111, 111f, 252–254, 254f
Peritoneal fluid analysis
    in cattle, 529t
    in horses, 530t
Phalanx, distal (third), radiography of, 386–387, 386f
Phenothiazine tranquilizers, 464t–465t
Physical examination, 70–123
    of cattle
        beef, 91–95, 93f, 94f
        dairy, 87–90, 89f
    complications of, 71
    equipment for, 71
    for health certificate, 120–121, 120f
    history taking for, 70–81. *See also* History taking
    of horses, 83–86
    for legal purposes, 119–123
    of llamas, 95–97, 95f
    patient observation in, 81
    of pigs, 98–101
    purpose of, 71
    reproductive. *See* Reproductive examination
Pig(s)
    anatomic and lay terms for, 534t
    anesthesia in. *See* Anesthesia
    basic husbandry periods for, 528t
    blood chemistry values in, 532t
    body condition score for, 99
    bottle feeding of, 276
    castration of, 330, 337–339, 338f
        restraint for, 46, 46f
    common names of, 533t
    complete blood count in, 531t
    drug administration in. *See* Drug administration
    ear notching in, 267–268, 269f
    handling of, 45
    history taking for, 80–81
    intravenous catheterization in, 242–243
    labor and delivery in, 289–290
        assisted vaginal delivery in, 352
        cesarean section in, 354t

    milk composition in, 532t
    nasal swabs in, 160, 160f
    physical examination of, 98–101
    piglets
        needle teeth clipping for, 295–296, 296f
        postpartum care for, 285, 289–290
    reproductive examination in, 109
    respiratory sampling in, 159–163
    restraint of, 45–48, 46f, 47f
    tooth eruption in, 528t
    vaccines for, 540t
    venipuncture in
        jugular, 136–138, 137f, 138f
        marginal ear vein, 139, 139f
    vital signs in, normal values for, 527t
Pig boards, 45–46
Pig handstand, 46, 46f
Pinging, of rumen, 89, 89f
Pink eye, in calves, 94
Pitchfork, 61, 61f
Placenta, passage of
    in cows, 288, 288f
    in horses, 287
    in llamas, 290
Play-Doh, 382
Porcine stress syndrome, 451, 482
Postmortem examination, 191–195. *See also* Necroscopy sampling
Postpartum care
    for calves, 285, 288, 288f
    complications of, 284
    for cria, 285, 290
    equipment for, 284–285
    for foals, 285–287, 286f
        enema in, 287, 293–295
        nasal oxygen in, 292–293
        parenteral nutrition in, 291–292
    immediate, 285
    for kids, 285f, 289
    for lambs, tail docking in, 298–300
    for piglets, 289–290
Post technique, for field anesthesia, 484, 485f
Pregnancy. *See also* Labor and delivery; Postpartum care
    examination in. *See also* Reproductive examination
        in cows, 106
        in mares, 103–104
    length of, 528t
    twin
        in cows, 106
        in mares, 104
Pregnancy gates, 31, 31f
Prepuce, of stallion
    examination of, 111, 111f
    washing of, 110–111, 111f
Prilocaine (Citanest), 502t
Procaine (Novocaine), 502t
Processing facilities, for cattle, 31, 31f
Prods, 39–40, 39f
Prolapse
    rectal, 370–372, 372f
    uterine, 357–359, 358f
    vaginal, 353–357, 355f
Proparacaine (ophthaine), 502t
Propofol, 474t

Pulse
   bounding, 440
   digital, in horses, 86
   femoral, in pigs, 100
   normal values for, 527t
   thready, 440
Pulse oximetry, 461–463, 462f
Pulse rate monitoring, in anesthesia
   equipment for, 462–463, 462f
   in horses, 440, 440f
   in pigs, 449, 449f
   in ruminants, 445

Q
Quality assurance, injections and, 202–203, 205
Quick-release knot, 5, 6f

R
Radiofrequency ear tags, 266–267
Radiography, 378–424
   of carpus
      dorsopalmar view in, 405t, 410–411, 411f
      extended/flexed lateral views in, 405t, 411–412, 412f, 413–414, 413f, 414f, 415f
      skyline views in, 405t, 415, 416f
   of coffin
      45-degree dorsopalmar view in, 385t, 387, 387f
      horizontal 0-degree dorsopalmar view in, 385t, 388, 388f
      lateral view in, 385t, 386–387, 386f
      oblique views in, 385t, 389, 389f
   directional terms in, 384, 385f, 385t
   equipment for, 379t
   of fetlock
      dorsopalmer view in, 399–400, 399t, 400f
      extended lateral view in, 399t, 401, 401f
      flexed lateral view in, 399t, 402, 402f
      flexor surface (skyline) view in, 399t, 404–405, 404f
      oblique views in, 399t, 403–404, 403f
   hoof preparation for, 381–382
   identification marks and labels in, 384
   of metacarpals/metatarsals, 405–409
      dorsopalmar view in, 405t, 406, 406f
      lateral view in, 405t, 407, 407f
      oblique views in, 405t, 408–409, 408f
   of navicular, 390–394, 390t
      lateral view in, 390t, 392, 392f
      oblique views in, 390t, 393–394, 394f
      palmarproximal (skyline) view in, 390–391, 390t, 391f
      65-degree dorsopalmar (upright pedal) view in, 390t, 392–393, 393f
   optional views in, 385t
   of pastern
      dorsopalmar view in, 395–396, 395t, 396f
      lateral view in, 395t, 396–397, 397f
      oblique views in, 395t, 397–398, 398f
   patient preparation for, 383–384
   safety guidelines for, 379–381
   standard views in, 385t
   of tarsus, 417–422, 417t
      dorsoplantar view in, 417–418, 417t, 418f
      flexed dorsoplantar view in, 417t, 422, 422f
      lateral/flexed views in, 417t, 418–419, 419f
      oblique views in, 417t, 420–421, 420f, 421f
   ultrasonography in, 423–424
Rectal endoscopy, in horses, 373–375
Rectal palpation
   in bulls, 115
   in cows, 105–106
   in mares, 103, 104
   in stallions, 111–112
Rectal prolapse, 370–372, 372f
Rectal temperature
   in cattle, 88
   in horses, 84, 86f
   in llamas, 96
Reflexes, in anesthesia monitoring
   in horses, 443
   in pigs, 450
   in ruminants, 448
Regional anesthesia, 500–524. See also Anesthesia, local and regional
Reimers emasculator, 330f, 331–332, 331f
Reproductive cycle, of cattle, 78
Reproductive examination
   female, 101–109
      complications of, 101, 104
      in cows, 105–106
      equipment for, 101
      in ewes, 107–108, 108f
      in goats, 107–108
      in llamas, 107–108
      in mares, 102–104, 102f
      purpose of, 101
      in sows, 109
   for health certificate, 120–121
   for insurance, 121
   male, 109–119
      in bulls, 113–117
      complications of, 109–110
      equipment for, 109–110
      purpose of, 109
      in stallions, 110–113, 111f
   prepurchase, 121–123
Reproductive treatments, 252–255
   intramammary infusion, 255
   sheath cleaning for, 220–222f, 252–254, 254f
Respiratory monitoring, in anesthesia
   in horses, 441
   in ruminants, 445–446
Respiratory rate, normal values for, 527t
Respiratory sampling, 159–163
   nasal swabs in, 160, 160f
   transtracheal wash in, 161–163, 162f
Restraint
   of cattle, 30–43
      casting in, 35, 36f
      flanking in, 36–37, 37f
      for foot examination, 37–38, 38f
      guidelines for, 30
      halter in, 12–14, 13f, 14f, 33, 33f
      hot shot in, 39–40, 39f
      nose rings in, 41
      nose tongs in, 40–41, 41f
      for reproductive examination, 105, 113
      tailing-up in, 34, 34f
      tail tie in, 9
      tail twisting in, 34
   of goats, 43–44, 45f
   of horses, 17–30
      chains in, 22–23, 23f
      for foals, 28–29, 28f, 29f
      guidelines for, 17–18
      halter in, 14–15, 15f, 20–21, 20f
      lead rope in, 21, 21f
      for reproductive examination, 102, 102f, 110
      rules for tying, 18
      in stock, 18–19, 19f
      tail tie in, 9–10, 10f
      twitches in, 23–26, 24f, 25f
   of llamas, 48–50, 49f, 50f
   of pigs, 45–48, 46f, 47f
      for castration, 46–47, 46f, 337–338, 338f
      snout snare in, 47–48, 47f
      rope halters for, 12–15, 13f–15f
   of sheep, halter in, 12–14, 13f, 14f
Reversal agents, in anesthesia, 468t
Ring block, 512–516
Ringworm, in cattle, 94
Rompun (xylazine), 465t–466t
Rope(s), 2–14
   braiding of, 10–11, 11f
   cotton, 3
   eye splice in, 10–11, 11f
   finishing/securing end of, 3–5, 4f
   knots in, 5–9. See also Knots
   lariat, 3
   lead. See Leading
   for restraint. See Restraint
   rump, 28, 28f
   sisal, 3
   synthetic, 3, 5
   for tail tie, 9–10, 10f
   tensile strength of, 3
   types of, 2
Rope halters, 12–15, 13f–15f. See also Halters
   for cattle or sheep, 12–14, 13f, 14f
   for horses, 14–15, 15f
Rubber mats, cleaning of, 62–63
Rugs, 57–59, 59t
Rumen
   auscultation of
      in beef cattle, 93
      in dairy cattle, 88–89
   ballottement of, 88
   pinging of, 89
Rumenocentesis, 158–159, 159f
Rumen tympany, 89, 89f
Rump rope, 28, 28f

S
Safety guidelines
   for cattle, 30
   for horses, 17–18, 29–30
   for llamas, 48
Salmonella, fecal sample for, 145

Salt blocks, 64
Sample collection, 126–191
    biopsy for, 163–174
    blood, 126–144
        arterial puncture for, 140–142
        for coagulation studies, 143–144
        venipuncture for, 126–140
    cultures and testing in, 176–184
    fecal, 144–145
    impression smears for, 175–176
    in necroscopy, 184–191
    respiratory, 159–163
    skin scrapes for, 174–175
    urine, 146–150
Sand evaluation, sample collection for, 144–145
Scrape-brush-disinfect method, for growth hormone implants, 221
Scrapes, 174–175
Scrotum
    examination of
        in bulls, 114, 116f
        in small ruminants, 118–119
        in stallions, 111, 111f
    measurement of, in bulls, 114, 116f
    washing of, 110–111, 111f, 252–254, 253f
Scrub, surgical
    for patient, 316–317, 317f
    by personnel, 305–307, 306f
Scrub nurse, 302
Sedatives
    in anesthesia, 463–471
    extra-label use of, 464
    withdrawal times for, 464
Semen collection
    from bulls, 115, 116f
    from llamas, 119
    from stallions, 112–113
Semen evaluation
    for bulls, 116
    for stallions, 113
Sevoflurane, 476t–477t
Shaving, preoperative, 316–317
Sheath cleaning, 252–254, 254f
    washing of, 110–111, 111f, 253–254, 254f
Sheep
    anatomic and lay terms for, 534t
    anesthesia in. See Anesthesia
    basic husbandry periods for, 528t
    blood chemistry values in, 532t
    common names of, 533t
    complete blood count in, 531t
    drug administration in. See Drug administration
    health certification for, 120–121
    intramammary infusion for, 255
    labor and delivery in
        assisted vaginal delivery in, 351–352
        cesarean section in, 353, 354t
    lambs, tail docking in, 298–300
    laparotomy for
        flank, 339, 343–345
        ventral, 339, 345–348
    milk composition in, 532t
    reproductive examination in
        in females, 107–108, 108f
        in males, 118
    rope halter for, 12–14, 13f, 14f
    tooth eruption in, 528t
    transport of, 120–121
    vital signs in, normal values for, 527t
Sheet blankets, 57–59, 59t
Simple continuous sutures, 328, 328f
Simple interrupted sutures, 327–328, 328f
Sisal rope, 3
Skin biopsy, 170–171
Skin cultures, 179–181
    bacterial, 179, 181
    dermatophyte, 180–181
Skin preparation, preoperative, 316–317, 317f
Skin scrapes, 174–175
Smears, impression, 175–176
Snout snare, 47–48, 47f
Society of Theriogenology form, for breeding soundness in bulls, 116, 117f
Solubility coefficient, of inhalant anesthetics, 474t
Solutions, parenteral nutrition, for foals, 291–292
Sows. See Pig(s)
Speculum, for orogastric intubation, 226, 226f
Sperm motility/morphology, in stallions, 113
Spinal nerve blocks, 503t, 523–524, 524f
Splicing, eye splice in, 10–11, 11f
Stable blankets, 57–59, 59t
Staining
    for dermatophytes, 181
    ocular, 257–258
    for skin scrapes, 176
Stall care
    cleaning and disinfection in, 60–63
    feed tubs and, 63–64
    hay nets and, 64–65, 65f
    salt blocks and, 64
    water buckets and, 63–64
Stallions. See also Horses
    breeding soundness examination for, 110–113, 111f
    handling of, 29–30
Stanchions, for goats, 44, 45f
State health certificate, 120–121, 120f
Steam sterilization, 311t, 312f, 313–315
Steers. See Cattle
Sterilization, 310–315
    cold, 311t, 312f
    gas (ethylene oxide), 311
    steam (autoclave), 311t, 312f, 313–315
Stethoscope. See Auscultation
Stocks
    for cattle, 31, 31f
    for horses, 18–19, 19f
    for llamas, 49–50, 50f
Stool sample collection, 144–145
Subcutaneous injections, 217–220
    in cows, 218
    in goats, 219
    of growth hormone implants, 220–221
    in llamas, 219
    in pigs, 219–220
Surcingles, blanket, 58
Surgeon's knot, 325
Surgery, 302–375
    aseptic technique in, 302. See also Sterilization
    castration, 329–339
    cushioning for, 320–321
    dehorning
        of calves, 360–364, 361f, 362f
        of goats, 297–298, 297f, 364–367, 366f
    draping for, 318–320
    endoscopic, in horses, 373–375, 375f
    eye enucleation, 367–370, 369f
    hand washing for, 305–307
    instrument pack for
        autoclaving of, 311t, 312f, 313–315, 314f
        preparation of, 307–309, 309f, 310f
    laparotomy
        flank, 339–345
        ventral, 339, 345–348, 347f
    obstetrical, 349–360. See also Obstetrical procedures
    patient preparation for, 316–321
    personal preparation for, 304–307
    suturing in, 322–329
Surgery team, 302–303
Surgical pack, preparation of, 304–307
Surgical scrub
    for patient, 316–317, 317f
    by personnel, 305–307, 306f
Sutures/suturing
    cruciate mattress, 328–329, 328f
    cutting of, 326, 326f
    equipment for, 324f
    Ford-interlocking, 328f, 329
    knot tying for, 323–326, 325f
    needles for, 323, 323f
    patterns of, 327–329, 328f
    simple continuous, 328, 328f
    simple interrupted, 327–328, 328f
    size of, 322–323
    technique of, 325–326, 325f, 326f
    types of, 322t
Swabs, nasal, 160, 160f
Sweep tubs, 31, 31f
Sweet feeds, 67
Syringe guns, 204, 205–206, 206f

T

Tachycardia, in anesthesia
    in horses, 439
    in ruminants, 445
Tachypnea, in anesthesia, in ruminants, 446
Tags, ear, 266–267, 267f
Tail combs, 53f
Tail docking, in lambs, 298–300
Tailing-up
    for cattle, 34, 34f
    for foals, 29, 29f
Tail tie, 9–10, 10f
Tail twisting, for cows, 34
Tail vein venipuncture, in cows, 132–133, 132f
Tail wraps, 250–251, 251f
Taper needles, 323, 323f
Tarsus, radiography of, 417–422, 417t
    dorsoplantar view in, 417–418, 417t, 418f
    flexed dorsoplantar view in, 417t, 422, 422f
    lateral/flexed views in, 417t, 418–419, 419f
    oblique views in, 417t, 420–421, 420f, 421f

Tattooing, ear, 264–266
Teat block, 512–513, 513f
Teat cannula method, for abdominocentesis, in horse, 151–153
Teeth
   eruption of, 528t
   needle, clipping of, 295–296, 296f
Telazol (tiletamine plus zolazepam), 472t
Temperature, normal values for, 527t
Temperature monitoring, in anesthesia
   in horses, 443–444
   in pigs, 448
   in ruminants, 448
Temperature taking
   in cattle, 88
   in horses, 84, 86f
   in llamas, 96
Tensile strength, of rope, 3
Testis, examination of
   in bulls, 114
   in stallions, 110–111, 111f
Tetanus antitoxin, postpartum, for mare, 287
Thiopental, 473t
Third phalanx, radiography of, 386–387, 386f
Thoracocentesis, 155–158
   for cattle, 156–157, 156f
   for horses, 156–157, 156f
   for llamas, 157–158
Thready pulse, 440
Through-the-needle catheter, 228t, 229f
   for horses, 236–237
Thumb forceps, 324f
Tiletamine plus zolazepam (Telazol), 472t
Tissue perfusion, in anesthesia monitoring
   in horses, 441–442
   in pigs, 449–450
   in ruminants, 446–447
Tolazoline, 468t
Tomfool knot, 6–7, 7f
Torbugesic (butorphanol), 470t
Total parenteral nutrition, for foals, 291–292
Towels, for draping, 318–320, 319f
Trailers, horse, loading of, 26–27, 27f
Tranquilizers
   in anesthesia, 463–471
      extra-label use of, 464
      withdrawal times for, 464
Transport
   health certification for, 120–121, 120f
   horse trailer loading in, 26–27, 27f
Transtracheal wash, 161–163, 162f
Turn-out blankets, 57–59, 59t

Twin pregnancy
   in cows, 106
   in mares, 104
Twitches, 23–26
   ear, in anesthesia monitoring, 443, 448
   hand, 24, 24f
   Kendal humane, 25–26
   mechanical, 25–26, 25f
Tympany
   cecal, in horses, 85
   rumen, 89, 89f

U

Ulcers, biopsy of, 170–171
Ultrasonography
   of lower limb, 423–424
   in reproductive examination
      in cows, 105, 106
      in mares, 102f, 103–104
      in small ruminants, 107–108
Umbilical cord care
   for calves, 288
   for cria, 290
   for foals, 286–287
   for kids, 289
   for piglets, 289
Units of measure, conversion chart for, 527t
Urethral blockage, in cattle, 95
Urinary catheterization
   in horses, 146–147
   in llamas, 149–150, 150f
Urine collection, 146–150
   from cattle, 148
   from horses, 146–148
   from llamas, 149–150, 150f
Uterine biopsy, 163–165
Uterine culture, 176–178
   for cows, 177–178
   for horses, 177
Uterine prolapse, 357–360, 358f

V

Vaccination
   of cattle, 536t–537t
   of goats, 538t
   history of, 72–73
   of horses, 535t
   of llamas, 539t
   of pigs, 540t
Vagina, artificial, 112
Vaginal examination. See Reproductive examination, female
Vaginal prolapse, 353–357, 355f

Valium (diazepam), 469t
Vapor pressure, of inhalant anesthetics, 474t
Venipuncture, 126–140. See also Intravenous injections
   alternative sites for, 140
   coccygeal, in cows, 132–133
   collection tubes for, 128, 128f
   complications of, 127
   equipment for, 127–128, 128f
   facial sinus, in horses, 130, 130f
   jugular
      in cows, 131–132, 131f
      in goats, 135–136, 136f
      in horses, 128–129, 129f
      in llamas, 133–134, 133f
      in pigs, 136–138
   marginal ear vein, in pigs, 139
   purpose of, 127
Ventilatory support, in anesthesia, in horses, 441
Ventral laparotomy, 345–348
Veterinarian, health certification by, 120–121, 120f
Veterinary anesthetist, roles of, 426
Vital signs, normal values for, 527t

W

Water buckets, care of, 63–64
Water withdrawal, preanesthesia
   in horses, 430
   in llamas, 434
   in pigs, 435
   in ruminants, 433
Weaning, timing of, 528t
Weight tape, 431, 431f
Wheat, 67, 67t
Whipping, of rope end, 4, 4f
Wounds, bandaging of, 247–249, 251. See also Wraps
Wraps
   abdominal, 252
   basic lower limb, 247–249, 249f
   tail, 250–251, 251f

X

Xylazine (Rompun, AnaSed), 465t–466t
Xylocaine (lidocaine), 502t

Y

Yohimbine (Yobine), 468t

Z

Zolazepam plus tiletamine (Telazol), 472t